Just Ordinary Robots

Automation from Love to War

OTHER TITLES FROM AUERBACH PUBLICATIONS AND CRC PRESS

Just Ordinary Robots

Automation from Love to War

Lambèr Royakkers

Rinie van Est

CRC Press
Taylor & Francis Group
Boca Raton   London   New York

CRC Press is an imprint of the
Taylor & Francis Group, an **informa** business

CRC Press
Taylor & Francis Group
6000 Broken Sound Parkway NW, Suite 300
Boca Raton, FL 33487-2742

© 2016 by Taylor & Francis Group, LLC
CRC Press is an imprint of Taylor & Francis Group, an Informa business

No claim to original U.S. Government works

Printed on acid-free paper
Version Date: 20150709

International Standard Book Number-13: 978-1-4822-6014-4 (Hardback)

Visit the Taylor & Francis Web site at
http://www.taylorandfrancis.com

and the CRC Press Web site at
http://www.crcpress.com

Contents

FOREWORD   XI
ACKNOWLEDGMENTS   XV
AUTHORS   XVII

CHAPTER 1   ROBOTS EVERYWHERE   1
     1.1   With Vision   3
     1.2   Technically Speaking   6
           1.2.1   From Automata to Robots   6
           1.2.2   Robot-Friendly Environments   8
           1.2.3   The Robot Body   10
                   1.2.3.1   Appearance Is Important   10
                   1.2.3.2   Opportunities for Physical Activity   12
                   1.2.3.3   Artificial Senses   13
           1.2.4   The Robot Brain   14
                   1.2.4.1   Strong and Weak Artificial Intelligence   14
                   1.2.4.2   Predictions from the Past   15
                   1.2.4.3   Through Trial and Error   16
                   1.2.4.4   Brute Computational Power   17
                   1.2.4.5   Artificial Social Intelligence   18
                   1.2.4.6   Artificial Morality   21
           1.2.5   Networked Robots and Human-Based
                   Computing   23
     1.3   Seen Socially   25
           1.3.1   Information Technology   25
           1.3.2   A Lifelike Appearance   26
           1.3.3   Level of Autonomy   26
           1.3.4   Robotization as Rationalization   27
                   1.3.4.1   Irrationality of Rationality   30

1.4   More Explorations   31
Interview with Luciano Floridi (Philosopher and Ethicist of
Information at Oxford University, United Kingdom)   32
References   38

CHAPTER 2   HOME IS WHERE THE ROBOT IS: MECHANOIDS,
             HUMANOIDS, AND ANDROIDS   43
2.1   Introduction   43
2.2   Mechanoid Robots: The Vacuum Cleaner   44
       2.2.1   Experiences of Early Adaptors:
                Roombarization   48
       2.2.2   Reducing the Complexity of Household Tasks   50
       2.2.3   Liability of Home Robots   53
2.3   Humanoid Robots: The Companion   55
       2.3.1   The Robot Body   56
       2.3.2   Human–Robot Interaction (HRI)   65
       2.3.3   Social De-Skilling   68
2.4   Android Robots: The Sex Robot   69
       2.4.1   Roxxxy   70
       2.4.2   Benefits of Sex Robots   72
       2.4.3   Social and Ethical Issues   74
                2.4.3.1   Cultural Acceptance of Sex Robots   74
                2.4.3.2   Dehumanization   75
                2.4.3.3   Sex with Child Robots   76
2.5   Observational Conclusions   78
       2.5.1   Household Robots   79
       2.5.2   Companion and Sex Robots   80
                2.5.2.1   Expectations   80
                2.5.2.2   Social, Ethical, and Regulatory
                           Issues   81
Interview with Kerstin Dautenhahn (Professor of Artificial
Intelligence, University of Hertfordshire, United Kingdom)   82
References   85

CHAPTER 3   TAKING CARE OF OUR PARENTS: THE ROLE
             OF DOMOTICS AND ROBOTS   91
3.1   Introduction   91
3.2   Domotics for the Care of the Elderly   94
       3.2.1   Paradigmatic Shift in Care   96
       3.2.2   Ethical Issues   97
                3.2.2.1   Privacy   97
                3.2.2.2   Human Contact and Quality of Care   99
                3.2.2.3   Competence of Caregivers and
                           Care Recipients   99
3.3   From Home Robotics to Robots in the Home   101
       3.3.1   Increasing the Pace of the Paradigmatic
                Shift in Care   101

3.3.2	General Ethical Issues Relating to Care Robots		105
	3.3.2.1	Safety	105
	3.3.2.2	Designing Care Robots	107
	3.3.2.3	Physical Appearance	108

3.4   Specific Ethical Issues with Regard to the Role of
Care Robots   109

3.4.1	The Robot as a Companion		110
	3.4.1.1	Deception	110
3.4.2	The Robot as a Cognitive Assistant for the Care Recipient		111
	3.4.2.1	Autonomy	111
3.4.3	The Robot as a (Supporter of the) Caregiver		115
	3.4.3.1	Dehumanization	115
	3.4.3.2	Quality of Care	116
	3.4.3.3	Human Contact	118

3.5   Observational Conclusions: The Long Term   118
Interview with Hans Rietman (Professor of Physical
Medicine and Rehabilitation, University of Twente,
the Netherlands)   121
References   125

CHAPTER 4   DRONES IN THE CITY: TOWARD A FLOATING
ROBOTIC PANOPTICON?   131

4.1	Introduction: Amazon Prime Air		131
4.2	Civil Applications of Drones		135
	4.2.1	Recreational Use	135
	4.2.2	Drone Journalism	138
	4.2.3	Precision Farming	140
4.3	Drones for Law Enforcement		144
	4.3.1	Robocops	144
	4.3.2	Tasks of Police Drones	147
	4.3.3	Examples of Police Drones	148
	4.3.4	Legal and Ethical Issues	152
4.4	Safety		154
	4.4.1	Aerial Safety	155
	4.4.2	Improper Operations	155
	4.4.3	Hacking of Drones	157
	4.4.4	Drone Hunting	158
4.5	Privacy		159
	4.5.1	Reasonable Expectations of Privacy	162
	4.5.2	Voyeurism	163
	4.5.3	Big Brother Drone Is Watching You	164
	4.5.4	The Chilling Effect	165
4.6	The Regulation of Drones		165
	4.6.1	Regulations in the United States	166
	4.6.1.1	The U.S. Federal Aviation Administration	166

		4.6.1.2	The U.S. Federal Government	168
		4.6.1.3	Local and State Governments	169
	4.6.2	Regulations in the European Union		170
		4.6.2.1	The European Commission	170
		4.6.2.2	European Countries	172
	4.6.3	Proliferation of Drone Regulations		173
4.7	Concluding Observations: Drones Create a Floating Robotic Panopticon			174
	Interview with Mark Wiebes (Innovation Manager with the Dutch National Police, the Netherlands)			178
	References			181

CHAPTER 5 WHO DRIVES THE CAR? 185

5.1	Introduction			185
5.2	Problems for Modern Road Traffic and the Costs			188
	5.2.1	Traffic Victims		188
	5.2.2	Traffic Congestion		189
	5.2.3	Pollution		190
5.3	Driver Assistance Systems (Levels 1 and 2)			191
	5.3.1	ABS and ESC		191
	5.3.2	Adaptive Cruise Control and Stop-and-Go Systems		194
	5.3.3	Pedestrian and Cyclist Airbags		195
	5.3.4	Pre-Crash Systems		196
5.4	Limited Self-Driving Automation (Level 3)			196
	5.4.1	Traffic Management		197
	5.4.2	Cooperative Systems		199
	5.4.3	Cooperative Driving		202
5.5	Autonomous Cars (Level 4)			207
	5.5.1	Google		210
	5.5.2	AutoNOMOS and the Remotely Controlled Community Taxi		213
5.6	Social and Ethical Issues Surrounding Car Robotization			216
	5.6.1	Acceptance		216
	5.6.2	Privacy		218
	5.6.3	Security and Safety		221
		5.6.3.1	Reliability	221
		5.6.3.2	Negative Behavioral Adaptation	222
		5.6.3.3	Cyber Security	223
	5.6.4	Better Drivers		223
	5.6.5	Liability		227
		5.6.5.1	Liability of the Manufacturers	228
		5.6.5.2	Liability of the Road Authorities	229
		5.6.5.3	Liability of the Driver	229
	5.6.6	Legislation for Limited and Full Self-Driving		231

5.7 Concluding Observations 232
    5.7.1 Short Term: Driver Assistance Systems
        (Levels 1 and 2) 233
        5.7.1.1 Expectations 233
        5.7.1.2 Social, Ethical, and Regulatory
                Issues 233
    5.7.2 Medium Term: Cooperative Systems (Level 3) 235
        5.7.2.1 Expectations 235
        5.7.2.2 Social, Ethical, and Regulatory
                Issues 236
    5.7.3 Long Term: Autonomous Cars (Level 4) 236
        5.7.3.1 Expectations 236
        5.7.3.2 Social, Ethical, and Regulatory Issues 238
Interview with Bryant Walker Smith (Assistant Professor of
Law, University of South Carolina, United States) 240
References 243

CHAPTER 6 ARMED MILITARY DRONES: THE ETHICS BEHIND
VARIOUS DEGREES OF AUTONOMY 249
6.1 Focus on Teleoperated and Autonomous Armed
    Military Robots 249
    6.1.1 Unarmed Military Robots 250
    6.1.2 Armed Military Robots 251
6.2 Autonomy of Military Robots Is High on the Agenda 254
6.3 Military Robots and International Humanitarian Law 259
    6.3.1 Tele-Led Drones 260
        6.3.1.1 Proportionality Principle 260
        6.3.1.2 Discrimination Principle 262
        6.3.1.3 Targeted Killing 263
    6.3.2 Autonomous Drones 265
6.4 Question of Responsibility 270
    6.4.1 Responsibility of the Manufacturers 271
    6.4.2 Responsibility of the Human Operators 273
    6.4.3 Responsibility of the Commanding Officer 276
6.5 Proliferation and Security 277
6.6 Concluding Remarks 280
    6.6.1 Social and Ethical Issues 281
        6.6.1.1 Proliferation and Abuse 281
        6.6.1.2 Counterproductive Nature 281
        6.6.1.3 Humanization versus
                Dehumanization 282
        6.6.1.4 Autonomy 282
    6.6.2 Regulation 283
        6.6.2.1 Work on an International Ban on
                Autonomous Armed Robots 283
        6.6.2.2 Curbing the Proliferation of
                Armed Military Robots 284

6.6.2.3   Broad International Debate on the
Consequences of Military Robotics   285
Interview with Jürgen Altmann (Physicist and Peace
Researcher at TU Dortmund University, Germany)   285
References   290

CHAPTER 7   AUTOMATION FROM LOVE TO WAR   297
7.1   Future Expectations and Technical Possibilities   298
7.1.1   Influential Strong AI Pipe Dream   299
7.1.2   Successful and Pragmatic Weak AI Approach   300
7.1.3   Exploring Artificial Social Intelligence   303
7.1.4   Exploring Artificial Moral Intelligence   304
7.2   Expected Social Gains   306
7.3   Robots as Information Technology   310
7.3.1   Monitoring and Privacy   310
7.3.2   Safety, Cyber Security, and Misuse   313
7.4   The Lifelike Appearance of Social Robots   314
7.5   Degree of Autonomy of Robots   317
7.5.1   Systems View on Responsibility and Liability   318
7.5.2   Man-in-the-Loop   318
7.5.3   Man-on-the-Loop   320
7.5.4   Man-out-of-the-Loop   321
7.6   Robot Systems as Dehumanizing Systems   322
7.6.1   Undermining Human Dignity   323
7.6.2   Undermining Human Sustainability   324
7.6.3   Current Relevance   326
7.7   Governance of Robotics in Society   328
7.7.1   Putting Users at the Center   328
7.7.2   Political and Regulatory Issues   329
7.7.3   Balancing Precaution and Proaction   332
7.8   Epilogue   333
References   336

INDEX   341

Foreword

American military drones fly above Afghanistan in search of terrorists. At the same moment, a robot with artificial lips plays the trumpet in a beautiful way. And in Europe, researchers are working on the
development of humanoid robots that are able to wash the behinds of
elderly people. Robots are thus no longer only used in factories but are
also rapidly becoming an integral part of our daily lives. Think about
human activities such as caring for the elderly and driving cars and
also about having sex and killing people. This book illustrates that
this new robotics is about literal automation from love to war. This is
driven by the ultimate engineering dream: developing an autonomous
and socially and morally capable machine.

Robotization, however, is not just about social humanoid robots
but is especially about the rise of all kinds of robotic systems in our
society. The Japanese Ministry of Economy, Trade and Industry
(METI) predicts the emergence of a *neo-mechatronic society*, in which
robots will routinely provide a number of services such as cleaning,
guarding buildings, providing recreational facilities, and caring for
the elderly. The United States foresees a development *from the Internet
to robotics* and strives for a leading position in the development of co-
robots: smart robotic systems that can cooperate with humans and
support them with tasks in health care, agriculture, energy, defense,

and space travel. The European Commission also has a lot of enthusiasm for the future of robotics and invests heavily in it.

Robotics will make our lives more pleasant. Telecare via domotics and care robots will enable people to live independently for a longer period of time. The robotization of car mobility will make our road traffic safer. And robots will gradually take over a lot of our current dirty, dangerous, and dull work. Robots are already used for dismantling explosives and will, according to some, eventually be used in the sex industry as a technological alternative for the often humiliating conditions many prostitutes find themselves in nowadays.

The new robotics will also make things more difficult, because we are forced to think, debate, and form an opinion about the many political, ethical, philosophical, judicial, and social issues that the rapid developments in the field of robotics raise. Are we capable of capturing the innovation opportunities offered by robotics? Have we thought about how to really shape that innovation in a responsible manner? How can we create the conditions for public trust in these new technologies? When can we tell is the best time to remove the legal barriers that hamper the introduction of beneficial robotic systems.

Have we already thought about the question of if and when we are morally obliged to use robots? Do we have a moral duty to make our traffic systems as safe as possible by means of the available robotic technologies? Does the use of tele-led armed military drones increase the emotional and, therefore, moral distance between the actions of drone operators and the ethical implications of those actions? Proponents think these robots might lead to less psychological suffering among military personnel, and eventually even to more rational decisions being made. Critics are afraid of the words of a young cubicle warrior who says about his job: "It's like a videogame. It can get a little bloodthirsty. But it's frickin' cool." A core challenge is therefore to prevent the potentially dehumanizing effects of robot systems.

Just Ordinary Robots: Automation from Love to War examines the social significance of the new generation of five types of robots: the home robot, the care robot, police and private drones, the car robot, and the military robot. The starting point is that innovation is only about developing technology. The challenge is to perceive and anticipate the chances and risks related to the new robotics in a timely way, because in the end, we humans have to decide how to shape the automation from love to war.

This book is the result of many years of research by the Rathenau Instituut, the Netherlands' key research and debating center for science, technology, and society. In 2012, this research led to the publication of the Dutch book *Overal Robots* (*Robots Everywhere*), written by Lambèr Royakkers, Floortje Daemen, and Rinie van Est (2012). This book is an updated and drastically revised version of that book.

Frans Brom
Head of Technology Assessment
Rathenau Instituut
The Hague, the Netherlands

Acknowledgments

The authors thank Floortje Daemen, the coauthor of the Dutch book *Overal Robots (Robots Everywhere)*. The authors also thank Gaston Dorren for conducting and writing the interviews, Beverley Sykes for the proofreading of the book, and the experts who have been interviewed: Jürgen Altmann, Kerstin Dautenhahn, Luciano Floridi, Hans Rietman, Bryant Walker Smith, and Mark Wiebes.

Authors

Lambèr Royakkers is an associate professor in ethics and technology at the Department School of Innovation Sciences of the Eindhoven University of Technology, Eindhoven, the Netherlands. He has studied mathematics, philosophy, and law. In 1996, he obtained his PhD on the logic of legal norms. During the past few years, he has done research and published in the following areas: military ethics, robo-ethics, deontic logic, and the moral responsibility in research networks. He was the project leader of the research program "Moral Fitness of Military Personnel in a Networked Operational Environment" (2009–2014) from the Netherlands Organisation for Scientific Research (NWO). His research has an interdisciplinary character and focuses on the interface between ethics, law, and technology. He is also involved in a European project, as chairman of the Ethics Advisory Board of the FP7-Project SUBCOP (SUicide Bomber COunteraction and Prevention, 2013–2016). Royakkers has authored and coauthored more than 10 books, including *Ethics, Engineering and Technology* (Wiley-Blackwell, 2011) and *Moral Responsibility and the Problem of Many Hands* (Taylor & Francis Group, 2015).

Rinie van Est is a research coordinator and "trendcatcher" with the Rathenau Instituut's Technology Assessment (TA) division, The Hague, the Netherlands. The Rathenau Instituut is Netherlands' key research and debating center for science, technology, and society. He has a background in applied physics and political science. At the Rathenau Instituut, he is primarily concerned with projects in the fields of sustainability and emerging technologies such as nanotechnology, cognitive sciences, persuasive technology, robotics, and synthetic biology. He has many years of hands-on experience with designing and applying methods to involve experts, stakeholders, and citizens in social debates on science and technology. In addition to his work for the Rathenau Instituut, he lectures on technology assessment and foresight at the School of Innovation Sciences of the Eindhoven University of Technology. Some of the studies he recently contributed to are "Check In/Check Out: The Public Space as an Internet of Things" (NAi Uitgevers, 2011), "European Governance Challenges in Bio-Engineering—Making Perfect Life: Bio-Engineering (in) the 21st Century" (STOA, 2012), "Energy in 2030: Busting the Myths" (Rathenau Instituut, 2013), "From Bio to NBIC: From Medical Practice to Daily Life" (Rathenau Instituut, 2014), "Intimate Technology: The Battle for Our Body and Behavior" (Rathenau Instituut, 2014), and "Sincere Support: The Rise of the E-coach" (Rathenau Instituut, 2015).

1

ROBOTS EVERYWHERE

The movie *I, Robot* starts with a fascinating scene. The camera slowly zooms in on a busy shopping street in Chicago in the year 2035. Slowly you realize that robots are walking among the shoppers. They are humanoid machines that provide public services. The robots take dogs out for walks, collect trash, and have a relaxed chat with the neighbors. These intelligent machines seem totally integrated into society. They are trusted. The basis of that public trust lies in the fact that robots act according to the three laws of Asimov (see Box 1.1), who wrote a number of short science fiction stories in 1950 under the title *I, Robot*. These robots are not just smart and handy, they are also moral machines. And then, of course, suddenly something goes very wrong.

In the meantime, the idea that robots can provide useful services in all sorts of places in our society is no longer science fiction. In recent decades, the robot has significantly changed the way goods are manufactured and thus factory work. However, according to many robotic experts, companies, and governments, the time has come for robot technology to be applied outside the factory. Bill Gates of Microsoft expects, for example, that in 2025 every household will be equipped with a smart mobile device (Gates, 2007). Automation is no longer limited to production processes. We are increasingly automating housekeeping, entertainment, transportation, caring for others, and waging war. Undoubtedly, this new robot revolution will have a major impact on our society. The first signs of this are already visible. Think of unmanned robotic aircraft, known as drones, deployed by the United States in the war in Afghanistan in order to detect and attack Taliban fighters. In his book *Wired for War*, Peter Singer (2009) sees this robotization of the army as the new "revolution in military affairs." A golden future is also predicted for robots used in medical care. A great deal of experimenting has been done with *hugging robots* that promise to sweeten the life of senile dementia victims in nursing homes. The Japanese believe strongly in a future in which humanoid robots,

BOX 1.1   ASIMOV'S THREE MORAL ROBOT LAWS (1950)

First Law

A robot may not injure a human being or, through inaction, allow a human being to come to harm.

Second Law

A robot must obey the orders given to it by human beings, except where such orders would be in conflict with the First Law.

Third Law

A robot must protect its own existence as long as such protection does not conflict with the First or Second Law.

designed to look and act like a human, will play an important role in caring for the elderly (Lau, Van't Hof, & Van Est, 2009).

These kinds of visions and developments often elicit many different emotional responses. We seem to accept the use of robots that perform dull, dirty, and dangerous jobs. But how far do we want to go with the automation of care for children and the elderly, of killing terrorists, or of making love? Would we not be setting human beings and our humanity aside if we accepted such automation? This book attempts to provide a socially involved but at the same time sober insight into the new robotics. Which robot technologies are coming? What are they capable of? And which ethical and regulatory questions will they consequently raise? We examine the social significance of the new generation of five types of robot: the home robot, the care robot, police and private drones, the car robot, and the military robot. We begin our quest close to home and end far from it, namely, on the battlefield.

This introduction first describes the emergence of the vision that robots will be broadly deployed in our society. We then identify some key technical characteristics of robotics. What kinds of machines are robots? What kinds of technologies are we discussing when we talk about robots, in particular social robots? The modern robot is used in social practices and thereby changes those practices. In order to provide the reader with a kind of compass for the following chapters, we identify four typical elements that influence how robotics intervenes in social practices: (1) robots as information technologies, (2) the life-like appearance of robots, (3) the level of autonomy of robots, and (4) robotization as rationalization. These four key characteristics of modern robots give rise to various social and ethical issues and thus raise questions of social acceptance. These four elements provide the reader thematic tools for reading the following five chapters.

1.1 With Vision

> Robots are at the dawn of a new era, turning them into ubiquitous helpers to improve our quality of life by delivering efficient services in our homes, offices, and public places.
>
> **European Robotics Technology Platform (EUROP) (2009, p. 7)**

Just after the World War II, it was not only writers like Asimov who fantasized about robots playing a role in many aspects of society. This was also true of the physicist Joseph F. Engelberger. In 1956, this "father of robotics" set up the first robot company. He developed the first industrial robot, the *Unimate* (see Figure 1.1), that is, the universal helper of

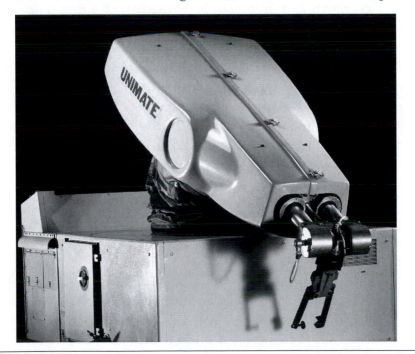

Figure 1.1 Unimate, the first industrial robot. (Photo courtesy of Science & Society Picture Library/Getty Images.)

humankind. This first industrial assembly robot was given a place in General Motors' automobile factory. Until now, the automotive industry has been the largest market for robots: more than half of all tasks have been automated. And the end is not yet in sight. For a long time, robots were primarily used for spot welding and spray painting. During the 1980s, robots were increasingly used not only to transport material (loading and unloading) within automobile factories but also for other industrial tasks. In addition, robots were developed for cleaning up environmental pollution and for exploration in outer space and the deep sea.

Robot technology progressively masters more and more complex operations. This is made possible by improved visibility (via 3D vision systems), better navigation and mobility, better voice recognition, and smarter interaction with people. The penetration of robots into the factory has always been highest in Japan (Mitchell et al., 2012, pp. 3–4). In 1980, the robot density, that is, the number of industrial robots per 10,000 manufacturing employees, was 10. In 2010, the robot density was above 300. Since the beginning of this century, various fully automated factories that require no human presence on site have been built in various parts of the world. For example, the Japanese robotics firm FANUC operates a so-called lights-out factory where robots are building 50 robots per day and can run unsupervised for a month. The deployment of robots for more complex operations has major consequences for employment and for the organization of the production process and the position of the people involved in it. In the 1970s, automation was a very sensitive topic of negotiations between employers and unions. Although it has been off the radar for decades, worries about the impact of computerization and robotics on employment have again become an important political issue. Opinions are heavily divided on this topic. The International Federation of Robotics (IFR) expects that robots will create 2–3.5 million new jobs worldwide between 2012 and 2020 (Gorle & Clive, 2013, p. 3). Guided by this vision, the European Union and the United States are banking on reindustrialization: robotics will bring manufacturing back to Europe and the United States. Studies by Frey and Osborne (2013) and Elliott (2014), however, picture a different scenario. Frey and Osborne analyzed what smart machines will be capable of doing in future and compared this to the human skills that jobs today require.

They found that in the next two decades almost half of all current jobs in the United States could potentially be replaced by computers or robots.

In the 1960s, Engelberger had already called for the use of robotic technology outside the factory, such as in services and health care. To demonstrate the capabilities of the Unimate to a wide audience, Engelberger appeared on television in 1966 with his robot friend and showed that it could also pour beer and conduct a band. It was only in the middle of the 1990s that the robot expert community embraced the idea that robots could play a role in offices, schools, hospitals, restaurants, and at home. Robots would become a mass product and we, ordinary people, the end user. People spoke of *service robots* or *serve-us robots*, and they were defined as "Machines that sense, think, and act to benefit or extend human capabilities and to increase human productivity" (Pransky, 1996, p. 4).

Meanwhile, the idea that in future robotics will play a central role in all spheres of our society has been vigorously propagated by many robot researchers and companies. They have been able to point to the significant technological advances achieved in recent years in the field of robotics. This has led to various applications in the areas of difficult production methods, logistics, and health care. In 2009, the cream of U.S. universities proposed a vision for the future of robotics, entitled *From Internet to Robotics* (Christensen, 2009). In this road map, robotics is seen as "one of the few technologies capable in the near term of building new companies and creating new jobs" (Christensen, 2009). The Obama Administration has taken this advice to heart and launched the *National Robotics Initiative* (NRI) in June 2011. The United States wants a leading position in the development of the latest generation of robots, the so-called collaborative robots or co-robots. This refers to intelligent robot systems capable of working together intensively with people in the fields of health, agriculture, the exploration of raw materials, defense, and aerospace. In the long term, robotic technology will become crucial in assisting the aging U.S. population. Moreover, according to the *European Robotics Technology Platform* (EUROP), their industry is entering an important phase. This platform predicts that from 2020 on, the robot market will rapidly turn into a consumer market (EUROP, 2009, p. 35). The European Commission, in the words of Commissioner Neelie Kroes (2011), expects this as well.

She claims that Europe is the leader in the field of research on robots and their industrial application. Since 2002, the European Commission research program has been investing heavily in cognitive systems and robotics. At present, around 100 research projects are funded by the FP7-ICT Cognitive Systems and Robotics Challenge.

In addition, various governments also support the vision of a robotic society. The South Korean Ministry of Information expects that by 2020 each South Korean household will own a robot. The Japanese government regards the application of robotic technologies in various social environments as an important growth market. The Ministry of Economy, Trade and Industry (METI) even foresees the rise of the so-called Neo Mechatronic Society (Lau et al., 2009). A major driving force behind this vision is the sharp drop in prices of all types of robotic technologies. METI predicts a home market for robot technology of ¥6000 billion (about U.S. $50 billion) in 2025. It is expected that robots will play a key role in combating the labor shortages in care as a result of an aging population. It is for this reason that Japan invests generously in the development of robotics for health care. The United States mainly invests in robotics for aerospace and the military.

1.2 Technically Speaking

The word *robot* was coined by the brother of the writer Karel Čapek. It stems from the Czech word *robota*, meaning "servitude" or "forced labor." Čapek's 1920 stage play *R.U.R.* (*Rossum's Universal Robot*) begins with a factory where artificial humans are created, the so-called robots. The artificial (re)production of living beings—humans and animals—has been continually imagined by philosophers and artisans since antiquity.

1.2.1 From Automata to Robots

The quest for artificial life emerged strongly in the eighteenth century. This effort focused on building automata. An automaton is a self-operating machine able to perform certain tasks without outside help. Automata are purely mechanical and often mimic animals or humans. They were built as luxury toys and as amusements for the nobility and the general public. The two most famous automaton makers may very well have been the Frenchman Jacques de Vaucanson and the Swiss-born watchmaker Pierre Jaquet-Droz. Vaucanson produced, among other things, a mechanical grocer and a life-size mechanical duck that was even equipped with a sort of digestive tract. Jaquet-Droz was famous for the three mechanical dolls he created from 1760 onward: the "artist" who could draw, the "musician" who could play a type of harpsichord, and the "writer" who could write with a pen (see Figure 1.2).

Such automata had an important symbolic and intellectual value. They showed the technical ingenuity of that time and also offered insight into a new way of manufacturing. In this way, they created the technical and conceptual foundation for the emerging Industrial Revolution. Toy attractions like these also played a central role in the intellectual debate at that time. Riskin sees, for example, that "Vaucanson's automata were philosophical experiments, attempts to discern which aspects of living creatures could be reproduced in machinery, and to what degree, and what such reproductions might reveal about their natural subjects" (Riskin, 2003, p. 601). The desire

Figure 1.2 Automaton of the "writer." (Photo courtesy of Bridgeman Art Library Ltd./Hollandse Hoogte.)

to recreate life by mechanical means thus lies at the heart of the Industrial Revolution. However, after 1760, with the advent of the Industrial Revolution, interest in the construction of automata and building artificial life gradually decreased. Since World War II, this interest has returned in full force, although it is no longer about the construction of mechanical automata, but the construction of intelligent electromechanical machines, the so-called robots.

The desire to build intelligent lifelike machines forms the basis of the current information revolution. This technological revolution is driven by a new engineering-led vision of life that emerged strongly after the World War II: life as an information system (Van Est, 2010). This is cybernetics, which suggests that both organic and mechanical processes can be controlled by feedback of information. This engineering vision considers and describes living and nonliving processes as equal. The founder of cybernetics, Norbert Wiener (1948), described both animals and machines as information processing systems that always carry out a particular purpose, check whether their actions have led to the desired result, and then adapt their behavior to achieve the required outcome. On the one hand, in this way, biological, cognitive, and social processes can be described in digital terms. On the other hand, machines can be built to act deliberately and show characteristics that we usually associate with living systems. Just like the mechanical automata of the eighteenth century, today's intelligent robots try to technically imitate various aspects of life. Robotics has, therefore, developed in close cooperation with cybernetics. In particular, the areas of artificial intelligence (AI) and human–machine interaction play a central role.

1.2.2 Robot-Friendly Environments

Robotics is an engineering field that deals with the development, design, creation, and operation of robots. To this end, robot engineers make use of a broad set of areas of expertise, such as electronics, computer science, AI, and mechatronics. The core of robotics is the integration of this broad field of technologies. Some technical skills are highly specific for robotics, with important key areas including navigation, perception, movement, and manipulation. Other technologies, such as batteries and fast processors, are also important in a number of other domains. The robot is consequently not only a mechanical form of technology but also an electronic technology and information technology (IT), and so forth.

The International Organization for Standardization (ISO) defines a conventional industrial robot as an "automatically controlled, reprogrammable, multipurpose, manipulator programmable in three or more axes, which may be either fixed in place or mobile for use in industrial automation applications" (ISO 8373:2012). In this book, we will look at robots employed outside the factory. These are generally described as service robots, which are robots that perform "useful tasks for humans or equipment excluding industrial automation applications" (ISO 8373:2012).

However, for our exploration of new robotics, this definition is insufficient. From the outset, the International Federation of Robotics (IFR) has defined the service robot as a machine that is useful to humans. This book is specifically searching for the social advantages and disadvantages of the service robot. That is why we are mentioning a number of the chief characteristics of the new robotics. They pertain to intelligent (usually networked) machines that perform physical actions with a certain degree of autonomy within a complex and, to a greater or lesser extent, unstructured environment and a dynamic social practice. This implies, among other things, that the interaction between environment and machine, and man and machine, plays an increasingly important role. To make this interaction possible, the robot employs sensors with which it can perceive the environment and human beings.

A major difference between industrial and service robots thus concerns the environment in which they must operate. The industrial robot usually works in a highly structured environment. Within this structured environment, robotics and humans are often strictly separated, and the people who interact with the robots are often specifically trained for this task. In this case, robotic operations can often be totally planned in advance, or, to put it more succinctly, preprogrammed. However, for service robots this is not usually possible. Their tasks are physical, in an ever-changing and much less structured environment, and they often interact with people to fully carry out their work. The use of robots in such environments therefore requires the strong rationalization of the environment. Think about the vacuum cleaner robot that cannot operate properly in a messy room and only comes into its own in a tidy room. Another route to the employment of service robots in a complex environment is the enhancement of the robot's learning and intelligent behavior. And this AI, according to Trevelyan (1999), is precisely what is characteristic of the twenty-first-century robot. He, therefore, defines robots as "intelligent machines that enhance human capabilities." Indeed, it is no longer only about cognitive intelligence. For interaction with ordinary people, robots should also possess social and emotional skills.

However, modern robots possess not only a robot brain (computer hardware and software) but also a robot body. There is a wide diversity of robots, both with regard to their behavioral repertoire, such as autonomy (robot brain) and appearance (robot body). In the following section, we want to do justice to the fact that intelligent robots exist in all shapes and sizes and have different cognitive and social skills. A robot is in fact a modular unit. What a robot is and can do is therefore determined by the capabilities of its "body" and "brain." The modern robot is not usually a self-sufficient system. In order to understand the possibilities and impossibilities of the new robotics, it is important to realize that the service robot is usually supported by a network of information technologies, as is, for example, the Internet. Thus, this implies, in particular, networked robots.

1.2.3 The Robot Body

We will briefly comment on three features of the robot body: the physical appearance, physical handling, and observation capabilities of the robot. These characteristics are often interrelated. A certain appearance, for example, two legs, makes a particular action, such as walking, possible and other actions impossible or very difficult, such as flying.

1.2.3.1 Appearance Is Important The physical appearance of the robot depends largely on its function. The robot can look like a machine. One speaks then of *mechanoids* (Walters et al., 2008). Just as with eighteenth-century automata, modern robots can also look like humans or animals. These animal-like and human-like robots

Figure 1.3   The android robot built on his own image of robotic scientist Ishiguro. (Photo courtesy of Rinie van Est.)

are called *humanoids*. The technical argument for designing robots to look like people is that such robots may well operate in human environments that are optimized for human use. One assumes that appearance is important for interaction between humans and robots. To investigate this, Japanese robot scientist Ishiguro builds humanoid robots as lifelike as possible (Minato, Shimada, Itakura, Lee, & Ishiguro, 2006) (see Figure 1.3). Humanoid robots that are built to aesthetically resemble humans are called *androids*.

The "uncanny valley" theory of Japanese robotic scientist Masahiro Mori (1970) has played an important role since the 1970s in thinking about the interaction between robots and humans (see Figure 1.4). Mechanoids elicit little emotional reaction in people. But the more a robot looks like a person or an animal, the more positive and empathetic feelings it will evoke in people. If robots resemble people very strongly, but their behavior is not human enough, then Mori predicts a strong sense of unease. In this case, the appearance is human-like, but there is very little familiarity. This is what Mori calls the *uncanny valley*. Mori recommends avoiding this valley by building

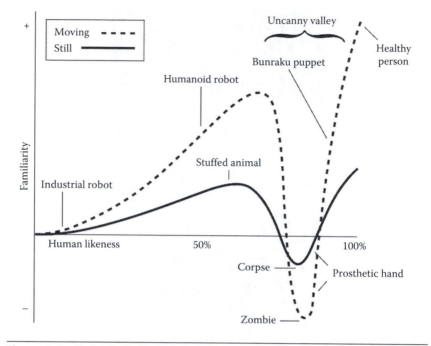

Figure 1.4 Mori's classic illustration of the "uncanny valley." (From MacDorman, K.F. and Ishiguro, H., *Interact. Studies, 7*, 297, 2006.)

robots that do not resemble people or animals too much, but are still human-like or animal-like in behavior. *Paro* is a good example of this. Paro is a well-known pet robot that looks like a baby seal. Initially, the Japanese engineers wanted to develop a cat robot. Test subjects reacted negatively to it, probably because a "cat" elicits a clear expectation that the robot cat did not deliver. With the seal robot, one did not come up against the uncanny valley effect. A second way to avoid the uncanny valley is to build robots that are so similar to humans (or animals) in appearance and behavior that they are indistinguishable.

1.2.3.2 Opportunities for Physical Activity Possibilities for physical activity are often determined by the robot's shape or physical body. We saw that the industrial robot is often a nonmobile robot. In addition, there are numerous mobile robots. Consider moving robots, robotic craft, and flying robots, such as unmanned drones (see Chapter 6) deployed by the U.S. military in the wars in Iraq and Afghanistan. Another example concerns humanoid robots, such as Honda's *Asimo* (see Figure 1.5) and Toyota's *Partner Robot* that

Figure 1.5   Humanoid robot Asimo. (Photo courtesy of Bart van Overbeeke.)

can walk at 6–7 km/h (or 3–4 mph). Or animal robots such as the four-legged *BigDog*, created in 2005 by Boston Dynamics in collaboration with the National Aeronautics and Space Administration (NASA) and Harvard University. In addition to moving, there are many other physical actions that robots can perform. The *RIBA II* care robot, developed by the Japanese RIKEN research institute, can lift patients weighing up to 80 pounds from the floor into a bed or a wheelchair. One important technical challenge concerns the energy source of mobile robots. The *Roomba* is a vacuum cleaner robot that goes looking for its own recharger when its battery begins to get low. In the United States, Robotic Technology Inc. and Cyclone Power Technologies Inc. developed the EATR (*Energetically Autonomous Robot*), which can look for food (biomass) on its own, and from this can create biofuel for its own energy needs.

1.2.3.3 Artificial Senses   People have five senses: ears to listen with, eyes to see with, skin to feel with, a nose to smell with, and a tongue to taste with. Robots can also be fitted with all kinds of artificial senses, or rather sensors. Think of electronic noses and taste sensors. Cameras with light sensors are used for facial and emotion recognition. The perception of robots can outperform human perception by a long way.

Some unmanned military aircraft, so-called drones, use infrared cameras for observation at night and use radar to be able to look through clouds. Researchers want to improve surgical robots by applying touch sensors. In this case, the robot communicates about the surgical procedure the surgeon is performing by exerting force on his hands. One speaks of *haptic feedback* or *haptic perception* (i.e., perception through the sense of touch).

1.2.4 The Robot Brain

> Artificial Intelligence is the science of making machines do things that would require intelligence if done by men.

Minsky (1968, p. v)

A robot is an IT-containing computer hardware and software. The robot contains no human intelligence, but AI. This AI determines the behavioral repertoire of the robot and its cognitive, social, and moral capabilities (Böhle, Coenen, Decker, & Rader, 2011). The assumption is that emotional intelligence, social behavior, and dynamic interaction with the environment are prerequisites for the individual and social behavior of robots in complex social practices.

1.2.4.1 Strong and Weak Artificial Intelligence   In the 1950s, the idea arose that all forms of intelligence and learning could be so precisely described that a machine would be able to mimic them. Some thought that human intelligence could be completely understood with the help of computers and that it is possible to make machines that act like humans and can think, reason, play chess, and show emotions. This attitude is called the *strong AI* thesis. In this vision, machines are ultimately smarter and morally more sensitive than humans. Supporters of the *weak AI* synthesis see computers as a tool in the study of the mind. They expect that machines can perform specific "intelligent" tasks to assist human users. Although the *weak AI* concept has the most followers, the *strong AI* vision receives the most media attention, partly because it has some very outspoken advocates, such as Marvin Minsky (1968), Hans Moravec (1988), and Ray Kurzweil (1990, 2005).

Minsky has been one of the main advocates of AI from its beginning (Noble, 1997). He was already at Dartmouth College in 1956
when the first meeting in the AI field took place. Minsky suggested
it is possible to build intelligent machines, because brains themselves
are machines. According to Minsky, steps in that direction have been
taken by machines that are able to look up information, recognize
patterns, have expert knowledge, and prove mathematical theorems.
He also thought about the advent of robotics. Minsky foresaw a fusion
between man and machine in the distant future. According to him,
thinking machines represent the next step in evolution. The *machina
sapiens* is a new species that will eventually surpass *Homo sapiens*. AI
was, therefore, seen as the ultimate turning point in human evolution.
The main contemporary spokesperson on this theme is Raymond
Kurzweil. He is a pioneer in the field of speech recognition, and the
inventor, in 1976, of a device that turned text into voice for the blind
reader. In his book and movie *The Singularity Is Near* (2005), he suggests that science and technology are developing exponentially. This
will inevitably lead, he believes, to a point at which AI will surpass
human intelligence. Vernor Vinge (1993) calls that moment "singularity." Kurzweil thinks that we will achieve this technical and cultural turning point before the middle of this century.

1.2.4.2 Predictions from the Past   Let us return to predictions from
the 1950s and 1960s. Norbert Wiener believed that computers would
come to play an important role in the production process, and spoke
of a forthcoming second Industrial Revolution (Umpleby, 2008). But
in addition to the use of AI for industrial tasks, all sorts of creative
and social tasks were foreseen for AI. Alan Turing thought that computers would be able to communicate with people, and invented the
so-called Turing test. In it, a person sends questions to both another
person and a computer located in another room. On the basis of their
replies, the interrogator must determine whether he or she is communicating with a human or a machine. Turing predicted that in
50 years (thus around now), computers would master this question-
and-answer game so well that the questioner would have a less than
70% chance of distinguishing, within 5 minutes, the computer from
the person. Marvin Minsky stated in 1958: "Our mind-engineering
skills could grow to the point of enabling us to construct artificial

accomplished scientists, artists, composers, and personal companions" (cited in Noble, 1997, p. 157). Simon and Newell (1958) predicted in the same year that within 10 years a computer would beat the world chess champion, discover and prove important new mathematical theorems, and compose beautiful music.

1.2.4.3 Through Trial and Error However, none of this went very quickly or smoothly. Over the years, the development of AI has witnessed many ups and downs. In this process, changes in thinking about AI go hand in hand with advances in brain and cognitive science (Böhle et al., 2011, pp. 129–132). In the 1960s, the AI community assumed that any form of intelligence could be mimicked by a computer code, for example. However, they slowly but surely ran into the counterintuitive situation in which computers had relatively little difficulty in solving geometric problems that are very difficult for most people. In contrast, the computer appeared to experience great difficulty in matters that are trivial for humans, such as recognizing faces. This setback led to a sharp decline in interest, in particular on the part of the U.S. government, in stimulating AI research. All of the 1970s were characterized by this so-called AI winter.

In the early 1980s, the emergence of expert systems led to new, high expectations. This type of system is based on the idea that experts make decisions based on a set of clear rules. At the same time, expert systems are dependent on large databases full of things that are common knowledge for people, such as words from various languages or the names of famous people. In the mid-1980s, neural networks became popular. Since the beginning of the 1990s, there have been systems on the market that can recognize characters and voices using neural networks. Such networks, however, need be trained. Skills are learned through reward and punishment. In this type of reinforcement learning, the robot is rewarded with points, and it is programmed so that it strives for as many points as possible. Robot researchers often have a hard time designing the appropriate reward and punishment system of robots. At Delft Technical University (TU Delft), they tried to teach a two-legged robot called *Leo* to walk (see Figure 1.6). Initially, the researchers punished Leo when he fell over. This meant, however, that Leo learned not to fall (Schuitema, Wisse, Ramakers, & Jonker, 2010), which he did by putting one leg on his neck. A reward for

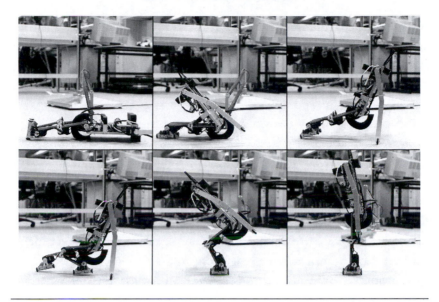

Figure 1.6   Leo, the two-legged robot developed by TU Delft. (Photo courtesy of Delft University of Technology, the Netherlands.)

good walking behavior appeared to work better, but led to all kinds of strange ways of walking. After a long time, the researchers discovered that when one rewards the robot for efficient energy use, it starts to learn the human way of walking.

Another important AI product is the "intelligent agent." This is "a computer system that is capable of flexible autonomous action in dynamic, unpredictable, typically multi-agent domains" (Luck, McBurney, Shehory, & Willmott, 2005, p. 11). These are computer programs that can "observe" their environment and can autonomously calculate and thus affect their environment. This approach is seen as an important new paradigm for software development. In the late 1980s, a new AI approach emerged, the situated or "embodied" AI. It assumed that intelligence is built from the ground up by trial and error and that, in addition, the computer really needs a "body" to actually get to know the world. This AI approach thus provides an additional motivation to build a robot.

1.2.4.4 Brute Computational Power   What has become of the expectations of the very first AI experts? The second Industrial Revolution that Wiener predicted has come and still rages on with the rise of the industrial robot. In the meantime, a computer has also beaten the world chess champion. That did not happen within the 10-year time span predicted by Simon and Newell (1958), but took nearly 40 years. In particular, the lack of computing power played tricks on AI for quite a while. A decade-long exponential increase in the speed and processing power of computers, the so-called *Moore's Law*, now makes much possible on the basis of brute computational power. In March 1997, the Deep Blue II chess computer defeated the then world chess champion Gary Kasparov. The match of six games ended with a score of 3.5–2.5. The chess Grandmaster complained afterward that there had been people playing the game. According to Kasparov, one particular move was clearly too stupid for a computer, while another had been too creative. Other chess computers, however, proved able to produce similar moves. One year later, Professor David Cope showed a program that analyzes the musical style of old masters such as Bach and Stravinsky.* On this basis, the program, EMI (Experiments in Musical Intelligence), creates synthetic classical music. A symphony in the style of Mozart, entitled *Austrian Composer's 41st*, has already been performed. Only real connoisseurs could distinguish the EMI symphony from a real Mozart piece. What is perhaps an even stronger example of AI or creativity was created in 2009. The universities of Aberystwyth and Cambridge designed an artificial scientist. The scientific Adam robot was the first robot to independently discover a number of new scientific findings. This robot discovered a gene in yeast that had been hunted for by researchers for decades (Ravilious, 2009). In the same year, researchers from Cornell University developed a computer program that deduced Newton's laws of motion from the motion of a pendulum (Keim, 2009). It is hoped that in the future, computers will be capable of discovering laws of nature that are as yet unknown.

1.2.4.5 Artificial Social Intelligence   Earlier, we indicated that robot experts want to develop robots with a human-like body, because our physical environment is adapted to our human dimensions. With respect to the robot's behavior, an identical argument has been in use since the mid-1990s. It is suggested that if robots are to operate in human environments, it is important that these machines are

---

* http://www.computerhistory.org/atchm/algorithmic-music-david-cope-and-emi/.

so programmed that they are able to interact socially with people and that they can also act in a moral way. The study of interactions between humans and robots is called *human–robot interaction* (HRI). In this multidisciplinary field of research, we see the emergence of concepts such as *social robots, artificial (robotic) companions,* and *artificial moral agents.*

According to sociologist Sherry Turkle (2011), many of us have been philosophically and emotionally prepared by now to "seriously consider machines as potential friends, confidants and even romantic partners" (p. 9). She claims that we stand on the verge of considering artificial human companions as a normal part of our lives, and coined this the "robotic moment" in human history. The technologies that are built to interact with humans via humans' social rules are referred to as social robots or artificial (robotic) companions. These machines are embodied artificial agents, either as virtual robots (avatars) or as physical robots. Floridi (2014) sees artificial companions evolving in various directions. Like pets, he believes that artificial companions will address social needs and the human desire for emotional bonds and playful interactions. Artificial companions will also provide information-based services in various contexts, such as communication, entertainment, education, training, health, and safety (pp. 154–156). Researchers also aim to develop artificial agents that are able to deal with human emotions. Floridi expects that artificial companions will act as "memory stewards," creating and managing a repository of information about their owners, even to the point that they—based on long-term life-logging—will be able to simulate a person.

Social robotics is still in an early phase of development. Fong, Nourbakhsh, and Dautenhahn (2003) define three classes of social robots or goals for social robot research and development: socially situated, socially embedded, and socially intelligent (p. 145). Socially situated robots can perceive and react to a certain social environment. For example, they are able to distinguish between other social agents and various objects in that environment. Socially embedded robots are structurally linked to a certain social environment. They are able to interact with agents and humans and are at least partially aware of human interactional structures (e.g., taking turns). These first two goals also play a central role in the so-called ambient intelligence

(AmI) vision.* AmI literally means that people are surrounded by intelligent equipment. A smart environment not only knows that people are present but also who is present and what characteristics, needs, emotions, and intentions they have. The founding fathers of AmI are Aarts and Marzano (2003) from Philips. At the beginning of this century, this vision was a way to shape the R&D agenda of Philips and the European Commission. Today, many world-leading IT firms, such as Microsoft, have embraced it and proclaimed that the "era of ambient intelligence" has begun (Sandoval, 2014).

The ultimate goal of social robotics is to develop socially intelligent robots, that is, "robots that show aspects of human style social intelligence, based on deep models of human cognition and social competence" (Fong et al., 2003, p. 145). Human social characteristics that engineers try to incorporate in machines are recognizing faces and emotions, expressing emotions, communicating using high-level dialogue, establishing/maintaining social relationships, using natural cues such as gaze and gestures, exhibiting distinctive personality and character, and having the ability to learn social competencies. Although developing these features will be very challenging, Fong et al. believe that modern technology will make it increasingly possible to interact with robots in a refined manner.

For a long time, we have seen the emergence of the social virtual robot (also called the *softbot*) or the *chatbot*. The chatbot is a so-called intelligent agent. On the IKEA website, you can put questions to Anna, the virtual assistant. Among the most developed chatbots are *Cleverbot* and *Eugene Goostman*. It is even claimed that the Cleverbot passed the Turing test during a technology festival in India in 2011 (Aron, 2011). Based on 5-minute online chats, 10 out of 30 judges at the Royal Society in London concluded that the program "Eugene" was a 13-year-old Ukrainian boy (Sample & Hern, 2014). Both computer programs make use of earlier answers to similar questions from people that can be found on the Internet when generating their answers. To get an impression of the state of this technology, it is instructive to watch a video on YouTube, in which a conversation between two Cleverbots can be seen (Labutov, Yosinski, & Lipson, 2011). The conversation is realistic, but at the same time it is also

---

* The resemblance between the abbreviations AmI and AI is not a coincidence.

very strange. It seems as if we are dealing with a kind of uncanny valley, in the sense that the language seems real, but not the social interaction. The conversation between these two computers is, however, not at all creepy, but is rather humorous.

Social interaction between people concerns not just verbal information, but, more importantly, nonverbal communication as well; think about posture or emotions that can be read from facial expressions. "Affective computing" deals with this area of the human–machine interaction process. According to one of the founders of this field, Rosalind Picard (1995) of MIT, we are dealing with "computing that relates to, arises from, or influences emotions" (p. 1). The goal is that computers learn to recognize human emotions and learn to adapt their behavior on that basis. To this end, affective computing analyzes aspects such as intonation of the voice, gestures that people make, bodily posture, and facial expression. For example, the Dutch company Noldus has developed *FaceReader*, which is used regularly by marketing researchers. This technology uses the *Facial Action Coding System* (FACS). This was developed by Paul Eckman, a renowned psychologist, who as far back as the 1970s suggested that there are six basic human emotions—anger, disgust, fear, happiness, sadness, and surprise—all of which can be read from the face within a millisecond. This coding can also be used so that avatars, softbots, or real robots can show emotions. It is expected that the user friendliness, and thus the acceptance, of such technologies will increase (Picard & Klein, 2002).

1.2.4.6 Artificial Morality The question of whether not only social behavior but also moral behavior can be programmed into computers is currently being discussed. This is quite clearly a very recent scientific field. At the beginning of this introductory chapter, there was a reference to the three Asimov ethical laws that Asimov robots are supposed to comply with. Especially in the field of military robots, there is reflection on the use of robots, which should behave according to international humanitarian law, as defined in the Geneva Convention. Ronald Arkin (2009) assumes that it is possible to develop robots that can make better decisions under combat conditions than human soldiers. He proposes not only that AI is independent of emotions, as it is only based on logic, but at the same time that the use of such moral machinery will limit the number of military casualties to a minimum.

Wallach and Allen (2009) are also in favor of the development of so-called *artificial moral agents*. They argue that "today's information systems are approaching a level of complexity that … requires the systems themselves to make moral decisions—to be programmed with 'ethical subroutines' to borrow a phrase from Star Trek" (2009, p. 4). They suggest that people are unable to oversee all the consequences of very complex interacting software systems, which could therefore possibly lead to catastrophes. For example, the automation of financial transactions, or *robotic trading*, led to the 2010 Flash Crash. On May 6th, the Dow Jones Industrial Average first plunged about 1000 points within minutes—about U.S. $1 trillion in market value vanished—and then recovered those losses in less than 3 minutes. It is telling that there still is no consensus on the exact root of the 2010 Flash Crash (Steiner, 2012, p. 4). As more and more of our world is placed under the control of multiple linked algorithms, it becomes more difficult to supervise or gain insight into who or what is causing undesired catastrophic events. In order to identify such catastrophes at an early stage and to be able to prevent them, Wallach and Allen hold that software bots should exhibit ethical behavior.

However, just as a robot with social intelligence is an engineering dream, a machine that exhibits full moral agency is also a futuristic vision. The idea of a moral robot functions as a spot on the horizon that may help to define pathways from current to more sophisticated technologies. James Moor (2006) defines four categories of machine ethics and/or artificial moral agents: (1) ethical impact agents, (2) implicit ethical agents, (3) explicit ethical agents, and (4) full ethical agents. Since, by its nature, computing technology is normative, all robots can be considered ethical impact agents that can be evaluated for their ethical consequences. At this first level, machine ethics is very close to mainstream computer ethics. At the second level are implicit ethical agents: machines that are designed so that they implicitly promote ethical behavior or at least do not have negative ethical effects. Over the past decade, the interest in addressing values in design has grown. In particular, *value sensitive design* has become a widely accepted approach to design (information) technology that accounts for human values in a comprehensive manner throughout the design process (Friedman, Kahn, & Borning, 2006). The next two levels concern the engineering challenge of incorporating ethical decision making into machines. Until this moment, it is not yet clear to what extent this might be feasible. At the third level come explicit ethical agents, which are machines that have human morality encoded in their software. Finally, Moor wonders whether a machine, like an average adult human, can be a full ethical agent that can make explicit ethical judgments and is generally competent to reasonably justify them.

1.2.5 Networked Robots and Human-Based Computing

In the analysis of robots, all eyes are often on the robot itself. However, the performance of many robots depends heavily on the support of other technologies plus human intelligence. The term *networked robots* indicates that these robots are supported by various information networks, without which they could not function. Military drones above Afghanistan, for instance, make use of 32 global positioning system (GPS) satellites, of which only 4 would be necessary for an unmanned aircraft to determine where it is located. In addition, smart machines often need human intelligence to become really clever systems. As Floridi (2014) points out, "[S]ometimes our ICTs need to understand and interpret what is happening, so they need semantic engines like us to do the job" (p. 146). This recent trend is known as *human-based computing*. We have already described how the Cleverbot generates answers based on earlier answers from people to similar questions that can be found on the Internet. In a similar vein, Google Translate searches through millions of documents that are translated by human translators to come up with a translation within a fraction of a second. Expert systems are dependent on large databases for their operation, and the Internet now provides a great deal of information in many areas, from shopping behavior to human faces. In our exploration of the new robotics, we therefore continue to pay attention to the networks and human effort behind the robot. If we did not do this, the significance of the modern robotics society would be incomprehensible.

The popular, but incomplete, image of the robot as an independent and self-sufficient machine is supported by contemporary visions. But it is probably also a relic of past thinking about the future. In the 1950s and 1960s of the last century, the robot was often depicted as an independently moving machine with legs and sensors. At the end of the 1960s, futurologists Kahn and Wiener (1967, pp. 86–98) thought that up to the year 2000 the influence of computers would primarily be through the automation of mechanical machines. This was true for production, and also, for example, for domestic work. Around the year 1984, the English mechanical engineer Thring already foresaw a robot "having no more feeling than a car, but having a memory for instructions and a limited degree of instructed or built-in adaptability according to the position in which it finds various types of objects. It will operate other more specialized machines, for example, the vacuum cleaner or clothes-washing machine" (quoted in Kahn & Wiener, 1967, p. 94). Kahn and Wiener saw the automation of information processes as the next, more difficult, step. They quote Lipetz (1966), who thought that by means of these developments "the geographical boundaries of traditional information storage and retrieval systems are beginning to evaporate. In their place are beginning to emerge vast networks of compatible communication devices linking users with many specialized and overlapping collections" (quoted in Kahn & Wiener, 1967, pp. 95–96).

In recent decades, however, the automation of machines and information processes has occurred simultaneously and often go hand in hand. This may also be explained from the perspective of the convergence of IT with other technologies. The effect of IT on various other technical fields—also known as *digitizing*—is often referred to by the term *convergence*. The automation of all kinds of production processes requires the convergence of mechanics and electronics, or mechatronics, which is the basis of the industrial robots. The emergence of the Internet depended on the convergence of information and communication technologies, labeled ICT. The high expectations of the new robotics are based on the convergence of the Internet and robotics that is expected during the coming years. The development of the new robotics is thus carried by the Internet, and this also continues to change it.

During the past decade, the Internet has penetrated the whole of society. Three technology trends are responsible for this (Van't Hof, Van Est, & Daemen, 2011, pp. 128–130). The first trend is the rise of digital devices in public spaces, from ATM terminals, cameras, gates, and navigation systems to smartphones. Digital convergence is the second trend. This trend means that the networks behind these devices are increasingly linked to the Internet—the mother of all networks. The third trend is that in recent years the Internet has become available in ever more places, especially through the smartphone. Seen from this perspective, service robots are a new kind of smart device that will populate our world. On the one hand, the new robotics is based on existing networks. Conversely, it thereby changes the nature of these networks. This is clearly expressed by the vision of robotics as developed in the United States in *From Internet to Robotics* (Christensen, 2009). This indicates that robotics builds on and uses the existing ICT infrastructure. But the message is that robotics is a further technical development of the existing information networks. The worldwide web has been extended with robotics, giving the Internet "senses and hands and feet."

1.3 Seen Socially

In this book, we want to investigate the social significance of robotics in the years to come. We do this by studying robotics developments in five different areas, surveying the home, long-term health care, police and private drones in cities, traffic, and the army. The central approach is that the use of robotics will affect these fields of applications and social practices in many ways. We will describe four key characteristics of modern service or social robots that produce various social and ethical issues and thus raise questions of social acceptance: (1) robots as IT, (2) the lifelike appearance of robots (body and brain), (3) the level of autonomy of robots, and (4) robotization as rationalization. This gives the reader a number of thematic tools for reading the following five chapters.

1.3.1 Information Technology

Robots are IT. This means that social issues such as privacy, cyber security, the digital divide (access to technology, computer skills), algorithmic transparency, and data ownership also play a role in robotics. The fact that within robotics great attention is given to improving the interface between machines and humans brings new questions with it, especially in the area of privacy. The vision of *affective computing* can, for example, only be realized if the robot is allowed to measure and store data about our facial expressions.

1.3.2 A Lifelike Appearance

In the previous section, robots were described as having a body and a brain. Robots with human or animal forms were developed in order to improve the interaction between humans and robots. This capitalizes on the ability of man to anthropomorphize technology: the ability of people to bridge the gap between humans and technology, allowing human attributes. Robotics makes explicit use of this option. Moreover, attempts are being made to develop social robots to engage with humans on an emotional level. This raises the question of the limits within which this social psychological phenomenon may be used. To what level do we want to deploy the emotional bond between people and machines? And how do we ensure that there is no abuse of the trust that is artificially built between man and machine? Sharkey and Sharkey (2012) consider the "embodiment" of the robot as an important approach for further ethical analysis of robot technologies, and say: "Robots and robotic technology require a new perspective on many ethical issues, because of their embodiment, and their often life-like appearance" (p. 37).

1.3.3 Level of Autonomy

The third characteristic concerns the degree of autonomy of the robot or, in other words, the degree of control that is delegated by the user to the machine. When man is in control of what the machine does, man is said to be "in-the-loop." When the robot acts autonomously, man is said to be "out-of-the-loop." When man and machine are both partly in control, man is said to be "on-the-loop." Each situation can raise various social and ethical issues.

When man is in-the-loop, the location of the operator and the robot may be different. The robot makes it possible for people to act remotely and thus allows social actors to have a presence elsewhere. Take, for example, the surgical robot, which provides the surgeon with the ability in the United States to perform operations in

Australia, or to an operator in Nevada (in the United States) who controls the drones flying in Afghanistan. These robots provide a form of telepresence that is mediated by IT. This shift toward man-on-the-loop raises the question of to what extent the user still receives enough information to make well-informed decisions. The delegation of control to robot systems also raises issues in the field of safety, responsibility, and liability.

In extreme cases, one could even give full control or autonomy to robots. Again, this raises the question of what decisions and actions we want to leave to a robotic machine. Do we want to have decisions and actions with a strong emotional or moral dimension, such as killing people or looking after children, taken solely by computers? A central question is what decisive position people should take in the control hierarchy. The European Robotics Technology Platform envisions mainly a supporting role for robots: "Robots should support, but not replace, human caretakers or teachers and should not imitate human form or behaviour" (EUROP, 2009, p. 9). But maybe other countries think otherwise in this area.

1.3.4 Robotization as Rationalization

When human robots are found, mechanical robots cannot be far behind. Once people are reduced to a few robot-like actions, it is a relatively easy step to replace them with mechanical robots.

George Ritzer (1983, p. 105)

At the beginning of the twentieth century, German social theorist Max Weber (1864–1920) created a theory of rationalization. He reflected on industrialization, urbanization, scientism, and capitalism and found that the modern Western world had become dominated by a belief in rationality. Weber saw the bureaucracy as the paradigm for the rationalization process in his day. It is well known that belief in efficiency led to the redesign of the factory and labor. Engineers not only mechanized separate actions but aimed to design the factory as one "great efficient machine." Rationalization, however, took place in many social practices. Historian Van den Boogaard (2010) explains that during the 1920s, the kitchen also became seen "as a factory that converted input (groceries) into out-put (meals) by means of specific activities, technologies, and spatial distances" (p. 137). In a similar vein, offices, airports, and cities were defined in terms of flows that could be designed and mechanized in an integrated manner. Weber discussed rationalization as a double-edged phenomenon. On the one hand, it can have many benefits, such as broader access to cheaper products and services with con-sistent quality. On the other hand, he was worried about the many irrationalities of rational systems. For example, bureaucracies can become inefficient because of too many regulations. Weber was most concerned about the so-called iron cage of rationality, the idea that an emphasis on rationalization can reduce the freedom and choices people have and lead to dehumanization.

Faith in rationalization implies that efficiency, predictability, cal-culability, and control through substituting technology for human judgment present dominant cultural values (Ritzer, 1983). Please note that in our information age, calculability is, most of the time, about programmability, and control often relies on digital control by means of algorithms. Rational systems aim for greater control over the uncertainties of life, in particular over people, who present a major source of uncertainty in social life. One way to limit the dependence on people is to replace them with machines. After all, robots and computers are easier to control than humans. We have witnessed this rationalization process in the factory. Originally, craftsmen ruled the production process. Then, the work was divided into many simple partial activities that could be performed by unskilled workers. This far-reaching simplification and specialization of the work paved the way for mechanization of various parts of the production process and finally made it possible during the second half of the twentieth cen-tury to automate and robotize such activities. Robotization thus pres-ents a way to rationalize social practices and reduce their dependence on people. As Ritzer (1983) argues: "With the coming of robots we have reached the ultimate stage in the replacement of humans with nonhuman technology" (p. 105).

As indicated earlier, this book is not about the use of robots in the clean factory, but about its use in places crowded with "unpre-dictable" people, such as the home, the city, or the battlefield. The use of robots in these messy circumstances is only possible when their environments are organized around their simple activities. One way of doing this is to buy a robot that is embedded in the required microenvironment: think of a dishwasher. For a long time, this type of stand-alone application of robots was the only way to exploit their limited capacities in complex social environments. Floridi (2014) argues that our world is rapidly becoming well adapted to the limited capacities of ICTs. He holds, "Nowadays, enveloping the environment into an ICT-friendly infosphere has started pervading all aspects of reality and is visible everywhere, on a daily basis. We have been enveloping the world around ICTs for decades without fully realizing it" (p. 144). In addition to technological changes, adapting the world to the limited capacities of robots requires social and cultural changes.

In *The McDonaldization of Society*, Ritzer (1983, 2004) argues that the processes of rationalization described by Weber have accelerated in our times and become globalized. He talks about McDonaldization, because he sees the fast-food restaurant as the paradigm for the rationalization of contemporary society. The entire food chain—from farm to fork—is geared toward efficiency. Besides production, the consumption side has also become rationalized. In the United States, cooking from scratch and family meals have become relatively rare nowadays. Many U.S. citizens opt for more efficient ways of eating, such as dining at fast-food restaurants or eating microwavable food in front of the TV.

Ritzer argues that no aspect of people's lives is immune to rationalization any more. Even social relationships and sex have become McDonaldized. IT is a major driver of this process. Technology is nestling itself within us and between us, collects a lot of information about us, and can sometimes even operate like us, that is, mimicking the facets of our individual behavior. In short, IT has become "intimate technology" (Van Est, 2014). The McDonaldization of sociability is evident in "rationalized online systems such as Facebook, where friendship is reduced to clicking an 'add' button and never needing to interact with that 'friend' on an individual basis ever again" (Flusty, 2009, p. 436). Also, sex has undergone substantial rationalization. Aoyoma, a relationship counselor in Tokyo, believes Japan is experiencing "a flight from human intimacy" (Haworth, 2013). Many Japanese young people have lost interest in conventional relationships and sex, because they find it "too troublesome" (*mendokuzai*). Dependence on "complicated" humans can be reduced by replacing them with convenient technologies, such as virtual-reality girlfriends or sex robots. Some people even believe that the future of relationships in Japan and the rest of the world will be largely technology driven (Haworth, 2013). This example shows that information technologies, like rational robotic systems, can have a profound effect on how we define ourselves and our relationships with other human beings.

1.3.4.1 Irrationality of Rationality Weber and Ritzer are concerned about the great costs associated with rationalization, which they group under the heading of the "irrationality of rationality." This label comprises all the negative aspects and effects of rationalization. More specifically, irrationality refers to the opposite of the dimensions of rationality: inefficiency, unpredictability, incalculability, and loss of control. Most specifically, irrationality implies that rational systems are unreasonable systems. Rationalization thus may lead to dehumanizing systems that may become antihuman or even destructive to human beings. For example, the kind of meals served at fast-food restaurants, and the fast-food culture in general, have detrimental effects on the health of many people. The rationalization of food consumption has also caused the loss of the communal meal for which families got together every day. Moreover, the more face-to-face and skin-to-skin intimacy is lost to technology; we should ask ourselves whether the rationalization of social relationships and sex may cause people to become physically and socially disconnected from each other. The ultimate irrationality is the possibility that the system (and/or the elite controlling it) would come to control us or even replace us.

Related to the latter, the theme "robots taking over society" has, for a long time, been center stage in the cultural imagination and public debate around robots. This relates to the hegemony of the strong AI and robotics vision in the media, which constantly stirs up the debate, causing anxiety among people. Some robotic technologists and scientists expect that such robots will soon become smarter than human beings (Brooks, 2002). Kurzweil (2005) describes a future in which even the "ordinary" man has no place other than to function as a kind of pet. The world will belong to people who have stepped up their intelligence (cyborgs) and to intelligent beings (robots). Bill Joy, nicknamed the "Edison of the Internet" for his revolutionary contribution to the development of the Internet, is worried about this development, and in 2000 he wrote a controversial article, "Why the Future Does Not Need Us" (Joy, 2000), in which he warns of the dangers of uncontrolled intelligent systems, stating that we may actually destroy ourselves if we continue with the further development of intelligent systems. Are these merely tall tales and wild speculation or should we take these kinds of future predictions seriously?

1.4 More Explorations

In the next five chapters, we will explore the types of robots in five major areas of application: home robots, health care robots, police and private drones, car robots, and military robots. Thus, this book begins close to home and then moves further and further away from our home situations. After examining health, traffic, and the use of drones in the cities by businesses, consumers, and the police, we end at the battlefield. Each chapter describes the social and ethical issues that emerge in these areas. We will explain how these issues are evoked by the four key characteristics of service and social robots that were introduced earlier. Each chapter provides an example of how and to what extent a certain social practice is expected to be rationalized by means of robotization and inquires into the potential positive and negative aspects of these rationalization processes. Lifelike features of robots are central in the chapters on home and care robots. For example, we will describe the use of mechanoid, humanoid, and android robots in the home situation and explain what kinds of ethical issues these different robots raise. The focus in the chapters on the use of police and private drones in cities is on robots as information technologies that raise all kinds of privacy issues. The level of autonomy of robots plays an essential role in our analysis of the social and ethical aspects related to car and military robots. By letting the four key characteristics guide our descriptions and reflections, we try to grasp the social significance of robotics for the medium and long term, and the political and administrative issues related to it.

In the following five chapters, these central questions are considered:

- What are the technical promises and societal expectations of robotics in the application field under study? What is possible right now in terms of robotic technologies and what might become possible in the short and the medium term?
- What social and normative questions loom, according to experts, in the shorter and the longer term?
- What regulatory issues are raised by these social and ethical issues? In other words, what points should be publicly discussed or put on the agenda by politicians and policy makers?

In the final chapter, some recommendations for politics and policy are drafted, based on the findings from Chapters 2 through 6.

Interview with Luciano Floridi (Philosopher and Ethicist of Information at Oxford University, United Kingdom)

"Those in power love to present technological developments as inevitable."

Factories were changed dramatically by the arrival of industrial robots, which required tidy and predictable *working envelopes*. Social robots will do something similar to the outside world. The process of adapting both our physical and mental living space to ICT-based agents has already begun, argues philosopher Luciano Floridi in his book *The Fourth Revolution*. "But let nobody tell you that these things are inevitable."

"I have the impression that we are moving towards a scenario where things are built in such a way that robots will have an easier life. Traffic lights, for instance, are colour-coded because that is convenient for most of us, humans. For artificial agents, such as driverless cars, it would be infinitely easier if an ultrasound signal were emitted instead of light."

But that would leave humans in the dark.

"In a context where there is both us and driverless cars driving, we would have hybrid systems, with lights and ultrasound. But look at the immense warehouses of Amazon: they are entirely built around the capacities of robots. Amazon can't easily replace them with humans, because they can't navigate these places."

Is a robot-friendly environment always less suitable for humans?

"My answer would be a cautious yes. Cautious in the sense that there is a very real risk that we overlook the inconvenience to humans. Let me give you some simple examples. A city council in England decided that for parking near the railway station, the only way of paying would be by mobile phone, not coins. People didn't like it and the city backtracked, 'oops, sorry, bad idea.' The classic case that we've all come across is where something has to be done in one particular way 'because that's how the computer system works.' So the ICT environment can force us to adapt to what the machine is able to do. I think there's a real danger of blindly moving in that direction."

Can we do something about this?

"Oh yes. I'm very distant from any deterministic view that says, 'Technology has its own course, taking us from A to B and we are not in charge.' First of all, that's not true. If I can't make a difference and you can't make a difference, it doesn't follow that we together can't make a difference. That's a fallacy, like saying that a lot of sand doesn't make a beach because a single grain of sand doesn't make a beach. More importantly, it's also ethically dangerous, indeed irresponsible. You will get this from people who want to disempower us: politicians, influencers, companies, anyone with power over large groups. They love to present their position as inevitable. If we can't do anything about technological development, they need not do anything about it, which suits them just fine. In reality, policy can do a lot about this, for good and for bad."

An essential term in your book is enveloping. *This refers to the reshaping of the world to make it suitable for robots, similar to the working spaces or "reach envelopes" of industrial robots. You extend that concept to include not only the material world around us, but even our inner lives. How is that?*

"Let me put it in two steps. To me, human life is mostly mental. We spend so much time with our thoughts, concerns, wishes, gossip, memories, hopes, and fears—that is where we live, it's where we are *present.* At this very moment, you are located in your study, I'm located in mine, but we are both *present* in our conversation. Now, information technology is currently reshaping this space of mental presence. We are constantly being invited to spend more and more time in a space of information, supported by technologies, where we can exchange our thoughts, hopes, and so on. It is almost like a house in which we are guests. And it is the owner of the house who decides which rooms and what facilities are available."

Are you saying that Facebook and Twitter and other social media are shaping our minds?

"They are shaping the space in which millions of people spend billions of hours. If I may use a bit of jargon, every technology has its own affordances and constraints. A stool is useful at a high kitchen table—that would be an affordance—but not so much for watching TV—there's a constraint. With an armchair, it's the other way round. If you have only a stool in your house, you won't watch much TV, if you only have an armchair, you will. These different technologies affect your behavior. The same is true for the information environment of social media. Therefore, the people shaping them have huge influence. They do not advertise their power, they don't carry a big stick, because they like to appear neutral, non-biased. Until a scandal breaks out, for instance about Facebook covertly involving hundreds of thousands of people in a social experiment to see how a change in the information feed can change their moods. Apply the resulting sort of knowledge to politics and it may be possible to influence people's voting behavior. So this is what social media can do and what they are like—not so neutral after all."

Social media are "locked" into computers and smartphones. Social robots will give ICT arms and legs, so to speak. What will they do to our lives?

"I think the effects will have a lot to do with a previous historical step, namely urbanization. Which is surprising: we thought that ICT would reverse that process, because there would be less need to meet face-to-face. But actually cities are still growing, because that's where opportunities are. By 2050, two-thirds of humanity will live in cities. The enveloping will mostly take place there; not in the green fields between Oxford and London, but *in* Oxford and *in* London. So if we assume that human lives will be more and more city-bound, then it is the structure, the architecture, the organization of cities that will be enveloped most around robots. Smart cities are already a huge step in that direction. Think Rio de Janeiro, a complex and dangerous city which had to prepare for the 2014 World Cup and the 2016 Olympics. They've done this basically with massive monitoring and data gathering from all possible sources, to run things efficiently: traffic, energy, security, and so on. Once you have a 'smart city' like that, it does not take much to adapt it to more robotic entities, such as driverless cars."

Apart from transport, social robots are also expected to meet other needs, in healthcare, in education, as companions, even as sex partners.

"Sex with robots, that's tabloid stuff. There are more than enough prostitutes in the world, unfortunately, and a robot is simply too expensive. But there are other markets where robots can make huge waves. The security market has already changed immensely. Today war is inconceivable without computers and robots. You and I grew up at a time when this would have been science fiction. Drones still have that sort of Azimov ring to them, but there you are.

Now imagine that robots will have a similar impact on health care or education. We might actually be on the verge of revolutions that redefine what we mean by taking care of people in a hospital, or educating people. I don't think it's happening yet, but I won't be surprised if the generation living now is going to see the same immense step forward in these areas that we have seen in conflict, security, war."

So you consider it a step forward?

"Oh, I'm not saying it's a good thing. It could be a step forward into the abyss! But again, it all depends on policy choices, on people voting, in a word, on what we are going to do with it. We have a responsibility of shaping these forces as they happen. And part of it is anticipating the problems and the potential benefits.

I do have a major worry though. Through the millennia, when we developed new technologies, there was always a chance of regretting and then correcting mistakes. Basically, time was on our side. We made horrific mistakes, such as exploding two atomic bombs on human beings. But we learned, we prevented even the worse scenarios. What I fear now is that the transformations we are seeing today are so profound and so fast-paced that there may not be time to say 'sorry, let's find a different solution.' Once you have recruited an army made of robots, we'll never go back to humans. Once you have decided that people can be taken care of by robots only… and so on.

We may even lose the skills. The recent British decision to rein-stall a nuclear energy programme highlights that. From the moment Britain decided to stop investing in nuclear plants some years ago, nuclear skills went into decline. Now the new plants will have to be built by France. You may or may not like nuclear energy, but this shows how some things are doable only when there are human beings able to do it."

One of the worries about social robots is that they are devoid of morality. But according to some, they can become moral agents in their own right, given the right software. Do you have a position in that debate?

"I was one of the first to write about this, in the late nineties. You can approach artificial agents in two ways. One is asking, who can be a moral agent? I argued that we should define 'moral agent' in an inclu-sive way, including for instance political parties and families, but also animals and machines, since there is a lot of agency in the world that is not based on human individuals. The other perspective is the one that your question is addressing. What happens when smart robots with a lot of autonomy, such as we might be building in 10 or 20 years, are in charge of actions with huge moral implications, in warfare, healthcare, education, social interaction?

There are three different strategies to deal with this. One says, 'We need to develop a morality for machines.' I think this is utter science fiction; a pipe dream, or a pipe nightmare. It can be fun to think about, but increasingly I think it is irresponsible to spend money and energy on this. Because it can't be done. Not in a very long time.

The second, more responsible strategy is to make sure that these machines have some safety measures implanted. To grasp the con-cept, think of microwave ovens: you cannot commit suicide by putting your head in them, because they won't work when the door is open. The machine is not moral in itself, but we can design standards that are morally good. In the case of drones, we could decide internation-ally that they cannot possibly fire under certain conditions or that, if they lose control, they will go to the nearest sea and drown them-selves. This is not as exciting as 'moral machines,' but on the plus side, it is doable.

The third strategy, which can go together with the second, is to have a human in control. Not *in*-the-loop—the robot can operate on its own—but *on*-the-loop, supervising—if something goes wrong, it can be stopped. It's like the Mars Rovers: they are autonomous, but NASA can intervene.

Unfortunately, quite a few people keep believing in the moral-machine option. Some think that machines can develop morality the way babies do, on the basis of feedback to their actions. But really, there is nothing in computer science at the moment to justify thinking that this is even remotely feasible. For let's face it, we don't even *have* artificial intelligence yet; we have artificial smartness. I remember how John McCarthy, who coined the word AI, was deeply disappointed with Deep Blue, the chess computer who beat Kasparov, but only because it had a database filled with matches played by humans. My wife, who is a neuroscientist here in Oxford, tells me we still know very little about human intelligence, so how can we build an artificial version? We have discovered this new continent, the human brain, but we've only just set foot on the beach. Can we reproduce a continent on the basis of a mere beach? Yet there is a whole church of believers in artificial intelligence and you can't talk them out of their faith. I've given up trying."

EXPERT INTERVIEWS AS A BONUS

At the end of each chapter, except for the conclusive one, the reader will find an elaborate interview with an internationally acknowledged expert. After Chapters 1 through 6, interviews can be read from, respectively: Luciano Floridi (Philosopher and Ethicist of Information at Oxford University in the United Kingdom), Kerstin Dautenbach (Professor of Artificial Intelligence at the University of Hertfordshire in the United Kingdom), Hans Rietman (Professor of Physical Medicine and Rehabilitation at Twente University of Technology in the Netherlands), Mark Wiebes (Innovation Manager with the Dutch National Police), Bryant Walker Smith (Assistant Professor of Law at the University of South Carolina in the United States), and Jürgen Altmann (Physicist and Peace Researcher at TU Dortmund University in Germany). All these interviews have been conducted and written by Gaston Dorren.

References

Aarts, E., & Marzano, S. (2003). *The new everyday: Views on ambient intelligence*. Rotterdam, the Netherlands: 010 Publishers.

Arkin, R. C. (2009). *Governing lethal in autonomous robots*. Boca Raton, FL: CRC Press.

Aron, J. (2011, September 6). Software tricks people into thinking it is human. *New Scientist Tech*. http://www.newscientist.com/article/dn20865-software-tricks-people-into-thinking-it-is-human.html?DCMP=OTC-rss&nsref=online-news (accessed October 26, 2014).

Asimov, I. (1950). *I, Robot*. New York: Gnome Press.

Böhle, K., Coenen, C., Decker, M., & Rader, M. (2011). Engineering of intelligent artefacts. In R. van Est, & D. Stemerding (Eds.), *Making perfect life: Bio-Engineering in the 21st century* (pp. 128–167). Brussels, Belgium: European Parliament, STOA.

Brooks, R. A. (2002). *Flesh and machines: How robots will change us*. New York: Pantheon Books.

Christensen, H. I. (Ed.). (2009). *From Internet to robotics: A roadmap for U.S. robotics*. Snowbird, UT: Computing Community Consortium (CCC). www.us-robotics.us/reports/CCC%20Report.pdf (accessed August 25, 2014).

Elliott, S. W. (2014). Anticipating a Luddite revival. *Issues in Science and Technology, 30*(3), 27–36.

European Robotics Technology Platform (EUROP). (2009). *Robotic visions to 2020 and beyond: The strategic research agenda for robotics in Europe, 07/2009*. Brussels, Belgium: European Robotics Technology Platform.

Floridi, L. (2014). *The fourth revolution: How the infosphere is reshaping human reality*. Oxford, UK: Oxford University Press.

Flusty, S. (2009). A review of "The McDonaldization of society 5". *Annals of the Association of American Geographers, 99*(2), 435–437.

Fong, T., Nourbakhsh, I., & Dautenhahn, K. (2003). A survey of socially interactive robots. *Robotics and Autonomous Systems, 42*(3–4), 143–166.

Frey, C. B., & Osborne, M. A. (2013). *The future of employment: How susceptible are jobs to computerization?* Oxford, UK: Oxford University Programme on the Impacts of Future Technology.

Friedman, B., Kahn, Jr., P. H., & Borning, A. (2006). Value sensitive design and information systems. In P. Zhang, & D. Galletta (Eds.), *Human–computer interaction and management information systems: Foundations* (pp. 248–372). New York: ME Sharpe.

Gates, B. (2007). A robot in every home. *Scientific American Magazine, 296*(1), 58–65.

Gorle, P., & Clive, A. (2013). *Positive impact of industrial robots on employment: Updated in January 2013 to take account of more recent data*. London, UK: Metra Martech/International Federation of Robotics (IFR).

Haworth, A. (2013, October 20). Why have young people in Japan stopped having sex? *The Guardian*. http://www.theguardian.com/world/2013/oct/20/young-people-japan-stopped-having-sex (accessed January 23, 2014).

Joy, B. (2000). Why the future doesn't need us. *Wired, 8*(4), 238–262.

Kahn, H., & Wiener, A. J. (1967). *The year 2000: A framework for speculation on the next thirty-three years*. New York: The Macmillan Company.

Keim, B. (2009, April 2). Computer program self-discovers laws of physics. *Wired* (online). http://www.wired.com/2009/04/newtonai.

Kroes, N. (2011). Commentary—Robots and other cognitive systems: Challenges and European responses. *Philosophy of Technology*, *24*(3), 355–357.

Kurzweil, R. (1990). *The age of intelligent machines*. Cambridge, MA: MIT Press.

Kurzweil, R. (2005). *The singularity is near: When humans transcend biology*. New York: Viking.

Labutov, I., Yosinski, J., & Lipson, H. (2011). *YouTube-film: "AI vs. AI: Two chatbots talking to each other"*. www.youtube.com/watch?v=WnzlbyTZsQY (accessed September 4, 2014).

Lau, Y. Y., Van't Hof, C., & Van Est, R. (2009). *Beyond the surface. An exploration in healthcare robotics in Japan*. The Hague, the Netherlands: Rathenau Institute.

Lipetz, B.-A. (1966). Information storage and retrieval. *Scientific American*, *215*(3), 224–242.

Luck, M., McBurney, P., Shehory, O., & Willmott, S. (2005). *Agent technology: Computing as interaction—A roadmap for agent based computing*. Southampton, England: Agentlink. http://agentlink.org/roadmap/index.html (accessed October 5, 2014).

MacDorman, K. F., & Ishiguro, H. (2006). The uncanny advantage of using androids in social and cognitive science research. *Interaction Studies*, *7*(3), 297–337.

Minato, T., Shimada, M., Itakura, S., Lee, K., & Ishiguro H. (2006). Evaluating the human likeness of an android by comparing gaze behaviors elicited by the android and a person. *Advanced Robotics*, *20*(10), 1147–1163.

Minsky, M. L. (Ed.). (1968). *Semantic information processing*. Cambridge, MA: MIT Press.

Mitchell, J., Clarke, C., Shaffer, J., Su, J., Chen, P., Kuroda, S., ... Savaris, B. (2012). *Global industrial automation—Connection series*. London, UK: Credit Suisse.

Moor, J. H. (2006). The nature, importance, and difficulty of machine ethics. *IEEE Intelligent Systems*, *21*(4), 18–21.

Moravec, H. (1988). *Mind children: The future of robotics and human intelligence*. Cambridge, MA: Harvard University Press.

Mori, M. (1970). (Translated by K. F. MacDorman, & T. Minato). The uncanny valley. *Energy*, *7*(4), 33–35.

Noble, D. F. (1997). *The religion of technology: The divinity of man and the spirit of invention*. New York: Alfred A. Knopf.

Picard, R. W. (1995). *Affective computing: MIT Media Laboratory Perceptual Computing Section* (Technical Report No. 321). Boston, MA: MIT.

Picard, R. W., & Klein, J. (2002). Computers that recognize and respond to user emotion: Theoretical and practical implications. *Interacting with Computers*, *14*(2), 141–169.

Pransky, J. (1996). Service robots: How should we define them? *Service Robot: An International Journal*, *2*(1), 4–5.

Ravilious, K. (2009, April 2). First robot scientist makes gene discovery. *National Geographic News*. http://news.nationalgeographic.com/news/2009/04/090402-robot-scientists.html (accessed March 17, 2014).

Riskin, J. (2003). The defecating duck; or, the ambiguous origins of artificial life. *Critical Inquiry*, *29*(4), 599–633.

Ritzer, G. (1983). The McDonaldization of society. *Journal of American Culture*, *6*(1), 100–107.

Ritzer, G. (2004). *The McDonaldization of society*. Thousand Oaks, CA: Pine Forge Press.

Sample, I., & Hern, A. (2014, June 9). Scientists dispute whether "Eugene Goostman" passed Turing test. *The Guardian*. http://www.theguardian.com/technology/2014/jun/09/scientists-disagree-over-whether-turing-test-has-been-passed (accessed August 25, 2014).

Sandoval, L. (2014, April 16). Microsoft CEO Satya Nadella talks ambient intelligence, announces big data analytics platform. *Tech Times*. http://www.techtimes.com/articles/5658/20140416/microsoft-ceo-satya-nadella-talks-ambient-intelligence-announces-big-data-analytics-platform.htm (accessed September 3, 2014).

Schuitema, E., Wisse, M., Ramakers, M. J. G., & Jonker, P. P. (2010). The design of LEO: A 2d bipedal walking robot for online autonomous reinforcement learning. *Proceedings of Intelligent Robots and Systems* (*IROS 2010*) (pp. 3238–3243). Taipei, Taiwan: IEEE.

Sharkey, A., & Sharkey, N. (2012). Granny and the robots: Ethical issues in robot care for the elderly. *Ethics and Information Technology*, *14*(1), 27–40.

Simon, H. A., & Newell, A. (1958). Heuristic problem solving: The next advance in operations research. *Operations Research*, *6*(1), 1–10.

Singer, P. W. (2009). *Wired for war: The robotics revolution and conflict in the twenty-first century*. New York: The Penguin Press.

Steiner, C. (2012). *Automate this: How algorithms came to rule the world*. New York: Portfolio/Penguin.

Trevelyan, J. (1999). Redefining robotics for the next millennium. *The International Journal of Robotics Research*, *18*(12), 1211–1223.

Turkle, S. (2011). *Alone together. Why we expect more from technology and less from each other*. New York: Basic Books.

Umpleby, S. A. (2008). Cybernetics. In S. R. Clegg, & J. R. Bailey (Eds.), *International encyclopedia of organizational studies* (pp. 350–354). Thousand Oaks, CA: Sage Publications.

Van't Hof, C., Van Est, R., & Daemen, F. (Eds.). (2011). *Check in/Check out: The public space as an Internet of things*. Rotterdam, the Netherlands: NAi Publishers.

Van den Boogaard, A. (2010). Site-specific innovation: The design of kitchens, offices, airports, and cities. In J. Schot, H. Lintsen & A. Rip (Eds.), *Technology and the making of the Netherlands: The age of contested modernization, 1880–1970* (pp. 124–177). Cambridge, MA: MIT Press.

Van Est, R. (2010). Rinie van Est on living technologies. In M. Bedau, P. G. Hansen, & E. Parke (Eds.), *Living technologies: 5 questions* (pp. 195–215). Milton Keynes, England: Automatic Press VIP.

Van Est, R. with assistance of Rerimassie, V., van Keulen, I., & Dorren, G. (2014). *Intimate technology: The battle for our body and behaviour.* The Hague, the Netherlands: Rathenau Institute.

Vinge, V. (1993). The coming technological singularity: How to survive in the post-human era? *Whole Earth Review* (Winter), *81,* 88–95.

Wallach, W., & Allen, C. (2009). *Moral machines: Teaching robots right from wrong.* Oxford, UK: Oxford University Press.

Walters, M. L., Syrdal, D. S., Dautenhahn, K., Boekhorsten, R. T., & Koay, K. L. (2008). Avoiding the uncanny valley: Robot appearance, personality and consistency of behavior in an attention-seeking home scenario for a robot companion. *Autonomous Robots*, *24*(2), 159–178.

Wiener, N. (1948). *Cybernetics: Or control and communication in the animal and machine.* New York: John Wiley & Sons.

2

HOME IS WHERE THE ROBOT IS

Mechanoids, Humanoids, and Androids

2.1 Introduction

Given the media attention, a lot is being expected of the home robot. During the past few years, popular magazines and newspapers have regularly reported on new robotics applications for domestic use. Headlines such as "A Robot in Every Home" (Gates, 2007), "Get Ready to Be Pampered by Household Robots in Next Decade,"* "Now, a Technology-Laden Robot That Does All the Household Work,"† "Are You Ready for Your First Home Robot?,"‡ "A Social Robot for Every Household?,"§ "Forget Babysitters: Nanny Bot Takes Care of Your Family,"¶ "Hooray for House,"** and "He Laughs, Learns, and Has Impeccable Manners. Oh, and He's a Robot"†† suggest that the robots have started their march and will appear on a large scale in the home.

According to a 2009 poll by the Institute of Electrical and Electronics Engineers (IEEE) asking respondents what specialty they would find most useful in a robot they might own one day, 39% of the respondents answered doing household chores and 37% answered sex. Some household robots already have a commercial application, especially the vacuum-cleaning robot. The sex robot

---

* ieet.org/index.php/IEET/more/pelletier20130723.
† articles.economictimes.indiatimes.com/2011-05-17/news/29552268_1_robotics-open-source-willow-garage.
‡ www.engadget.com/2014/06/12/home-robot-pepper/.
§ www.asme.org/engineering-topics/articles/robotics/social-robot-for-every-household.
¶ gajitz.com/forget-babysitters-nanny-bot-takes-care-of-your-family/.
** www.northernexpress.com/michigan/article-4313-hooray-for-house-robots.html.
†† fortune.com/2014/08/10/JIBO-robot-emotion-hotel/.

was introduced into the market some years ago but is not (yet) a commercial success. These two types of robots are very different with respect to lifelike appearance, to their physical appearance as well as the way they act. Household robots are mechanoids, robots that do not resemble a human being. Their physical appearance depends on the task for which they have been developed. Sex robots are androids, robots that look like a human being. They are designed with anthropomorphism in mind, which attributes human characteristics and behaviors to nonhuman subjects. In other words, their performance simulates human performance. This is in contrast to most household robots. Their performance is not based on simulating human behavior, but on doing the tasks in an efficient way to relieve humans from the routine chores of the home. To perform these tasks in an efficient way, we will see that the complex household tasks have to be redefined. In Section 2.2, we will discuss the vacuum-cleaning robots as an illustration for the household robots. The sex robot will be discussed in Section 2.4. The sex robot is actually a companion robot with a focus on the physical aspect of the relationship. In Section 2.3, we will discuss the companion robot, with the focus on the social interaction with human beings. These companion robots do not look perfectly like human beings, but are humanoids. A humanoid robot is a robot with body parts that resemble those of the human or animal body. Companion robots mostly have heads that have been designed to replicate human facial features such as eyes and mouths to make the social interaction more smooth. Especially with regard to future developments and expectations of companion and sex robots, these robots will raise many ethical issues, especially with regard to dehumanization, such as social de-skilling. We will end with some observational conclusions in Section 2.5.

2.2 Mechanoid Robots: The Vacuum Cleaner

According to some, the next revolution will enter the household in the coming years. Static household appliances that need to be operated by residents will be replaced by moving devices that govern themselves, that is, household robots. The idea is not entirely new, and in 2007, Bill Gates was certainly not the first with his prediction of "a robot in every home by 2015." In 1999, the European Commission, in a report on the market for industrial robots and service robots, predicted that the household robot would conquer the field between 2009 and 2014.* Much was especially expected of the vacuum robot.

Although robots were first used practically in industry, the earliest fantasies about robots are persistently associated with delegating household chores in exchange for more efficiency, time saving, and comfort. In 1921, this image was presented by the introduction of the word "robot" in the play *Rossum's Universal Robots* (*RUR*) by the Czech writer Karl Čapek. In this play, Rossum's business places an ad with the message that everyone should buy their own robot to increase their comfort.

In 1964, Medith Wooldridge Thring predicted that by around 1984 a robot would be developed that would take over most household tasks:

> The great majority of the housewives will wish to be relieved completely from the routine operations of the home such as scrubbing the floors or the bath or the cooker, or washing the clothes or washing up, or dusting or sweeping, or making beds. By far the most logical step to allow this variety of human homes and still relieve the housewife of routine is to provide a robot slave which can be trained to the requirements of a particular home and can be programmed to carry out half a dozen or more standard operations. ... There are no problems in the production of such a domestic robot to which we do not have already the glimmering of a solution. It is therefore likely that, with a strong programme of research, such a robot could be produced in ten years. If we assume that it also takes ten years before industry and government are sufficiently interested to find the sum required for such development (which is of the order of £1 million), then we could still have it in 1984.
>
> **Thring (1964, p. 38)**

Most historical household robot predictions were not very future-proof. Despite a multitude of investment, the multifunctional home robot is still not within reach. During the last 10 years, the first robots have made their entry into the household, but they are all "one trick ponies" or *monomaniacal*—specialized machines that can perform only one task. According to Bill Gates (2007), monomaniacal household robots really are on our doorstep. According to Internet celebrities, we are currently witnessing a technological turning point, just as with the PC in the 1980s, with the invasion of households by robots: "We may be on the verge of a new era, when the PC will get up from the desktop and allow us to see, hear, touch and manipulate objects in places where we are not physically present" (Gates, 2007, p. 62).

At the moment, the vacuum-cleaning robot and the robot lawn-mower are the two most successful household robots. Most leading brands regularly provide new and improved models of these robots. Robots are also commercially available for cleaning swimming pools, but of course this is only interesting for a much smaller number of households.

Around 1980, it dawned on robot developers that service robots would have a potentially large market in the household even if only a fraction of the domestic tasks could be automated. Initially, they contemplated cleaning robots for cleaning floors and windows. It still took until 2002 before the first cleaning robot, the vacuum-cleaning robot, was launched. More than 20 years passed between the initial concept and the launch, as the technical challenges proved greater than expected (Prassler & Kosuge, 2008).

Roomba, a vacuum-cleaning robot (see Figure 2.1), developed by the company iRobot, was the first home robot to appear on the commercial market. New applications that are being developed for the commercial market are window-cleaning robots and others that can do the ironing and cleaning and are able to fold the laundry. Based on the introduction of the most popular household robot, the vacuum-cleaning robot, we will show that the rise of household robots has not been entirely unsuccessful, mainly because of the apparently high complexity of household tasks.

The Roomba vacuum-cleaning robot is also the most widely sold robot, and more than 8 million have been sold worldwide.* Although this is an impressive number, the vacuum-cleaning robot sales do not beat those of the regular vacuum cleaner. Prassler and Kosuge (2008) estimated the ratio of sales of vacuum-cleaning robots to sales

---

* www.irobot.com/en/us/Company/About.aspx?pageid=79.

Figure 2.1   Roomba, the vacuum-cleaning robot. (Photo courtesy of Rinie van Est.)

of regular vacuum cleaners to be between 1:40 and 1:400. On the basis of recent sales figures of vacuum-cleaning robots and regular vacuum cleaners, the estimation by Prassler and Kosuge still seems correct. We can state that the vacuum-cleaning robot is still widely regarded as a highly technical gadget and that the mass consumers have not been reached.

The vacuum-cleaning robot consists of two parts: the fixed base and the robot. At its base, the robot can recharge itself. By pressing a button, the robot will start vacuuming and moving around the room. The robot can also be programmed to start vacuuming at a set time.

The vacuum-cleaning robot has a small dust collection container, which needs to be emptied daily. For the robot, very thick carpet or laminate or parquet with grooves present a problem. The vacuum-cleaning robot has too little power capacity for deep suction. It also fails to operate on black-colored surfaces, because the cliff sensors that recognize an edge or a stair view this color as an edge and will not clean over it. Most vacuum-cleaning robots come with two virtual walls. These are small appliances that provide an infrared beam of 2.5 meters (about 8 feet), and the cleaner cannot move beyond these virtual walls. In this way, the cleaner is kept away from items such as power cords and bowls of pet food. The cost of a vacuum-cleaning robot ranges from U.S. $300 to $900.

2.2.1 Experiences of Early Adaptors: Roombarization

Research by Sung, Grinter, Christensen, and Guo (2008) shows that people who buy a vacuum-cleaning robot belong to the group of early adopters, namely curious and well-educated young people, and that the general public has not been reached. From the study by Sung et al., it is clear how these early adopters experienced the vacuum-cleaning robot and how they used it. In general, they are positive about the robot. Users with young children or pets were significantly positive. This is a different use of the vacuum-cleaning robot for them: the robot invites babies or very young children to learn to crawl or walk, because they want to follow the robot and the robot chases the pets, or the pets (especially birds) are driven around the room on top of the robot.

A striking fact that emerges from research is that people who own vacuum-cleaning robots tidy their home more often. A path must be cleared in order to make room for the vacuum-cleaning robot. In addition, small items, for example, small parts of toys, must be removed from the floor; otherwise, the robot sucks them up. As a result, for most people this leads to promoting household order, as they will go through the house with a cleaning cloth. In addition, a large majority still continues to use the regular vacuum cleaner for several reasons. The robot version does not always manage to get the floor completely dust-free, and pet hair in particular is sometimes left behind. Sometimes speedy cleaning is desired, and since the vacuum robot is unable to walk up stairs, reach into corners or other difficult places, such as vacuuming a round cables or removing cobwebs, people will still use the regular vacuum cleaner.

Here we see a tendency comparable to what occurred with the introduction of the regular vacuum cleaner. It was expected then that housecleaning could be done significantly more swiftly, so housewives would gain time for other activities. However, it turned out that hygiene standards were soon raised considerably, which ultimately required more time for cleaning than before the era of the vacuum cleaner (Cowan, 1983). Thus, domestic labor does not necessarily get lighter because of the introduction of the vacuum-cleaning robot, as the advertisement promises us: "Think of your most tedious housecleaning task. Now think about never having to do it again. Indoors and out, our robots are engineered for cleaning performance and convenience, bringing the latest robotic technology to real-world homes."*

The study by Sung, Grinter et al. (2008) showed that almost all users of a robotic vacuum cleaner made changes to the organization of their home and their home furniture. The more tidy and less furnished the household is, the easier it is to make use of that vacuum-cleaning robot. This process of rationalizing the environment so that the vacuum-cleaning robot can do its job better is known as *roombarization* (Sung, Guo, Grinter, & Christensen, 2007), referring to the Roomba. Typical modifications are moving or hiding cables and cords, removing deep pile carpet, removing lightweight objects from the floor, and moving furniture. When purchasing new furniture or floors, one should take the capabilities of the vacuum-cleaning robot into account. An inhibiting factor for the rise of the commercial vacuum-cleaning robot probably lies in this need for a structured environment. The history of technology research shows that the interest in new devices quickly decreases when existing practices require too many changes (Oldenziel, 2001). That this probably also holds for vacuum-cleaning robots follows from the study by Vaussard et al. (2014). In this study, most households stopped using the robot after a while, because they became disappointed as they actually assessed the robot's relevance within their own ecosystem; in other words, they assessed how well the robot integrates inside the user's space and perception and considered the fact that the robot does not actually decrease the amount of work for the user.

Presently, the huge potential for selling vacuum-cleaning robots has not yet been fulfilled. In a Dutch magazine for computer technology, 24 different vacuum-cleaning robots were tested, including the most popular, such as the Roomba, and its conclusion is telling in that, economically, there is currently no convincing argument to buy any of the tested robots.† Most of the reviews indicate that vacuum-cleaning robots cannot entirely replace the traditional human-driven vacuum cleaners.‡ Acceptance of useful household appliances is difficult if

---

* www.irobot.com/.
† C't Magazine voor Computertechniek 2011, 11, 106. See also http://robot-vacuum-review.toptenreviews.com/.
‡ http://www.cnet.com/news/robot-vacuum-roundup/.

those devices only work partially, and the regular household appliance remains an indispensable device in the household.

A historical technological argument to the detriment of the vacuum-cleaning robot is that devices that can save time and labor lose out to devices used for relaxation. In the 1930s, many households bought a radio and then often had no money left for a washing machine (Bowden & Offer, 1996). We see a similar tendency in relation to the vacuum-cleaning robot. Despite the economic crisis, people are flocking to buy expensive game systems for entertainment, such as the Nintendo Wii and the Microsoft Xbox 360. More than 100 million Wiis have been sold since its introduction in 2006.*

The experiences and problems outlined above with the vacuum-cleaning robot also occur with the lawnmower robot and the mop robot. With those, a form of *roombarization* will also take place. For the lawnmower robot, the owner must have a power socket outside for charging and a flat lawn without water features. There should be no trees in the yard from which fruit or branches might drop, because the robot will have problems with those objects, and loose items should not be strewn on the lawn as the robot mower tends to mow much more frequently than a regular human-operated mower. Additionally, for the robotic lawnmower, it is recommended that children are not playing in the yard when the robot is working, because of safety issues and also because a robot can be too attractive as a toy (see Section 2.4.2 regarding liability). Although the lawnmower robot operates on very low electric power, it can sever fingers or toes when these come into contact with the rotating blades. An integrated safety circuit therefore switches the blades off automatically when the machine is lifted.

2.2.2 Reducing the Complexity of Household Tasks

It appears that household tasks require very complex decisions of a robot, because they are comparable to solving the so-called frame problem. In such a problem, a clear criterion is missing for checking whether a proposed solution is acceptable. In a domestic task, often unconsciously, we search, organize, and select all kinds of relevant information and we will take a decision on the basis of that information.

---

* www.nintendo.co.jp/ir/library/historical_data/pdf/consolidated_sales_e1406.pdf.

The environment of the household tasks is not static, but changes constantly. For example, the contents of the laundry basket, clothes that need folding, change each time we use it. Different clothes, some even inside out, will be in the basket. The problem of forcing a robot to adapt to these changes, which humans often consider to be common sense, is the basis of the frame problem in artificial intelligence, and it is very hard to represent (see also Floridi, 2014).

The degree of difficulty is shown by research from the University of California at Berkeley, which aims to develop a robot able to fold laundry.* The main result that emerged is how difficult it really is to fold textiles mechanically, which is a relatively simple task for humans. Eventually, a robot was developed that took nearly 25 minutes to fold one towel.† Also, for the first prototypes of cooking robots, it shows that simply breaking an egg into a pan forms an insurmountable obstacle. The prospect that we will not have to do any cooking in the near future is still very far away. Nevertheless, there are already cooking robots in highly structured environments where cooking is reduced to a few basic sequences of actions, where the vegetables are precut and washed for the robot and the ingredients have already been put in scales at fixed locations. For a particular dish, the robot will take the necessary ingredients for the recipe from the dishes by following preprogrammed instructions, put them into a wok or pan and stir them together. The cooking robot may not be suitable for a household, but it might well be useful in school cafeterias and military camps, for example, in terms of mass production and where a prefabricated kitchen is fitted.‡

To deal with the complexity of cooking, the environment is therefore rationalized: a closed microenvironment is created within an unstructured environment. A good example is one of the most successful household devices—the dishwasher. By redefining the process of washing the dishes by spraying hot water at the dishes, instead of manual dishwashing, which relies largely on physical scrubbing to remove soiling, a closed microenvironment is created to deal with the complex task of washing the dishes in a dynamic and

---

* newscenter.berkeley.edu/2010/04/02/robot/.
† www.youtube.com/watch?v=Thpjk69h9P8.
‡ www.twanetwerk.nl/default.ashx?DocumentID=11839.

unstructured environment. Another recent household device, *Tubie* the ironing robot, has become available for U.S. $1100.* With Tubie, the process of ironing is also redefined. Tubie is in fact a doll over which the consumer pulls the garment. Then a motor inflates the doll, and hot air is blown all through the garment. It thus takes about 5 minutes per garment before the ironing program is over. In contrast with the dishwasher, although we do not have to iron, we must be constantly present during the process to pull the clothing on and off the doll and guide the deflation process. Therefore, Tubie does not fully take over the job, and consequently ironing in this way takes even longer than if we had used the old-fashioned steam iron.

The process of rationalizing the environment also applies to the vacuum-cleaning robot, as we have already seen, which we call *room-barization*. Before using the robot, the environment needs to be prepared so that the robot can do the task properly without, for example, sucking up a sales receipt. When cleaning up the room, we find all kinds of stuff that we may or may not throw out or move—a loose Lego brick is thrown back into the Lego box, a sales receipt on the floor is perhaps not just vacuumed up, but we first check whether we still need to keep it for guarantee purposes. These are all decisions that are impossible (especially for the time being) for a vacuum-cleaning robot to make. To enable them to make these decisions, engineers are trying to make robots smarter by connecting them to a cognitive control system. The international robotics project "Web-Enabled and Experience-Based Cognitive Robots That Learn Complex Everyday Manipulation," from 2011,[†] funded by the European Commission, has had this aim. Such cognitive controls allow robots to pick up information available on the Internet and transfer it into *proper* behavior. Think of web instructions, online dictionaries, and online encyclopedias. In this way, a household robot—as is expected—could even learn to cook simple dishes based on recipes found on the web. In addition, numerous robotic technology–based domotics solutions (see Chapter 3) and ambient intelligence (Aarts & Encarnação, 2006)[‡]

---

* www.ironing-machines-tubie.com/.
[†] ias.cs.tum.edu/research/robohow.
[‡] In ambient intelligence, the emphasis is more on increasing comfort for humans (creating smart homes, etc.) than on a robot's specific actions. Further equipping robots with intelligent functionality we can perhaps file under the label ambient intelligence.

find their way into households. These days there is so much research into smarter household robots, but thus far robotic developers always come up against the fact that household tasks are still harder than we originally thought.

Due to the complexity of the household, the household robots are monomaniacal: they can only take over a portion of one specific household task. For a successful launch in the market, these robots will have to be able to fully take over and complete certain tasks, in order to actually make household work lighter. The *non-monomaniacal* weekly cleaner will probably remain more efficient for far longer than a household robot.

2.2.3 Liability of Home Robots

The use of household robots raises no urgent ethical issues. The main ethical issue concerns safety in the deployment of these devices in the household—developing robust devices that do not walk in people's way and have a much-reduced risk of accidents. With regards to the safety of using a household robot, the degree of its autonomy affects the outcome of any act performed by the robot and the possible legal consequences. Asaro (2012) sees no legal issues relating to liability as long as the robot does not act of its own accord—either unexpectedly or according to decisions it reaches independently of any person. The potential harms posed by such a robot will be covered by civil laws governing product liability, since the robot, available for the commercial market nowadays and in the short term, can be treated as a tool, such as a toy, weapon, or car and so on.

This completely changes when the robot performs independent actions or decisions, and although there is a pressing need to deal with questions of liability, there has been little discussion about the legal aspects of human–robot interaction in the literature. No legislation is yet available to deal with questions of liability in accidents involving these robots. Schaerer, Kelley, and Nicolescu (2009) propose a creative framework for liability in human–robot interaction based on the principle "robots as animals" for semiautonomous robots programmed to act within a predictable range of behavior.

On this basis, our negligence rationale becomes clear. The owner of a semi-autonomous machine should be held liable for the negligent supervision of that machine, much like the owner of a domesticated animal is held liable for the negligent supervision of that animal. Semi-autonomous machines, like domesticated animals, are more predictable than wild animals, but remain occasionally prone to sporadic behaviour – even in the absence of manufacture or design defects, and even despite adequate warnings, e.g., "Do not leave this product unattended."

Schaerer et al. (2009, p. 75)

This framework involves a two-step analysis: (1) making a decision about whether the product was defective. If so, then the manufacturer is held liable and the analysis ends. If not, then the second step is (2) to decide whether the owner was negligent. If so, the owner is liable. If not, the victim alone bears the cost of the accident.

For example, if a vacuum-cleaning robot knocks over and seriously injures a baby, the manufacturer might be held liable because of a defect of the vacuum-cleaning robot. If the robot was not defective, we must ask whether the owner of this robot was negligent in causing the accident because the baby was a victim within the foreseeable zone of danger. In other words, that a reasonable person would believe that the negligent supervision caused the accident and that the accident was therefore foreseeable.

The proposed framework of Schaere, Kelly, and Nicolescu has been developed for semiautonomous robots with a level of autonomy well below that of humans. In the future, we may be able to create robots that are capable of moral autonomy or even legal responsibility (see Wallach & Allen, 2009). Then many questions about legal responsibility, legal liability, and legal action become relevant, because the robot may operate beyond the control of a user. Robots, however, have not reached this state, and it seems likely that it will be decades before this is realized (on this subject, see Richards, 2002). Therefore, there is little point in talking about law or judicial rulings, because such statements can only be based on expectations about the development of robot technology and not on specific legal issues.

In 2013, the International Electrotechnical Commission (IEC) prepared an International Standard that covers the safety aspects of vacuum-cleaning robots and their washing peers, IEC 60335-2-2, since the Commission states that the safety of cleaning robots, like that of all household appliances, is essential.* The International Standard provides reliable guidance for ensuring safety in the design and construction of vacuum-cleaning robots as well as their integration, installation, and use throughout their life cycle.

2.3 Humanoid Robots: The Companion

In this section, the focus is on companion robots. These robots interact with humans, which requires social intelligence based on deep models of human cognition and social competence. The companion robot is designed with anthropomorphism in mind, which means attributing human characteristics and behaviors to nonhuman subjects, in our case socially interactive robots. In robotic social interaction, an activity shared between human and machine is the goal. The crucial purpose of a companion robot is to work with the user to activate, entertain, and engage the user. Companion robots thus enter into social interactions with people at home. A major design goal for producers of companion robots is that they will also be able to communicate on a nonverbal level and will be able to transfer emotions because they are embodied (Breazeal, Takanski, & Kobayashi, 2008). Because of the design of the robot's body, its bodily communication can be greatly broadened; body messages can run parallel with spoken language, for example, through gestures and touch, or warmth and texture, or by exchanging facial expressions. The role of embodiment will be discussed in Section 2.3.1. Due to their mechanical outer body, robots can also combine communicative functions to perform other tasks, such as household chores, keeping track of calendars and shopping lists, and reading news from the newspaper and so on. But the robot's behavior can also have a purely social purpose, such as companionship or entertainment. According to the creators, the private living space is the habitat of choice for these new social robots.

Although the companion robots that are currently available communicate on a very primitive level and the predictability of their

---

* http://www.iec.ch/etech/2013/etech_0113/tech-2.htm.

response is high, this also motivates researchers to proceed in order to reach a more efficient and effective interaction (Breazeal, 2003; Heerink, Kröse, Wielinga, & Evers, 2009). Not only is a lot of knowledge missing about the mechanisms that encourage communication between humans and robots, but knowledge about how behavior occurs between humans and robots, and even how the interaction between people actually works, is also absent. This knowledge is critical to the design of robots, because the success of the social robot depends on successful interaction (Breazeal et al., 2008; Dautenhahn, 1995). This research is still in its infancy and is being studied in the discipline of human–robot interaction. In Section 2.3.2, we will briefly describe the state-of-the-art human–robot interaction and the focal points of this discipline today.

Since users are strongly inclined toward anthropomorphism, these robots quickly generate feelings (Duffy, 2003). The mechanization (rationalization) of the interaction with a human being therefore evokes ethical questions with respect to dehumanization. This issue will be detailed in Section 2.3.3.

2.3.1 The Robot Body

In a study by Ray, Mondada, and Siegwart (2008), it was found that people do not like the idea that a household robot looks like a human being, and that people prefer the robot to look like a small machine. This machine should also match the interior of the house, so that householders feel more comfortable adopting it (Sung, Christensen, & Grinter, 2008). For companion robots, however, the humanoid shape enables the robot to perform its task, that is, mainly the interaction with a human being, in a better way. If the companion robot has, for example, a face, so that it is able to use his eyes, mouth, ears, and eyebrows to express all kinds of emotions, it will be capable of intuitive communications with human beings. Companion robots have, however, one remarkable thing in common and that is they do not resemble humans in appearance in the slightest. They usually express childish or abstract life forms (see Boxes 2.1 and 2.2). For most designers of companion robots, the key is about some kind of communication via verbal and nonverbal means, but the appearance of the robot matters less. The nonverbal means, however, can

BOX 2.1   FURBY

The *Furby* (Figure 2.2) was designed in Japan by Tiger Electronics and was introduced in 1998. Since its introduction, there has been a lot of hype generated for this intelligent stuffed animal. Within a year, 1.8 million items were sold, and the following year 14 million were sold; in total over 40 million Furbies were sold.* In 2005, yet another improved Furby by Hasbro was introduced, but 2 years later, the Furby was taken out of production. As usual with a craze, interest slowly ebbed and disappeared, leaving Furbies in the closet or even in the garbage bin. In 2012, there was a revival. Hasbro released a new Furby, *Furby 2012*, with more expressive liquid crystal display (LCD) eyes, a wider range of emotions, its own iOS and Android app and the ability to adapt its personality in reaction to user behavior.† A year later, a new Furby was released: *Furby Boom* that "has more than twice as many responses as the previous Furby, remembers the name you give it, has 5 new personalities to discover and, on top of all that, it's hatching a new generation!"‡ Furby Boom

Figure 2.2   Furby with one of the authors. (Photo courtesy of Rinie van Est.)

---

* www.time.com/time/specials/packages/article/0,28804,2049243_2048661_2049232,
   00.html.
† www.dvice.com/archives/2012/07/furby_returns_f.php.
‡ www.hasbro.com/furby/en_us/shop/furby-boom.cfm.

was the best-selling toy of 2013.* To prevent Furby 2012 and Furby Boom from soon becoming tiresome, just like the previous Furbies, these Furbies are connected to a smart device, such as an iPad, which enhances the user's experience.

With the Furby, a fantasy animal with cuddly fur, for the first time a popular, commercially available form of artificial life was created. We can consider the Furby as the successor of *Tamagotchi*, a virtual pet, which one could take care of (see Figure 2.3). The Tamagotchi is an egg-shaped plastic device that contains a small computer with an LCD screen on which the "creature" is displayed. It was introduced in 1996 and within a few years more than 40 million items were sold.† The Furby, a robot pet, was the next craze in which the virtual aspect was embodied. By means of sensors, Furby could respond to touch and could talk in *Furbish*, a language with a limited vocabulary of 200 short words. Many speculated that

Figure 2.3   Tamagotchi. (Photo courtesy of Rinie van Est.)

---

* www.toynews-online.biz/news/read/furby-boom-named-best-selling-toy-of-2013/041659.
† www.technology.org/2013/11/28/tamagotchi-virtual-pets-return-europe-north-america/.

Furbies could actually "learn" words that were spoken to it and could even record classified information, which would compromise national security. This belief led, in 1999, to the National Security Agency (NSA) officials banning the Furby from its premises in Fort Meade, Maryland.* Because it is a cuddly toy, Furby unconsciously plays on the belief of its owner that it can communicate with people and is able to simulate emotions. The Furby is also fairly responsive; the more cuddles it gets, the more it produces a particular behavior and the more often it will exhibit this behavior. However, the owner of the Furby does not have it completely under control, because it was programmed to sometimes deviate from predictable patterns, just as people tend to do.

Due to the success of the Furby, the *Furby Baby* was introduced in 1999 along with new Furbies with other programs and characteristics. The second generation Furby, *Emoto-Tronic Furby*, was introduced in October 2005. This new generation was able to express more emotions through facial feature recognition and voice recognition technology and technology that enabled it to simulate breathing.† As stated, this Furby was discontinued in 2007.

_____

* news.bbc.co.uk/2/hi/americas/254094.stm.
† electronica.infonu.nl/diversen/1390-furby-het-emotronische-vriendje.html.

be expressed by the robot body: the exchange of facial expressions, gazing, body postures, gestures, and showing emotions (Kirby, 2010; Mutlu, Kanda, Forlizzi, Hodgins, & Ishiguro, 2012; Torta, 2014). According to Fong, Nourbakhsh, and Dautenhahn (2003), companion robots do not have to look human or animal-like. They must, however, be so designed that users feel challenged to interact with them. The experience of authenticity is important. The toy seal *Paro* (see Chapter 3), for example, only became a runaway success when the outer design was changed. Paro was first developed in a cat shape, but initial results showed that people did not find this toy cat's reactions realistic enough. By redesigning the model into a seal, and most of us have never had a real seal on our lap, the Paro robot was by contrast experienced as real.

BOX 2.2   AIBO, NAO, AND JIBO

AIBO* is a robot companion shaped like a dog that was developed by Sony and has been marketed since 1999 (Figure 2.4). AIBO stands for *Artificial Intelligence Robot* and also the Japanese word for friend or mate. Because of the high price— over U.S. $1100—very few were sold, around 150,000 in total. In 2006, the robot dog was taken out of production because it was not a profitable product[†]; it was a victim of practicality and, some would say, of the innovation-for-innovation's-sake mentality.[‡]

The robot dog can move reasonably smoothly, but slowly, and knows all sorts of tricks and dances. He can also express six emotions, which are surprise, happiness, unhappiness, anger, sadness, and fear. AIBO has a number of "instincts" built in: curiosity, hunger (as in the need for its battery to be charged), sleep, love, and being hyperactive with a lot of extra movement. He is able to play with a ball, for example, as he observes the brightly colored ball from the color-sensitive camera situated within his nose. Just like the Furby, the AIBO can learn tricks.

Figure 2.4   AIBO. (Photo courtesy of Junko Kimura/Getty Images.)

---

* support.sony-europe.com/aibo/ and http://www.sonyaibo.net/aibostory.htm.
† news.cnet.com/Sony-puts-Aibo-to-sleep/2100-1041_3-6031649.html.
‡ www.theverge.com/2014/2/4/5378874/sonys-new-aibo-is-a-french-bulldog-named-boss.

The AIBO gradually became more sophisticated. Later on, AIBO was equipped with speech recognition, speech synthesis, a built-in camera, and a higher number of improved touch-sensitive sensors. It also responded to an increasing number of commands, in the latest version to more than a hundred commands, and when "hungry," AIBO can return to its charging station independently.

The AIBO has been widely used for research into human–robot interaction, and is also used as a platform for all kinds of robotics experiments. An example is the use of these robot dogs until 2008 at the RoboCup, an international football competition for robots.* In 2008, the two-legged Aldebaran Robotics humanoid *NAO* (see Figure 2.5) was chosen as the replacement competitor for the discontinued Sony AIBO.†️ NAO is mainly used by universities and laboratories for research and education purposes with regard to human–robot interaction. With 25 degrees of freedom, NAO is capable of executing a wide range of movements (walking, sitting, standing up, dancing, avoiding obstacles, kicking, seizing objects, etc.). Equipped as standard with an embedded computer and Wi-Fi connection, NAO is fully autonomous and

Figure 2.5   NAO. (Photo courtesy of Bart van Overbeeke.)

---

* The organizers of RoboCup are working toward having a champion team of robots in 2050 that is able to beat the human FIFA world champions.
† www.robots-dreams.com/2007/08/breaking-news-n.html.

can establish a secure connection to the Internet to download and broadcast content. NAO is equipped with a voice recognition system that locates where the sound is coming from, so that he can turn his head towards the origin of the sound. He has a text-to-speech function so he can interact with his environment verbally by pronouncing any text sent to him.* NAO costs about U.S. $6500.

JIBO, THE FAMILY ROBOT

JIBO (see Figure 2.6), the world's first family robot, is only 28 centimeters tall with a 15 centimeters base, and weighs about 6 pounds. It sports a color camera, microphones with sound localization, full-body touch sensors, and a touchscreen face. It looks like a white eye that swivels in two parts to greet you, take your picture, read to the children, check e-mail, and send you appointment reminders, among other things. JIBO is the result of years of work and study by the MIT robotics

Figure 2.6 JIBO, the family robot, with Cynthia Breazeal. (Photo courtesy of Fardad Faridi, Jibo, Inc.)

---

* http://www.active-robots.com/aldebaran-robotics-nao-evol-humanoid-robot.

researcher Cynthia Breazeal. She explains that JIBO is "there to help families connect, coordinate, to play together, to try to help with organisation, and to help with a new form of telepresence or video call."* Breazeal previously created Kismet, the robot head that was designed for interacting with children (see Section 2.3.3). Similar to JIBO, Kismet also mimicked human emotions, and had facial features such as eyes and a mouth, but was not commercially available.

JIBO can sing and dance, but cannot walk around. Its core competence is about enlivening content and engaging people, but JIBO's potential extends far beyond engaging in casual conversation and completing daily tasks. JIBO is connected to the Internet via Wi-Fi and benefits from a cloud-based updating system. The software development kit will allow third parties to build new tools that work with JIBO and utilize some of its scripts to build custom interactions.† According to Breazeal, JIBO is about human empowerment and can play a big role in educating children, health care management and aiding the elderly.‡

The family robot will be ready for mass production in 2016, and the price will be about U.S. $500.

---

* www.cbc.ca/news/business/JIBO-robot-can-do-just-about-anything-in-the-home-1.2733450.

† mashable.com/2014/07/16/JIBO-worlds-first-family-robot/.

‡ www.indiegogo.com/projects/JIBO-the-world-s-first-family-robot.

However, research shows that only robots that look exactly like a human being, called androids (see the next section), will trigger reactions in people that are similar to the way people react to each other (Minato et al., 2006). When robots seem to be more human, our sense of familiarity increases, so the communication is facilitated and people stay enthralled for longer. By contrast, the robot that seems almost—but not quite—human will repel us in the same way that watching a zombie does. This effect is indicated by the term *uncanny valley* (Mori, 1970). The limitations of the *uncanny valley* have recently been bridged successfully by a Japanese research group, led by Professor Hiroshi Ishiguro (MacDorman & Ishiguro, 2006).

This research group developed *Repliee Q2*,* one of the most advanced robots of our time (see Figure 2.7). It is a female android. Her skin is made of soft silicone; she blinks her eyes and even looks like she's breathing. She also exhibits unconscious human movements, such as briefly yawning. Through lip-synching, it seems as if she really speaks. However, she can only recite prerecorded messages. She cannot respond to any signal, so no social interaction can occur. Currently, the researchers are working on her natural, human walking movement. This has not turned out to be a simple process, and will require years of study (Wairatpanij, Patel, Cravens, & MacDorman, 2009).

This Repliee Q2 robot is used for research on human behavior to find out which gestures and facial movements are minimally required in order to create the impression that the robot is a real person. The challenge for researchers is to eventually create an android indistinguishable in appearance, movements, and gestures from a human being. In an interview, Ishiguro states that he is under no illusions about the near future, for he says that it might take a 30-years-plus commitment to develop specific applications, such as a receptionist or a guide and the like. Theorizing about an android that may eventually be completely indistinguishable from humans in appearance and behavior, he is skeptical. He expects it will take about a hundred years for such a robot to be realized (Epstein, 2006).

Figure 2.7   Female android Repliee Q2. (Photo courtesy of Rinie van Est.)

---

* www.is.sys.es.osaka-u.ac.jp/development/0006/index.en.html.

2.3.2 Human–Robot Interaction (HRI)

Due to the wealth and ambiguity of human language and behavior, and the complexity of social practices, there are many major and minor design issues in facilitating interaction between robots and humans. How can a robot attract the attention of the human user, and vice versa? How can a robot be regulated so that the robot and the human user each speak in turn, as in a regular conversation? How can a human and a robot start making mutual eye contact just before a face-to-face conversation as a signal that a conversation can begin? How can a robot tell from the intonation of a person's speech whether the sentence is a question? How can a robot distinguish between different types of touch, such as hitting, pushing, and stroking, and react appropriately, perhaps with a spoken response? The discipline of human–robot interaction is all about these subtle aspects of interaction.*

According to experts in the field of artificial intelligence, the interaction between man and machine would be greatly simplified if the machine could use natural language (Gardner, Kornhaber, & Wake, 1996). In addition, Turing (1950) already suggested that natural language is one of the most essential characteristics of intelligence. The still-famous chatbox is *Eliza*, which was developed in 1966 by Joseph Weizenbaum. The computer had typed communications with people, in which it assumed the role of a Rogerian psychotherapist (Weizenbaum, 1966). People who talked with Eliza often did not realize that they were communicating with a computer instead of a human being. Yet we cannot call Eliza intelligent. A conversation with Eliza takes a ridiculous turn when the topic is outside Eliza's domain, because its sentence structures are based on fixed rules for transformation (Gardner et al., 1996). The use of natural language by itself does not appear to be sufficient for a machine to earn the label "intelligent." This is because this machine must also be able to associate and possess common sense, for which people living in its domain often need only half a word. This type of communication still turns out to be very difficult for machines. Solutions are sought in neuroscience, where research takes place into the field of learning.

---

* For an overview of this field, we refer to Fong et al. (2003) and Breazeal et al. (2008).

The basic idea is to teach a language from scratch to a machine—just as we humans learn it—through evolution (see, for example, Marocco & Nolfi, 2007). Despite these attempts, until now communication through natural language with machines, and thus also with robots, has only progressed with difficulty.

The value of communication with robots is not only embedded in spoken language, but also, for example, in the exchange of facial expressions, gazing, body postures, gestures, and the showing of emotions (Kirby, 2010; Mutlu et al., 2012; Torta, 2014). One of the focal points of the human–robot interaction discipline is to implement emotional skills in robots, which is required for efficient interaction. In her book *Affective Computing*, Rosalind Picard (1997) states that three emotional abilities play a role in communicating human emotions: the ability to *recognize* the emotions of others and to *express* one's own emotions and to actually *have* emotions. The most famous researcher in this area is from the United States—HRI expert Cynthia Breazeal. She is experimenting with "empathic" robots such as *Kismet*, who is able, using his eyes, mouth, ears, and eyebrows, to express all kinds of emotions and act accordingly. Kismet can recognize some emotions, but recognition of the full range of human emotions still seems far beyond reach. However, in a study carried out by Du, Tao, and Martinez (2014), a computational model of face perception has been developed, and they claim that this model can achieve an accuracy rate of 96.9% in the identification of the six basic emotions, and 76.9% in the case of the compound emotions.* The responses of a robot to the emotions of a human are largely based on a mirror mechanism: when one person laughs, the feeling that comes with the laugh is raised in the other (Breazeal et al., 2008). Although this research is still at an experimental stage, it seems technically possible that

---

* Andrade (2014) fears that robots will not only recognize faces and voices in the future, but also truths and lies, which would constitute a serious limitation on the individual's autonomy: "At the individual level, the freedom to not tell the truth is an essential prerogative of our autonomy as human beings. What this technology jeopardizes is something that goes beyond the simple impossibility of lying without being caught. This technology represents an assault on our right to privacy, our right to identity, and our right to freedom of expression—which encompasses not just what we choose to say, but what we choose to keep to ourselves."

in the future a robot will be able to react to several emotions that it recognizes. The third ability, actually *having* emotions, is still very far away and is a subject for philosophers: will a robot ever be able to experience emotions? However, for social interaction between robots and humans that question is irrelevant, because having emotions can also be simulated, as in the robot Kismet. Brian Duffy calls this proper simulation of emotions "the power of the fake" (Duffy, 2006).

As the field of HRI evolves, it is important to understand how users interact with robots over long periods. Leite, Martinho, and Paiva (2013) found that the limited capabilities of the commercial robots cause a gap between people's initial expectations and what they really experience after the initial interactions with the robot. On the other hand, "even when the novelty effect fades away people allow robots to come closer to them in later interactions. This result highlights that there is potential for social robots in the home environment, as long as they are capable of engaging users over extended periods of time. To do so, the robot's functionality (in other words, how it can assist users in their home routines), appears to be a major determinant" (Leite et al., 2013, p. 302).

When people interact with another human or an object on a regular basis, they start creating bonds with that human or object. So going one step further means that we bond in friendships and possibly even in intimate relationships with an android (Levy's second prediction). According to Levy (2007a), this still requires a number of technological breakthroughs, especially in the field of human–robot interaction and in engaging in conversations. Within the human–robot interaction community, opening up the possibility of friendship with a robot is a kind of scientific goal: "At the pinnacle of achievement, they could befriend us, as we could them" (Breazeal, 2003). The question that arises is whether such a friendly reciprocity between human and robot is possible. According to Dylan Evans (2010), a real friendship between robots and humans is impossible. Crucial to the feeling of friendship is that friendship is *not* unconditional. Intimate friendship is a kind of paradox: on the one hand, we want a friend to be reliable and not to let us down, but we do not want absolute devotion by the friend because we then lose interest. In addition, Evans argues that we can only really care about a robot when the robot can actually suffer. If robots cannot experience pain, we will just consider them to be dolls. This raises further ethical questions, such as whether we should develop robots that can suffer and whether we should grant rights to robots (see Levy, 2009).

2.3.3 Social De-Skilling

The mechanization of the interaction with a human being by human–robot interaction evokes ethical questions with respect to dehumanization. This dehumanization mainly involves social de-skilling: an erosion of the skills necessary to maintain human relationships.

People assign robots a psychological status, and even a moral status, which we previously only attributed to living humans (Melson, Kahn, Beck, & Friedman, 2009). Research shows that young children are much more attached to toy robots than to dolls or teddy bears, and consider them as *friends* (Tanaka, Cicourel, & Movellan, 2007). This raises issues such as the effect on the social development of children, as children often come into contact with social robots. British robot technologists Sharkey and Sharkey (2010) continue to question especially nanny robots for kids, as they think the robots will damage their emotional and social development and will lead to bonding problems in children. U.S. engineering psychologist Sherry Turkle (2011) is concerned that children will get used to perfect friendships with perfectly programmed positive robots, so children will not learn to deal with real-life people with all their problems and bad habits.

Ethicist Robert Sparrow (2002) has already expressed worries about companion robots remaining as "simulacra" for true social interaction. Emotions expressed by robots promise an emotional connection, which they can never give, for emotions displayed by robots are indeed merely imitations, and therein lays the danger, according to Sparrow (2002, p. 317): "imitation is likely to involve the real ethical danger that we will mistake our creations for what they are not." For Turkle (2011), the advancement of the robot for social purposes is worrying, and she fears that people will lose their social skills and become more socially isolated. Even if people are happy with their companion robot, the reduction in human contact may make them less socially able and, as a consequence, not as effective as citizens. Turkle therefore sees the robot as a symbol of a great danger, namely that the robot's influence stops us from being willing to exert the necessary *effort* required for regular human relations: "Dependence on a robot presents itself as risk free. But when one becomes accustomed to 'companionship' without demands, life with people may seem overwhelming. Dependence on a person is risky—because it makes us the subject of rejection—but it also opens us to deeply knowing another" (p. 65). The company of robots may be preferred to humans because we as human beings feel that we are in control in a human–robot relationship and that robots will not judge us like humans do. These ideas, however, remain speculations, because there has been only limited research on the actual effects of the impact of companion robots on children and adults (Tanaka & Kimura, 2009).

2.4 Android Robots: The Sex Robot

The sex robot has been a feature of science fiction for decades. In the 1973 dystopian thriller *Westworld*, for example, wealthy tourists indulge themselves at a theme park staffed by robots who cater to their every desire, including sex.* Sex is often the driving force behind the acceptance of new technologies, such as the Internet, video cassette recorders (VCRs), cable TV, and CD-ROMs (Johnson, 1996). For example, VCRs were an obscure technical tour de force until the porn industry technology hijacked them and made them a great success. People from the porn industry had the bright idea that people with a home video player could watch porn anonymously for pleasure. Sex robots could also become the engine for the further development of socially interactive androids and a growing acceptance of social androids with multiple functionalities, such as childcare, keeping calendars and shopping lists, supplying certain information found on the Internet, and so on.

With the robots in the previous section, entertainment was central; these robots invite people to learn things or invite caring behavior. In this section, we discuss another type of companion robot, the sex

---

* www.youtube.com/watch?v=LcL3eP0Hfy4.

robot, built for intimate physical interaction between man or woman and machine. The main feature of the sex robot is its body. In contrast to the companion robots discussed in the previous section, sex robots look and feel like a human being in order to simulate the human performance of having sex. This rationalization of sex has some benefits (see Section 2.4.2), but also raises some serious ethical question with regard to acceptance, dehumanization, and sex with child robots (see Section 2.4.3). We start with the introduction of the first female sex robot: Roxxxy.

2.4.1 Roxxxy

In 2007, the 62-year-old Scottish chess master David Levy, then living in London, received his PhD on "Intimate Relationships with Artificial Partners," a dissertation on the topic of love and sex between humans and robots. He finished that dissertation in Maastricht, in the Netherlands, because the subject is taboo in Britain. In his dissertation, and in the commercially available edition *Love + Sex with Robots*, Levy (2007a) says that by 2050 it will be considered quite normal that people and robots tie the knot. Levy also expects the first sex robots will be introduced to the market within a few years, and that sex with a robot will inevitably become a popular human activity. The first prediction has already come true, because sex robots are now available over the Internet. In January 2010, at the *AVN Adult Entertainment Expo* in Las Vegas, the first female sex robot was presented that can even talk: *Roxxxy* (see Figure 2.8).*

Roxxxy is 1.70 meters tall, weighs 55 kilograms (121 pounds) (the newer ones are lighter, around 25 kilograms or 55 pounds), and embodies five personalities: the extravagant and adventurous Wild Wendy; the very reserved, shy Frigid Farah; the motherly Mature Martha; SM-Susan; and inexperienced Yoko. Consumers can further customize the sex robot itself, in terms of skin color, hair color, hairstyle, and breast size. The anatomically correct Roxxxy has synthetic skin, shows a heartbeat, can move and responds to touch. In addition,

---

* www.foxnews.com/scitech/2010/01/11/worlds-life-size-robot-girlfriend/.

Figure 2.8   Sex robot Roxxxy. (Photo courtesy of ANP.)

she can conduct simple conversations, which can be programmed so that she speaks about matters that interest the client.*

Roxxxy has strategically placed sensors that ensure, for example, that she responds to your touch. The robot is wirelessly connected to the Internet for updates to expand her capabilities and vocabulary on how to care for Roxxxy. Roxxxy was made to help improve the sex life of individuals and couples. The robot also provides companionship and unconditional love for lonely people or those who want to turn their fantasies into reality. Roxxxy's inventor, Douglas Hines of True Companion, wanted to create a doll that could talk with the owner and establish a bond. But whereas real love is free, you have to pay about U.S. $5500 for this sex robot.

Also, to cater to women, a male version of the robot, called *Rocky*, is already being developed. According to Douglas Hines, Roxxxy garnered about 4000 preorders shortly after its reveal in 2010. However, to date, no actual customers have ever surfaced for a Roxxxy doll, and it is doubtful whether any commercial Roxxxy dolls have ever been produced.† One reason could be the issue of the uncanny valley (see Chapter 1). The sex robots look normal and vaguely human, whereas their behavior is not human enough. This leads to a strong sense of unease, and people probably consider these dolls to be creepy.

* www.truecompanion.com/home.html.
† en.wikipedia.org/wiki/Roxxxy.

2.4.2 Benefits of Sex Robots

According to Levy (2007a), in the near future the sex robot will be a better sexual lover than humans. The "brain" of the future sex robot holds a complete sex encyclopedia: it is stuffed with tips and tricks from the world of porn, sex therapy, and the *Kama Sutra*. A major advantage is that the sex robot can patiently teach people the art of making love, so that men who are insecure about their performance in bed can take unlimited private lessons from their sex robot teacher. The second advantage that Levy mentions is that many relationships that suffer from an extinguished sex life will be saved, because no one will be stuck with poor or mediocre sex any more. Another advantage is that people who feel insecure about their own sexual orientation can easily experiment with male and female sex robots. In some cases, these robots could satisfy people's desire for illicit sexual practices, such as paedophilia, and could be used in therapy to remedy the underlying problem.

For society as a whole, sex robots do have benefits. According to Levy, there will be fewer cases of unwanted pregnancies, sexually transmitted diseases (STDs), and AIDS (see also Yeoman & Mars, 2011); however, if sex robots are reused by several clients, poor sanitation could carry a risk of infection. Sex robots can also help put an end to the major problem of prostitution and sex trafficking. That the problem of prostitution is huge is clear from the findings of a study of nine countries (Canada, Colombia, Germany, Mexico, South Africa, Thailand, Turkey, the United States, and Zambia): about 90% of the women in prostitution wanted to escape prostitution; 60%–75% were raped in prostitution; and 70%–95% were physically assaulted in prostitution (Farley et al., 2004). A conservative estimate of the number of victims of sex trafficking globally is 4,500,000 (International Labour Organisation, 2012). From a study by Raymond (2008), it follows that legalization or decriminalization of prostitution, such as in the Netherlands, does not make things better for women involved in prostitution or sex trafficking. However, a ban on visiting prostitutes, such as in Sweden, Norway, and Finland, where it is not the prostitutes who are punished, but those who pay for sex, seems a promising solution. The ultimate goal of this ban was "to protect the women in prostitution by addressing the root cause of prostitution and trafficking: the men who assume the right to purchase female beings and sexually exploit them" (Ekberg, 2004, p. 1210). In these countries, the number of men who buy prostituted women has decreased and the local prostitution markets have become less lucrative, with the result that traffickers choose other and more profitable destinations (Ekberg, 2004; Raymond, 2008). Sex robots could become popular in these countries—like sex dolls, which are now popular in South Korea due to the enforced ban on prostitution. But in countries where prostitution has been legalized, sex robots could also contribute to solving the problem of sex slavery and the trafficking of women as the quality of their sexual services is comparable to those of their human counterparts, and this could contribute to a decrease in the demand for human prostitutes.

Yeoman and Mars (2011) present a futuristic scenario about sex tourism in Amsterdam. In 2050, Amsterdam's well-known red-light district will be all about android prostitutes who are clean of sexually transmitted infections (STIs), not about prostitutes who are smuggled in from Eastern Europe and forced into slavery, and the city council will have direct control over android sex workers, controlling prices, hours of operations, and sexual services. They conclude that if "such a proposition came true Amsterdam would probably be the safest and best sex tourism destination in world and all the social problems associated with sex tourism would disappear overnight" (Yeoman & Mars, 2011, p. 370).

According to Levy, the benefits of sex robots create their own ethical justification, and the majority of the population will also come to see sex robots as "the ethical choice" because of their positive effects. Yet Levy also sees a downside to this development of sex robots, namely the forced resignation of redundant human prostitutes: "This problem, the compulsory redundancy of sex workers, is an important ethical issue, since in many cases those who turn to prostitution as their occupation do so because they have literally no other way to earn the money they need" (Levy, 2007b). Danaher (2014) has some doubts about whether prostitution is just as vulnerable to technological unemployment as other forms of human labor. He argues that there is a compelling case for the alternative hypothesis that the demand for and supply of human sexual labor is likely to remain competitive in the face of sex robots. His argument is based on his belief that "the preference for human sex partners could be a significant factor when it comes to the future of human prostitution, despite the alleged advantages of sex robots" (Danaher, 2014, p. 123), and is coupled with his thesis that technological unemployment in other industries is likely to increase the supply of human prostitutes.

2.4.3 Social and Ethical Issues

2.4.3.1 Cultural Acceptance of Sex Robots The demand for sex robots will probably be relatively small in the short term because of the current high price, although people seem to be interested: according to a 2009 poll by IEEE,* 37% of the respondents answered that they would prefer a sex robot to the question: "If you had a personal robot that could do only one thing, which ability would you prefer it to have?" Levy (2007b) expects that the sex robot will soon be introduced in the prostitution industry, because for most people renting a sex robot is the only option if they want to experiment with it. Levy predicts the expected success of the sex robot from the earlier successes of renting sex dolls in Japan and Korea.† Especially in South Korea, sex doll brothels are a big hit, says Levy, because there all human prostitution is prohibited. An hour with a doll costs about U.S. $25. Levy concludes that if renting sex dolls has been a success, then this will also apply to sex robots too. In response to this prospect, Amanda Kloer even sees robots as perfect prostitutes, but she inserts a note:

> In a way, robots would be the perfect prostitutes. They have no shame, feel no pain, and have no emotional or physical fall-out from the trauma which prostitution often causes. As machines, they can't be victims of human trafficking. It would certainly end the prostitution/human trafficking debate. But despite all the arguments I can think of for this being a good idea, I've gotta admit it creeps me out a little bit. Have we devalued sex so much that is doesn't even matter if what we have sex with isn't human? Has the commercial sex industry made sex so mechanical that it will inevitably become … mechanical?‡

---

* http://ieet.org/index.php/IEET/print/3573.

† Japanese advertisements for these sex dolls even go as far as to claim that customers will never want a real girlfriend again (see http://www.huffingtonpost.co.uk/2014/08/13/dutch-wives-sex-doll-toy-japan_n_5674461.html).

‡ http://news.change.org/stories/are-robots-the-future-of-prostitution.

After their first introduction in the prostitution industry, the price of sex robots will drop, and they will become available to a wider audience. In an interview in 2008, Levy pictures hordes of men with a lack of social contact who find it difficult to enter into a relationship with a woman. These men are desperate for a sex partner who transcends the level of the inflatable doll, he says. If the media then get wind of this new trend, according to Levy, the acceptance of sex robots will proceed at a fast pace.* One key question, however, is whether people really would want to have sex with a robot. A survey conducted among 2000 members of the British public by Middlesex University in 2014 found that 17% of people are prepared to "have sex with an android," whereas 29% said they had no problem with machines being used in this way. Forty-one percent of respondents said that they found the idea uncomfortable, and 14% said that in their view robots should not be used for sex.†

The possibility of having sex with robots may reduce the incidence of cheating on a partner and adultery. There is still the question of whether having robot sex would be considered as being unfaithful, or whether robot sex would become just as humdrum and innocent as the use of a vibrator nowadays (Maines, 1999). The main difference between a sex robot and a vibrator is that the idea of a sex robot is that it would be a stand-in for a human being (and therefore would be too much like really having sex with another human), in contrast to a vibrator, which is not meant to be a replacement for a human being but it is meant to create a certain kind of pleasure. According to Greta Christina (2014), the whole idea of a sex robot would be that it would be like having sex with a person, except without all that pesky business of it having desires and limits of its own. This last remark leads to the ethical issue of dehumanization.

2.4.3.2 Dehumanization   The rationalization of sex with sex robots leads to the fact, as Amanda Koer expressed it, that sex is made mechanical. Levy only sees the advantages of mechanical sex with robots, since robots "behave in ways that one finds empathetic,

---

* http://www.youtube.com/watch?v=K7FENChv6v4#t=185.
† www.theguardian.com/technology/2014/may/06/third-of-britons-fear-rise-of-robots-poll.

always being loyal and having a combination of social, emotional, and intellectual skills that far exceeds the characteristics likely to be found in a human friend" (Levy, 2007a, p. 107). According to Levy, it is almost a moral imperative that we work to make these theoretical robotic companions a reality, because these robots could add so much love and happiness to our world. Unlike Levy, Turkle (2011) states that the use of sex robots, rather than leading to social-ization, will result in de-socialization (social de-skilling, see also the previous section). Sex will become purely a physical act without commitment and even caring rather than a loving act. She describes a trend toward rejecting authentic human relationships for socia-ble, human-like robots, and asks herself the question what kind of people we are becoming as we develop increasingly intimate rela-tionships with machines. Sullins (2012, p. 408) responds that Levy ignores the deep and nuanced notions of love and the concord of true friendship:

> While it is given that robots can be built that people find sexually attractive, it is unlikely that a machine can be built that will be capable of building an erotic relationship between itself and the user. Instead, with these technologies as they are currently evolving, we have an engineering scheme that would only satisfy, but not truly satisfy, our physical and emotional needs, while doing nothing for our moral growth.

Furthermore, sex robots are "simulacra" for sex with a human being, with the danger that we may mistake sex with a robot for true love, which robots can never give (cf. Sparrow, 2002).

2.4.3.3 Sex with Child Robots   Finally, the issue of sex with child-like robots needs to be addressed. This is a new issue for lawyers and ethicists. In contrast to child pornography, which is illegal in most countries, the use of child robots does not involve any direct harm to children. Child pornography causes harm in its production and/or its dissemination. We can compare sex with a child robot with sex with a child avatar. In both cases, no direct harm to chil-dren is involved. In 2002, the Supreme Court in the *Ashcroft v. Free Speech Coalition* case ruled that virtual child pornography (in which young adults or computer-generated characters play the parts of children) is protected by the First Amendment and cannot be criminalized. In 2003, however, the PROTECT Act was passed into law to attack pandering to and solicitation of child pornography regardless of whether the material consisted of computer-generated images or even of adults who looked like children, or even if the material was fraudulent or did not exist at all. In the United Kingdom, "visual depictions featuring child abuse which appear to be photographic have been deemed illegal by both statute (the Children Act 1978) and courts (*R. v Bowden* 2001 QB 88, 2000 2 All ER 418) using the phrase pseudo-photographs" (Adams, 2010, p. 64), which means an image, whether made by computer graphics or in any other way whatsoever, which appears to be a photograph, including film. Also, in the Netherlands in 2011, a court held that possession of virtual child pornography is punishable.* The reasoning was that virtual child pornography could become part of a subculture that promotes sexual abuse of children. Based on this verdict, a judge in the Netherlands will probably criminalize sex with child robots at some point in the future, but the problem here will be that the law would be overstretched. According to the current legislation in most countries, having sex with child robots is not regulated (Bamps, 2012). The legislature probably could not have foreseen the development of robotics that enables sex with robots and therefore also sex with body robots that are built like children. The question is how the legislature would envisage dealing with those who have sex with child robotics. If we wish to suppress a subculture that promotes sexual abuse of children, there should be a legal framework for child–robot pornography. If law is seen as a public expression of morality, then such a legal framework is necessary, since 90% of the respondents of a 2010 questionnaire were against the development of child robotics for the satisfaction of sexual needs (Bamps, 2012).

Robot ethicists have also interfered with the issue of sex with child robots. In 2014, there was an event called Our Robot Future at the University of California where robot experts also discussed this topic. According to Ronald Arkin, who does not approve of child

---

* Rb. Rotterdam, March 31, 2011, LJN BP 9776, www.rechtspraak.nl/Organisatie/ Rechtbanken/Rotterdam/Nieuws/Pages/Bezit-virtuele-kinderporno-strafbaar.aspx.

sex robots for recreational use, "[c]hild-like robots could be used for paedophiles the way methadone is used to treat drug addicts, there are no presumptions that this will assuredly yield positive results—I only believe it is worth investigating in a controlled way to possibly provide better protection to society from recidivism in sex offenders, if we can save some children, I think it's a worthwhile project."* The robotics expert Ben Way also supports Arkin's idea. Speaking to the *MailOnline*, Way says: "Will child sex bots lead to some people acting out their dark and disgusting desires on real children? Yes, but I suspect having child sex bots will significantly reduce the number of people overall who abuse children. ... As repugnant as it may seem society should support this technology and do proper research into its effects before making a snap decision based on social norms, the most important thing we can do as a society is reduce harm to children whatever way we can do it."† It is clear that research is needed, since there is also the concern that a child sex robot could encourage pedophiles to act on their impulses instead of serving as a safe outlet for them. As Pardes states: "But we're not going to get anywhere with rehabilitating paedophiles if we treat them like monsters by encouraging them to go at it with weird, childlike sex bots."‡

2.5 Observational Conclusions

In this chapter, we have discussed three types of personal robots: the mechanoid functional household robot, the humanoid companion robot, and the android sex robot. The last two types of robots are socially interactive robots, where the robot body and the human–robot interaction is essential for the anthropomorphism, that is, attributing human characteristics and behaviors to nonhuman subjects in order to interact in a smooth way with humans. This anthropomorphism especially raises ethical questions.

---

* www.forbes.com/sites/kashmirhill/2014/07/14/are-child-sex-robots-inevitable/.
† www.dailymail.co.uk/sciencetech/article-2695010/Could-child-sex-robots-used-treat-paedophiles-Researchers-say-sexbots-inevitable-used-like-methadone-drug-addicts.html.
‡ www.bbc.com/news/blogs-echochambers-28353238.

In this final section, we will briefly discuss the expectations of these three types of robots in the short and the medium term, summarize the main ethical issues, and indicate some regulatory issues.

2.5.1 Household Robots

In relation to the household robots, we see a gap between the high expectations concerning multifunctional robots that completely take over housework and the actual performance of the currently available robots. Few of the current developments point toward the introduction of multifunctional robots that will do all the cleaning for us. Despite all the technological developments, for the time being household robots have quite a high science fiction level and are still very much in their infancy.*

It is unlikely that the *monomaniacal* simple cleaning robots such as vacuum-cleaning robots and robots that clean windows will turn up in large numbers in our households. These *monomaniacal* robots turn out not to be efficient because they cannot perform a specific entire household task, and they also force the user to adapt and streamline part of their environment. As we have seen in the vacuum-cleaning robot and the ironing robot, only parts of vacuuming and ironing jobs can be taken over. The household chores appear to be far more complex than previously thought, which leads to major challenges for the rationalization of household tasks so that they can be performed by a robot, since this rationalization encounters fundamental limitations. Many situations in which a household task must be performed do require a lot of decisions, which are largely based on common sense, for which no fixed algorithms exist, that is, this is the frame problem. Robots for these decisions cannot really be constructed.

Having "a robot in every home" in the coming years, in our opinion, is highly unlikely. We expect that this cannot yet be realized in the short or the medium term. Many technical challenges must be overcome before the home robot can convince the public that it can take over household chores completely and efficiently.

---

* See, for example, http://teresaescrig.com/service-robotics-is-still-very-much-in-its-infancy/, http://www.washingtonpost.com/wp-dyn/content/article/2010/12/20/AR2010122004781.html.

2.5.2 Companion and Sex Robots

2.5.2.1 Expectations Compared to the household robot, expectations concerning the companion robot are much less predefined. The goals are just communicating, playing, and relaxing. The need is not set, but arises in the interaction. We see an age-old dream come true: devices that resemble humans or animals, and with which we can interact. Examples are the dog AIBO, the fluffy cuddly toy Furby, JIBO, and the sex robot Roxxxy—all four invite us to play out social and/or physical interaction. People become attached to the robot and give it human attributes.

Nevertheless, we certainly cannot speak of a success story. The socially interactive robots that are currently available are very limited in their social interaction and are very predictable, so consumers will not remain fascinated for long and will forget about them after a short while. At this time—and probably within the next 10 years—we should therefore consider commercially available socially interactive robots such as AIBO, Furby, and JIBO as fads and gadgets whose lustre soon fades, rather than as kinds of family friends. How the sex robot will develop is still unknown, but the sex industry and some robot technologists see a great future for this robot and consider the sex robot to be a driving force behind the development of social robots and human–robot interaction research. Human–robot interaction science is still in its infancy but is one of the spearheads of ICT research with a long-term goal of the participation of the robot in human spheres. However, in order to let the robot interact with humans in a successful manner, many obstacles must still be overcome in order for there to be a social robot, which has the properties defined by Fong et al. (2003): it can express and observe feelings; is able to communicate via a high-level dialogue; has the ability to learn social skills, to maintain social relationships, to provide natural cues such as looks and gestures; and has (or simulates) a certain personality and character. It will take decades before a social robot has matured enough to incorporate these properties, but modern technology will make it increasingly possible to interact with robots in a refined manner. This will turn out to be a very gradual process.

2.5.2.2 Social, Ethical, and Regulatory Issues The humanoid companion robot is modeled with anthropomorphism in mind. The anthropomorphism is accomplished by the rationalization of social interaction by research in the field of human–robot interaction. This rationalization raises all sorts of social and ethical questions, particularly related to dehumanization, such as what influence do companion robots have on the development of children and on our human relationships? Fears range from damaging social development, especially in children, to no longer being able to cope with other people with all their problems and their bad habits and to an unwillingness to invest in long-term relationships. Relationships with robots are much less binding, making people less empathetic because intimacy is avoided. There is a fear of social de-skilling as people become attached to robots. Little research has been done in this field, but it is important that we think about boundaries: where and when do companion robots have a positive socializing effect and where do we expect de-socialization? Attention to the influence of robotics on our social capital should therefore be a major item on the public issues agenda.

The android sex robot, the ultimate symbol of the rationalization of lust, may in future contribute to solving the general problem of sex slavery and the trafficking of women, an issue which is currently ignored by many politicians. As long as policy makers in countries are unwilling to ban visits to prostitutes, as happens for example in Sweden, Norway, and Finland, they should stimulate the introduction of robotic brothels. Imperative to this is whether the sex robot is actually an alternative to a human prostitute for prostitute clients. Research in this field is still lacking.

In addition to the ethical issues concerning the dehumanization of the human companion robot, which also are of importance in relation to sex robots, sex robots put forward the issue of sex with child robots and the associated question of whether child–robot sex should be punishable. The questions that now arise are whether this child–robot sex contributes to a subculture that promotes sexual abuse of children, or whether it reduces the sexual abuse of children. The current national legislations do not establish that sex with child robots is a criminal offence. National legislators (or, for example, the European Commission) will have to create a legal framework if this behavior is to be prohibited.

Interview with Kerstin Dautenhahn (Professor of Artificial Intelligence, University of Hertfordshire, United Kingdom)

"I don't need a robot to respond to my every smile and frown."
"Domestic and care robots are getting better and better at human-like communication. But will they ever really get the hang of it? And perhaps more importantly, why should we want them to in the first place?" asks Professor Kerstin Dautenhahn.

The skills of social robots have advanced a great deal. Ten years ago, they could only tell people apart who looked directly into the camera under good lighting conditions. Nowadays, face recognition software is much more flexible.

Speech recognition too has advanced, though it still has serious limitations. Typically, it has to be trained by the individual user before it will do a good job, and then only in certain domains, such as dictation of letters. It can't do this important thing that we do all the time: guess what the other person is saying. Listening is much more than figuring out how phonemes combine into words and sentences. It's also about prediction and anticipation, based on who is talking and about what. We'll see more progress, I'm sure, but we tend to underestimate how hard it is. The AI pioneers in the 1950s, such as Marvin Minsky, thought they could solve 'the speech problem' in two months. I don't expect we will ever solve it, because it covers the whole of AI: cognition, intelligence, personality, the question of embodiment, theory of mind. Human language cannot be interpreted on the basis of the words alone. Things such as metaphor, humor and irony also enter into it. The fact that highly intelligent people with autism have trouble understanding these illustrates how complicated they are.

Another skill that robots have become better at is interpreting the facial expressions that come with strong emotions. However, in daily life, we don't experience much ecstatic happiness or deep sadness. We just smile, or we are somewhat annoyed. For a robotic personal assistant to be useful, it should be able to recognize the subtler emotions. Which implies that it should also notice our scratching our heads or rubbing our chins, for human communication is multimodal. It's not just the face, not just the posture, not just the words or the voice or the gaze, it's all of that and more. Robots have not yet integrated all these modes. To get there, I think all these separate fields of research need to be integrated first.

For robots that can interpret human behavior, the next challenge will be to respond meaningfully. And as so often, Hollywood is creating unrealistic ideas. The robot in the film *Her*, which interprets and responds to a person's emotions, is far beyond anything that is possible any time soon. It requires way more intelligence than robots have. The AI pioneers thought robot intelligence would be a matter of giving them access to a huge database, with all the information about the world: the weather, Denmark, you and me, everything. But it doesn't work like that. Having a robot communicate in a well-defined context, say a cash withdrawal machine, where the exchange is limited and largely predictable, that's easy. But for a personal robot assistant in the house to have a meaningful conversation, that's a very different challenge.

Another issue is movement. The human–robot interaction discipline is beginning to recognize that it's not only iconic gestures that matter in human interaction. People unconsciously move their bodies all the time, and expect others to do likewise. Even our breathing is noticed by others.

Sometimes I think that it might just be too difficult to build a robot that behaves like a human being. There are such fundamental differences. We do a lot of things intuitively, on a hunch, without high-level cognition kicking in, and then we make up stories to post-rationalize our actions. I can't see robots doing that. They don't have the experience of being a sentient animal, feeling real pain and fear and happiness and frustration and the rest.

Moreover, why should we want robots like that in the first place? If I broke my leg and needed a caregiver for some weeks, would I mind having to tell it what I need, or even type in commands? No! Do I need a robot that responds to my every little smile and frown? Does it have to think, "Oh, maybe she's thirsty now, or bored?" No, I just want it to help me. But there is this AI vision, which goes a long way back in human civilization, that we want to build things in our own image: from clay figurines and marble statues to Frankenstein's monster and, nowadays, humanoid robots.

Don't get me wrong: I like robots a lot. I find them fascinating and useful. In education, we have reached the stage where we are making things happen. We are putting the systems out, each built for specific purposes. Here in Hertfordshire, our pioneering work has resulted in Kaspar, a robot that can teach children with autism about social and tactile interaction: how will other children respond if you touch or otherwise treat them in certain aggressive or friendly ways? In Istanbul, Hatice Kose is working on a robot that teaches sign language to hearing-impaired children. Other robots help in learning maths or English or other subjects. We need more research to find out how robots can best fit the curriculum and how teachers are best trained. But they can teach children stuff. And children like them; that helps.

In care, too, social robots are making headway. At this stage, we have only prototypes, which require special environments, places that are tidy or even seriously modified. Advocates of these systems argue that when we build new housing for assisted living, we might as well create a robot-friendly environment right away, with lots of integrated sensors and with no rugs on the floor. To my mind, the challenge should be to have robots work in the sort of place where humans actually like to live, including the clutter that makes a house a home.

And again, I wonder if we really need our care robot to be an all-rounder. From a purely robotics point of view, it is very hard to build something that can effectively and safely help in the kitchen *and* in the living room *and* in the bathroom. Why not have dedicated stationary systems? In the bathroom, you might want a big strong arm attached to the wall, which you can use to lean on and put your towel on and which can help you come out of the shower. Why stick to this age-old vision of one complete system that does everything and is similar to a human? Surely not because that's what Hollywood makes us expect?

Human-like robots raise another issue: how do you stop users from considering them real friends, in the human sense? [MIT psychologist] Sherry Turkle, who draws a strict line between authentic and inauthentic interaction, is worried about the effects of humanoid robots on children. Some reply that robots are not all that different from pets. But I think she has a point when she says that systems built to manipulate people's behavior and emotions, what she calls relational artifacts, are a real game-changer. If a robot looks human and somehow responds to your emotions and behaves as if it too had emotions—well, you and I would realize it's programmed, but children or people with dementia might not. They might well think, "This machine is my friend" or perhaps even, "He's so much better than my other friends, who don't have time and are sometimes unkind." I think Turkle's worries are well-founded here.

As a robotics researcher, I feel we have a lot of responsibility in deciding what kind of shape and voice we give a robot, how we present it, what we tell people they can do with it. And here's a dilemma I've been wrestling with recently. I've always considered it wrong to build a robot companion for elderly people in care homes that is primarily there to socially interact with them, because interaction ought to be done by real people. Robots should be there to help with physical routine tasks, not to provide artificial interaction. But now a colleague of mine here in Hertfordshire has gone and interviewed elderly people in a care home. And she found that what they wanted was a robot to talk to, and that will talk to them. They don't get visitors that often and many of them don't particularly like their fellow residents. The caregivers are too busy, and listening to the stories of the residents, perhaps over and over again, is for them not the most interesting task.

So, if some elderly people say they would benefit from having a robot to talk to that will patiently listen to the repetitive stories of the residents and will even talk back a bit ... (sighs)—is that something we should build? Or should the ethical consideration that it's not authentic interaction prevail? With children, of course, the answer is easy: they have to learn how to live and deal with real people. But is that also true for the elderly, or even for the terminally ill? I don't have an answer. What I've learned is that it's not enough to address these questions from an intellectual point of view. We also have to ask people.

References

Aarts, E. H. L., & Encarnação, J. L. (Eds.). (2006). *True visions. The emergence of ambient intelligence*. New York: Springer Verlag.

Adams, A. A. (2010). Virtual sex with child avatars. In C. Wankel & S. Malleck (Eds.), *Emerging ethical issues of life in virtual worlds* (pp. 55–72). Charlotte, NC: Information Age Publishing.

Andrade, N. (2014, June 9). Computers are getting better than humans at facial recognition. *The Atlantic*, 2014. http://www.theatlantic.com/technology/archive/2014/06/bad-news-computers-are-getting-better-than-we-are-at-facial-recognition/372377/ (accessed October 14, 2014).

Asaro, P. M. (2012). A body to kick, but still no soul to damn: Legal perspectives on robotics. In P. Lin, K. Abney, & A. Bekey (Eds.), *Robot ethics. The ethical and social implications of robotics* (pp. 169–186). Cambridge, MA: The MIT Press.

Bamps, D. (2012). Sex with a child robot: Psychological, ethical and legal arguments. In H. Nelen & J. Claessen (Eds.), *Beyond the death penalty: Reflections on punishment* (pp. 161–182). Mortsel, Belgium: Intersentia.

Bowden, S., & Offer, A. (1996). The technological revolution that never was. Gender, class, and the diffusion of household appliances in interwar England. In V. de Grazia & E. Furlough (Eds.), *Sex and things: Gender and consumption in historical perspective* (pp. 224–275). Berkeley, CA: University of California Press.

Breazeal, C. (2003). Toward sociable robots. *Robotics and Autonomous Systems*, *42*(3–4), 167–175.

Breazeal, C., Takanski, A., & Kobayashi, T. (2008). Social robots that interact with people. In B. Siciliano & O. Khatib (Eds.), *Springer handbook of robotics* (pp. 1349–1369). Berlin, Germany: Springer.

Christina, G. (2014, January 16). Would you have sex with a robot? io9. http://io9.com/would-you-have-sex-with-a-robot-1502351287 (accessed August 1, 2014).

Cowan, R. S. (1983). *More work for mother. The ironies of household technology from the open hearth to the microwave.* New York: Basic Books.

Danaher, J. (2014). Sex work, technological unemployment and the basic income guarantee. *Journal of Evolution and Technology*, *24*(1), 113–130.

Dautenhahn, K. (1995). Getting to know each other. Artificial social intelligence for autonomous robots. *Robotics and Autonomous Systems*, *16*(2–4), 333–356.

Du, S., Tao, Y., & Martinez, A. M. (2014). Compound facial expressions of emotion. *Proceedings of the National Academy of Sciences of the United States of America*, *111*(15), 1454–1462.

Duffy, B. R. (2003). Anthropomorphism and the social robot. *Robotics and Autonomous Systems*, *42*(3–4), 170–190.

Duffy, B. R. (2006). Fundamental issues in social robotics. *International Review of Information Ethics*, *6*(12), 31–36.

Ekberg, G. (2004). The Swedish law that prohibits the purchase of sexual services. Best practices for prevention of prostitution and trafficking in human beings. *Violence against Women*, *10*(10), 1187–1218.

Epstein, R. (2006). My date with a robot. *Scientific American Mind*, *17*, 68–73.

Evans, D. (2010). Wanting the impossible. The dilemma at the heart of intimate human–robot relationships. In Y. Wilks (Ed.), *Close engagements with artificial companions. Key social, psychological, ethical and design issues* (pp. 75–88). Amsterdam, the Netherlands: John Benjamins Publishing.

Farley, M., Cotton, A., Lynne, J., Zumbeck, S., Spiwak, F., Reyes, M. E., ..., Sezgin, U. (2004). Prostitution and trafficking in nine countries. *Journal of Trauma Practice*, *2*(3–4), 33–74.

Floridi, L. (2014). *The fourth revolution: How the infosphere is reshaping human reality.* Oxford, UK: Oxford University Press.

Fong, T., Nourbakhsh, I., & Dautenhahn, K. (2003). A survey of socially interactive robots. *Robotics and Autonomous Systems*, *42*(3–4), 143–166.

Gardner, H., Kornhaber, M., & Wake, W. (1996). *Intelligence: Multiple perspectives*. Fort Worth, TX: Harcourt Brace.

Gates, B. (2007). A robot in every home. *Scientific American Magazine*, *296*(1), 58–65.

Heerink, M., Kröse, B. J. A., Wielinga, B. J., & Evers, V. (2009). Influence of social presence on acceptance of an assistive social robot and screen agent by elderly users. *Advanced Robotics*, *23*(14), 1909–1923.

Johnson, P. (1996). Pornography drives technology: Why not to censor the Internet. *Federal Communications Law Journal*, *49*(1), 216–227.

International Labour Organisation. (2012). *ILO global estimate of forced labour. Results and methodology*. Geneva, Switzerland: International Labour Organization. http://www.ilo.org/global/topics/forced-labour/publications/WCMS_182004/lang—en/index.htm (accessed October 26, 2014).

Kirby, R. (2010). *Social robot navigation* (PhD thesis). Pittsburgh, PA: Carnegie Mellon University.

Leite, I., Martinho, C., & Paiva, A. (2013). Social robots for long-term interaction: A survey. *International Journal of Social Robotics*, *5*(2), 291–308.

Levy, D. (2007a). *Love + sex with robots. The evolution of human–robot relationships*. New York: HarperCollins Publishers.

Levy, D. (2007b). Robot prostitutes as alternatives to human sex workers. Presented at *ICRA'07*, Rome, Italy, April 14. www.roboethics.org/icra2007/contributions/LEVY%20Robot%20Prostitutes%20as%20Alternatives%20to%20Human%20Sex%20Workers.pdf (accessed September 3, 2014).

Levy, D. (2009). The ethical treatment of artificially conscious robots. *International Journal of Social Robotics*, *1*(3), 209–216.

MacDorman, K. F., & Ishiguro, H. (2006). The uncanny advantage of using androids in social and cognitive science research. *Interaction Studies*, *7*(3), 297–337.

Maines, R. P. (1999). *The technology of orgasm: 'Hysteria,' the vibrator, and women's sexual satisfaction*. Baltimore, MD: Johns Hopkins University Press.

Marocco, D., & Nolfi, S. (2007). Emergence of communication in teams of embodied and situated agents. *Connection Science*, *19*(1), 53–74.

Melson, G. F., Kahn, P. H., Beck, A., & Friedman, B. (2009). Robotic pets in human lives: Implications for the human–animal bond and for human relationships with personified technologies. *Journal of Social Issues*, *65*(3), 545–569.

Minato, T., MacDorman, K. F., Shimada, M., Itakura, S., Lee, K., & Ishiguro, H. (2006). Evaluating human likeness by comparing responses elicited by an android and a person. *Advanced Robotics*, *20*(10), 1147–1163.

Mori, M. (1970). Bukimi no tani [the uncanny valley]. *Energy*, *7*(4), 33–35.

Mutlu, B., Kanda, T., Forlizzi, J., Hodgins, J., & Ishiguro, H. (2012). Conversational gaze mechanisms for humanlike robots. *ACM Transactions on Interactive Intelligent Systems*, *1*(2), 12:1–12:33.

Oldenziel, R. (2001). Epiloog. In J. W. Schot et al. (Eds.), *Techniek in Nederland in de twintigste eeuw IV* (pp. 147–151). Zutphen, the Netherlands: Walburg Pers.

Picard, R. (1997). *Affective computing.* Cambridge, UK: MIT Press.

Prassler, E., & Kosuge, K. (2008). Domestic robots. In B. Siciliano & O. Khatib (Eds.), *Springer Handbook of Robotics* (pp. 1253–1281). Berlin, Germany: Springer.

Ray, C., Mondada, F., & Siegwart, R. (2008). What do people expect from robots? *Proceedings of the EEE/RSJ 2008 international conference on intelligent robots and systems (IROS'08)* (pp. 3816–3821), Nice, France, September 22–26. http://infoscience.epfl.ch/record/125291/files/iros08_ray_final-10.pdf (accessed March 17, 2014).

Raymond, J. G. (2008). Ten reasons for not legalizing prostitution and a legal response to the demand for prostitution. *Journal of Trauma Practice*, *2*(3–4), 315–332.

Richards, J. W. (Ed.) (2002). *Are we spiritual machines? Ray Kurzweil vs. the critics of strong A.I.* Seattle, WA: Discovery Institute.

Schaerer, E., Kelley, R., & Nicolescu, M. (2009). Robots as animals: A framework for liability and responsibility in human–robot interactions. *Proceedings of RO-MAN 2009: The 18th IEEE international symposium on robot and human interactive communication*, September 27–October 2 (pp. 72–77), Toyama, Japan. http://papers.ssrn.com/sol3/papers.cfm?abstract_id = 2271466.

Sharkey, A., & Sharkey, N. (2012). Granny and the robots: Ethical issues in robot care for the elderly. *Ethics and Information Technology*, *14*(1), 27–40.

Sparrow, R. (2002). The march of the robot dogs. *Ethics and Information Technology*, *4*(4), 305–318.

Sullins, J. P. (2012). Robots, love and sex: The ethics of building love machines. *Affective Computing*, *3*(4), 398–409.

Sung, J.-Y., Christensen, H. I., & Grinter, R. E. (2008). Sketching the future: Assessing user needs for domestic robots. *Proceedings of the 18th IEEE international symposium on robot and human interactive communication (RO-MAN 2009)* (pp. 153–158), Toyama, Japan.

Sung, J.-Y., Grinter, R. E., Christensen, H. I., & Guo, L. (2008). Housewives or technophiles?: Understanding domestic robot owners. *Proceedings of the third ACM/IEEE intelligent conference human robot interaction*, March 12–15 (pp. 128–136), Amsterdam, the Netherlands. Atlanta, GA: ACM.

Sung, J.-Y., Guo, L., Grinter, R. E., & Christensen, H. I. (2007). 'My Roomba is Rambo': Intimate home appliances. In J. Krumm et al. (Eds.), *LNCS*: Vol. 4717. *UbiComp 2007* (pp. 145–162), Innsbruck, Austria, September 16–19. Berlin, Germany: Springer Verlag.

Tanaka, F., Cicourel, A., & Movellan, J. R. (2007). Socialization between toddlers and robots at an early childhood education center. *Proceedings of the National Academy of Sciences of the United States of America 104*(46), 17954–17958.

Tanaka, F., & Kimura, T. (2009). The use of robots in early education: A scenario based on ethical consideration. *Proceedings of the 18th IEEE international symposium on robot and human interactive communication (RO-MAN 2009)* (pp. 558–560), Toyama, Japan.

Thring, M. W. (1964). Robot about the house. In N. Calder (Ed.), *The World in 1984* (Vol. 2) (pp. 38–42), Baltimore, MD: Penguin Books.

Torta, E. (2014). *Approaching independent living with robots* (PhD thesis). Eindhoven, the Netherlands: Eindhoven University of Technology.

Turing, A. M. (1950). Computing machinery and intelligence. *Mind, 59*(236), 433–460.

Turkle, S. (2011). *Alone together. Why we expect more from technology and less from each other.* New York: Basic Books.

Vaussard, F., Fink, J., Bauwens, V., Retornaz, P., Hamel, D., Dillenbourg, P., & Mondada, F. (2014). Lessons learned from robotic vacuum cleaners entering in the home ecosystem. *Robotics and Autonomous Systems, 62*(3), 376–391.

Wairatpanij, S., Patel, H., Cravens, G., & MacDorman, K. F. (2009). Baby steps: A design proposal for more believable motion in an infant-sized android. In K. Dautenhahn (Ed.), *Proceedings of the new frontiers in human–robot interaction* (pp. 139–144), Edinburgh, UK.

Wallach, W., & Allen C. (2009). *Moral machines: Teaching robots right from wrong.* Oxford, UK: Oxford University Press.

Weizenbaum, J. (1966). ELIZA—A computer program for the study of natural language communication between man and machine. *Communications of the Association for Computing Machinery, 9*(1), 36–45.

Yeoman, I., & Mars, M. (2011). Robots, men and sex tourism. *Futures, 44*(4), 365–371.

3

TAKING CARE OF OUR PARENTS

The Role of Domotics and Robots

3.1 Introduction

The use of care robots feeds our fears of future nursing homes without human staff: we see robots at work at the bedside, efficiently and effectively executing the much-needed physical caring tasks and ignoring the social and emotional needs of the human being lying on that bed. Against a background of an aging population, robots can, however, offer additional "hands-on services" at the bedside. For robots do not suffer from stress, are deployable 24 hours a day, and never forget about providing medication. This image is disliked by a lot of people, because for them the concept of care rubs up against the concept of "technology." Care stands for attention, warmth, kindness, reciprocity, empathy, and helpfulness. By contrast, technology is associated with effectiveness, efficiency, distance, and impersonality. However, the trends in long-term care indicate that technological innovations are necessary to meet the expected demand in health care. A staffing shortage—due to future aging (see Box 3.1)—is often invoked as an argument for deploying robotics in long-term care. In particular, the growth of the very oldest group (people over the age of 80) will put pressure on care services and will also result in an increase in the demand for various services for the elderly: assisting the elderly or their caregivers in daily tasks, helping to monitor their behavior and health, and providing companionship (Sharkey & Sharkey, 2012).

As a consequence, there will be a huge demand for care staff. There will even be a double-care gap: both a quantitative and a qualitative one. There will be a quantitative gap between what is desirable and possible in terms of money and what is possible in terms of personnel.

BOX 3.1   AGING

Aging is defined as an increase in the number of persons aged 65 years and over compared with the rest of the population. According to the European Commission (2012), the proportion of those aged 65 years and over is projected to rise from 17% in 2010 to 30% in 2060, with the peak occurring around 2040. Moreover, it is expected that people will be living longer: life expectancy at birth is projected to increase from 76.6 years in 2010 to 84.6 in 2060 for males and from 82.5 to 89.1 for females. One out of ten people aged 65 years and over will be older than 80. In Japan, the country with the highest proportion of elderly citizens, the population is also rapidly aging; 23% of the population was already older than 65 years in 2010, predicted to rise to 31% by 2030,[*] and in the United States, 13% were over the age of 65 in 2009, expected to rise to around 19% by 2030.[†]

---

[*] http://pardee.du.edu/.
[†] http://www.aoa.gov/Aging_Statistics/.

Additionally, there will be a gap between the type of care needed and the type of care that is available: people prefer more and more to live in their own homes as long as possible instead of being institutionalized in sheltered homes or nursing homes when problems related to aging arise (Broekens, Heerink, & Rosendal, 2009). This also keeps political circles busy. The European Commission (2012), for example, foresees rising demand for care, and has opted for innovation in health care by encouraging investment in labor-saving technologies, such as robotics. The expectation is that the use of robots will improve the quality of, access to, and efficiency of health care for everyone: more care will be delivered, the care recipient will exert more influence over their own care process, and a lower number of professionals will realize a higher workload than is currently the case. The government in Japan also wants to address the aging problem with the use of care robots (Lau, van't Hof, & van Est, 2009).

The use of robotics in care starts with home automation (domotics). Domotics literally combines the words domestic and robotics, that is, home robotics or, in other words, home automation technology.

These smart technologies, which at present are being incorporated widely into our environment, are the prelude to a future home with care robots. Care domotics enables the provision of various care tasks remotely: monitoring the functioning of the elderly, supporting people with disabilities in their functioning, or facilitating patients with tele-communication from home, for example, by sending the data on their physical condition to a hospital. Most of the tasks concerning care, however, require the physical presence of a human caregiver. These physical tasks cannot be taken over by home robotics. It is expected that home robotics will one day become supportive of the application of care robots that support or fully replace human caregivers. The presumption is that with the help of robots it will be easier for caregivers to provide care and care recipients will be able to exercise more control over the care they receive.

In this chapter, we will focus on robotics (both home robotics and robots in the home) specifically designed to carry out tasks—partly or wholly independently—in long-term care practice. Long-term care is focused on preserving and promoting the quality of life by alleviating the burden of illness or disability with the help of medical and nonmedical resources. The subject is care in nursing homes or in homes for the elderly and care for people with a disability or chronic illness. This study wants to give an impetus to the debate on the social and ethical consequences of deploying robotics in care for the elderly. The field of the ethics relating to care robots has emerged in recent years with ever more rapid advances in robotics and the urgent need to address the aging (Borenstein & Pearson, 2010; Nagai et al., 2010). In 2006, Sparrow and Sparrow noted their thoughts about the use of robots in care for senior citizens. Their message is utterly clear—they believe that the use of care robots is unethical: "We see the idea that we can solve the 'problem' of caring for an ageing population, by employing robots to do it, as essentially continuous with a number of other attitudes and social practices which evidence a profound disrespect for older persons" (Sparrow & Sparrow, 2006, p. 143). They emphasize that robots are unable to provide the care, companionship, or affection needed by aging people. They see the use of robots by the elderly as an expression of a profound lack of respect for senior citizens. Other ethicists do not immediately reject the care robot, as they see opportunities for care robots, if used in certain conditions and with particular qualifications (Borenstein & Pearson, 2010; Coeckelbergh, 2010; Decker, 2008; Pearson & Borenstein, 2013). The manner in which robots are deployed proves to be a crucial point. For this reason, we will look at the different roles robots can play in care.

We start with care domotics for both technological and ethical reasons. Doing this agrees with the technological state of the art plus the fact that care domotics will provide the necessary socio-technological ground layer for the employment of care robots. Furthermore, the ethical aspects raised by domotics also play a role in the application of care robots. Section 3.3 deals with the envisioned introduction of robots in care for the elderly. Moreover, some general ethical aspects of care robots related to their safe use, design, and physical appearance will be discussed. In the next section, we will deepen our reflection on the use of health care robots for the elderly by distinguishing between three different roles: (1) the robot as companion for the care recipient; (2) the robot as cognitive assistant for the care recipient; and (3) the robot as (supporter of the) caregiver. We will show that each of these roles gives rise to various specific ethical issues. We will end this chapter with some observations about the use of both home robotics and robots in care for the elderly in the long term.

3.2 Domotics for the Care of the Elderly

Fewer and fewer senior citizens spend their old age in domiciliary care, residential care, or nursing homes. In recent decades, due to housing policy aimed at maintaining people's independence and to people's desire to live independently of extensive forms of care, the number of over-65s in these institutions has already significantly decreased (Meulendijk, van de Wijngaert, Brinkkemper, & Leenstra, 2011). This implies that people live independently at home for longer and will increasingly receive care at home. Using domotics provides opportunities for people to live independently at home even longer and still receive the necessary care.

Domotics are devices and infrastructures in and around homes that provide electronic information for measuring, programming, and controlling functions for the benefit of residents and the providers of services (Aldrich, 2003), such as

- Personal alarm systems, including systems that need to be actively triggered by the client and systems automatically triggering an alarm in case of an emergency as well as dedicated fire alarm systems
- Systems enabling teleconsultations and remote monitoring, including video-based systems requiring a broadband connection and systems enabling remote access to care records by professional staff and/or clients
- Home automation systems directed toward enabling the older person to control the immediate home environment, such as automatic door-opening systems, intercoms, and control systems relating to home appliances
- Systems enabling access to on-demand support in relation to daily living activities, such as meals on wheels and home care as well as social integration
- IT systems supporting human resource planning, logistics, and general administrative functions concerning health/care-related service provision
- Assistive devices such as panels with large buttons for people with dexterity problems and large screens for people with visual restrictions

With domotics, people will receive care in their homes. This is called ambient-assisted living, telecare, or remote care. As a consequence, patients will have shorter stays in care institutions and rehabilitation will take place more and more at home. Because of developments such as home diagnostics and teleconsultation, the patient will also be less likely to visit hospital. The patient will be able to receive home care and the caregiver will not even have to travel to the patient's home. Medical data such as blood glucose levels or electrocardiograms (ECGs) can be uploaded to the doctor or hospital.

The expectations surrounding domotics are higher quality through intensive care, better self-management by the client, and efficient use of the relatively scarce time of professionals. An example of a project about domotics automation with telemonitoring is KOALA (*looking at a distance, a logical alternative*) (M&ICT, 2010). The aim of this project was to increase patient autonomy and to provide care more efficiently. Through their TV or computer screen in their living room, patients could contact doctors, assistants, and district nurses 24 hours a day, 7 days a week. It also allowed them to forward data relating to medical test measurements, such as blood glucose and blood pressure, through the system. The parties involved, including the information and communication technology (ICT) service provider, concluded that there is lot of potential in this form of care. Yet, the results are not conclusive in terms of cost savings and a reduction in personnel, because this remote care functioned not as a substitute for but rather as a supplement to the care otherwise provided.

3.2.1 Paradigmatic Shift in Care

A combination of domotics and ambient intelligence (AmI) could significantly improve the functionality of domotics (Rodriguez, Favula, Preciado, & Aurora, 2005). AmI consists of a vision in which technology is integrated into and aware of environments and is able to make reasoned decisions (Aarts & Marzano, 2003). People are empowered through a digital environment that is aware of their presence and context and is sensitive, adaptive, and responsive to their needs, habits, gestures, and emotions (Riva, 2003). AmI could considerably enhance the effectiveness of standard domotics for care for the elderly and enable new solutions. For example, simple alarm buttons worn by users could be enhanced to measure physical discomfort and automatically alert emergency services (Meulendijk et al., 2011).

This intelligent technology brings about a paradigmatic shift in the traditionally low technology social practice of care. The shift is going to occur in the way we live with domotics: the *handing over of control* that increases the potential functionality of domotics—the house itself can be empowered to perform a greater range of tasks relating to the occupants' comfort, convenience, security, and entertainment. By using domotics, care practice can be enveloped in an ICT-friendly infosphere in order to make telecare possible. This rationalization of care offers remote care of elderly and physically less able people, providing the care and reassurance needed to allow them to remain living in their own homes. Telecare implies a new kind of man-in-the-loop for caregivers, since it reshapes the role of caregivers: caregivers will become teleoperators for all kinds of care tasks.

According to Oudshoorn (2008), there are increasing possibilities for telemonitoring and telediagnosing, or even treating patients, using domotics and ambient intelligence. This also implies that various technologies collect a lot of data about the people who receive remote care; sensors at home capture when someone is lying in bed, is opening the fridge, or goes for a walk outside; sensors in clothing may capture the physical condition of people and transmit information, and cameras can record what someone does around the house or even watch the care recipient 24 hours a day. These data provide a great deal of detailed information about the daily life of the care recipients and thereby raise questions about the privacy of these care recipients. The issue of proper treatment of these data is a sensitive matter: who is responsible when data are lost; which data are stored and for how long are they stored; and are the care recipients aware of the fact that information is being collected about them? These questions will play an important role in the near future with the growth of remote care. The ever-increasing scale of providing remote care requires reflection on the degree of the invasion of privacy. According to Borenstein and Pearson (2010), the degree of control by the care recipient of the information collected is important. When someone has actual control over the information collected, this enhances the autonomy of that person.

3.2.2 Ethical Issues

With the implementation of domotics, ethical values such as privacy, human contact, quality of care, and the competences of caregivers and care recipients appear to be at stake. In this section, we will briefly reflect on these ethical issues.

3.2.2.1 Privacy   Monitoring and taking control of these senior citizens may involve a breach of privacy. Care recipients will not like being recorded when they are not properly dressed or when they are taking a bath. The use of cameras for real-time observation could, therefore, cause the most serious feelings of violations of privacy (Mihailidis, Cockburn, Longley, & Boger, 2008), especially if the users cannot turn these cameras off. A solution could be a two-way communication system that uses a videophone on demand between the caregiver and care. This system creates a feeling of social security. It does not have the privacy restrictions of passive camera monitoring, but at the same time it improves demand-driven care. People can connect to the system whenever they feel like it (Kort, 2005). A risk of such a system is that there is the potential for improper use or overuse by the care recipients. Care recipients who feel lonely could overwhelm the doctor's surgery, for example, with unnecessary inquiries to generate attention (Center for Practice Improvement and Innovation, 2008).

The privacy issue becomes more complex in relation to dementia in old age: to what extent can someone with dementia indicate that they are aware of the presence of a technology that captures their daily lives? Can we use track-and-trace systems to cope with the problems accompanying dementia, such as wandering, a potentially lethal behavior? It is argued that a slight loss of liberty is acceptable in order to increase safety (Van Hoof, Kort, Markopoulos, & Soede, 2007).

Another issue with respect to privacy concerns the collected data. The questions are who exactly should own or control the data and how can safe data storage be guaranteed (Bharucha et al., 2009). Currently, no protocol is available for controlling and storing these data.

These privacy questions should be taken into consideration by developers and politicians when they are deploying home automation as well as care robots. In this era of rapid developments in ambient intelligence and robotics, where privacy issues are at stake, there is an urgent need for total transparency and clear definitions, otherwise the issue of privacy could block innovations in these areas. This requires developers to consider the consequences right from the beginning of the design process of their use of robotic technologies in terms of privacy. It should become a balance between some kind of protection of privacy on the one hand and the need to keep on living at home independently on the other, since systems that are designed to promote independence require varying degrees of violations of privacy (Bharucha et al., 2009).

Qualitative studies, however, have shown that most elderly people do not feel as if their privacy is violated (Courtney, 2008). The need for monitoring overrules any possible privacy concerns if there is an appropriate balance between this need and privacy. For example, some older adults prefer to be monitored in the bathroom while they are

naked, rather than running the risk of a fall or of their lying unconscious on the floor being unnoticed (Van Hoof et al., 2007).

3.2.2.2 Human Contact and Quality of Care Domotics is focused on
giving elderly people more independence; however, this may involve
a threat of a reduction in social contact. Many authors worry that
the use of domotics devices might lead to loss of human contact and
humane care (see, e.g., Boissy, Corriveau, Michaud, Labonte, &
Royer, 2007; Mihailidis et al., 2008), which may lead to elderly people
becoming socially isolated, since for some people the visits of caregivers are the only social contact available to them. Furthermore, health
professionals feel that good care is linked to genuine relationships and
social interaction (Sävenstedt, Sandman, & Zingmark, 2006), including aspects such as reciprocity, empathy, and warmth. Domotics will
create long-distance care relations instead of a personal and intimate
care relation (Boissy et al., 2007). However, a study by Pols (2010)
shows that the use of domotics could mean that care becomes even
closer to the care recipient, instead of being "care at a distance," and
that domotics could turn out to be more, rather than less, intensive.
Care by telecommunication devices appears to be more intense and
frequent than face-to-face contact. For example, telecommunication
using a webcam makes a conversation seem even closer to each person
than real encounters, since one is more focused on the image of the
face of the other person (Pols, 2011). Although the domotics devices
allow for more frequent consultations, not "everything" can be seen,
as in the case of caregivers who provide care on the home, because
different variables cannot be taken into account for diagnoses, such as
physical examinations.

The use of domotics requires a vision of care practice, and the discussion should be about what exactly we mean by "care," and what
the role of technology is in "care." More research is needed to investigate the changes to clinical practices in care for the elderly made by
domotics in order to create a clear image of what these changes will
mean for the notion of "care" and how we can use domotics.

3.2.2.3 Competence of Caregivers and Care Recipients The introduction of domotics devices has an impact not only on care recipients but also on caregivers. As shown in the previously mentioned

domotics project KOALA, the use of remote care at a distance not only requires acceptance by users, but caregivers have to adjust their activities as well. According to Van Oost and Reed (2011), the socio-technical network in which domotics is being implemented must be closely observed, since the use of domotics creates a new care practice. In this new practice, health care professionals get new roles with a different division of responsibilities (Akrich, 1992; Oudshoorn, 2008) that requires them to have new skills. Through mediation of technology, caregivers receive different information, and the possibilities for intervention are therefore changed. The use of telecommunication, for example, requires a different method of working by health care professionals or caregivers, such as being able to diagnose, monitor, and reassure people via a computer or TV screen.

New skills are also expected of the care recipient. The care recipient should be able to deal with teleconferencing and with forwarding messages containing data to a doctor. More serious is the use of medical devices by the care recipients. If a care recipient with diabetes, for example, has responsibility for monitoring their blood glucose levels, there is no longer a professional involved in the process. Even during training programs, patients make errors when calibrating and using the device (Mykityshyn, Fisk, & Rogers, 2002). Well-designed training and instructional programs for older adults, as well as care professionals instructing patients about the technology and familiarizing them with its use, are needed for the successful use of medical devices (Czaja & Sharit, 2012). Dealing with robotic technology therefore opens a new chapter in the training of caregivers as well as care recipients so that both of these groups can easily cope with it and can anticipate the possibilities and limitations of robotic technologies (see also Shaw-Garlock, 2011).

To conclude this section, it seems true to say that there seems to be little ethical objection against domotics by elderly people. Zwijsen, Niemeijer, and Hertogh (2011) state that this might be because most elderly people are eager to avoid the prospect of living in a nursing home. Living at home is valued very highly, and is traded off for any potential loss of human contact and privacy. That there is little ethical objection does not automatically mean that domotics is an ethically sound solution. As it is the prospect of living in a nursing home that propels many elderly people to opt for domotics, there is a risk that this could be an "adaptive preference" (Zwijsen et al., 2011, p. 424).

3.3 From Home Robotics to Robots in the Home

3.3.1 Increasing the Pace of the Paradigmatic Shift in Care

A new development within domotics which could increase the pace of the paradigmatic shift is the care robot, since robots can use their ability to move, and can perform physical tasks to support elderly people, in contrast to ambient intelligence, which is an embedded technology. Whereas ambient intelligence reshapes the role of care-givers so that the caregiver was in-the-loop, the use of care robots may create the nightmare scenario of a completely robotized care, in which human caregivers are cut out-of-the-loop. This scenario remains bleak for the time being, but the deployment of care robots evokes a lot of ethical questions, which we will discuss in the next subsection and Section 3.4.

A development such as the Japanese lifting robot *RIBA II* (see Box 3.2) has brought the use of care robots a step closer. This robot supports human caregivers in lifting their clients. Apart from robots that support caregivers, there are already robots that allow people to live longer independently at home. Possible functions of this robot are keeping a calendar and shopping lists, playing music, remembering to take medication or keep appointments, and give a warning when the gas is left on or a window is left open.

In recent years, many initiatives have been started in Europe to integrate robots into domotics.

Some large projects in which the development of care robots played a major role were *Mobiserv, CompanionAble, KSERA*, and *HOBBIT* (see Box 3.3). A special quality of these European projects was that both the needs and the acceptance of users lead within the development of this technology and not vice versa. Much attention was focused on the interaction between human beings and robots and on the acceptance of robots by senior citizens. In all these projects, a user-centered design approach, in which models and guidelines are used as fundamentals to build from users' expectations instead of technology's capabilities, was adopted. The literature stresses that

BOX 3.2   RIBA II: THE WORLD'S FIRST ROBOT THAT CAN LIFT UP A HUMAN IN ITS ARMS

The Japanese companies RIKEN and Tokai Rubber Industries (TRI), in their collaborative project *RIKEN-TRI Collaboration Center for Human-Interactive Robot Research*, have developed the lifting robot RIBA II (see Figure 3.1).* RIBA stands for "Robot for Interactive Body Assistance." This robot can lift people who weigh up to 80 kilograms (or 175 lb), if and when guided by a caregiver. RIBA II looks like a huge teddy bear on wheels, with two arms that can reach almost to ground level. The arms and hands of RIBA II contain 218 haptic sensors. They register anything the robot touches and must be controlled by a supervisor when lifting the patient. When the supervisor pushes the arm—toward the right, for example—the robot will respond to this. RIBA II requires the supervision of an attendant in order to optimally position the patient so that the robot can easily start lifting. The legs of the patient must be positioned in the right way, and the attendant's help is also needed to position the patient's back before lifting, so that the robot can start lifting the patient properly. RIBA II is covered with soft materials, so that the interaction between humans and robots is not only safe but also comfortable for

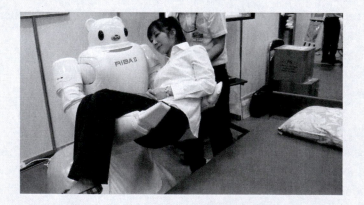

Figure 3.1   RIBA II. (Photo courtesy of Sumitomo Riko.)

* www.riken.go.jp/engn/r-world/info/release/press/2011/110802_2/index.html.

the patient. The appearance of the robot, a teddy bear, should also contribute to the ease of the patient. The eyes of RIBA II are functional by way of a digital camera. In addition, the lifting robot is equipped with two microphones, enabling RIBA II to record sound. Thus, through cameras and microphones, RIBA II can detect people: the robot hears where a sound is coming from and thus perceives the person as being in a given position. When the supervisor is standing within the field of view of the cameras, RIBA II determines the distance to the attendant and can roll toward the attendant.*

With this robot, RIKEN intends to support care staff in one of the most difficult and demanding health care jobs: lifting people. According to RIKEN, caregivers lift patients about 40 times a day. RIBA II takes over this strenuous job. This robot can move people from a bed into a wheelchair or lift them off the ground and into a bed. It is a prototype and not yet commercially available. RIKEN is striving to ensure that in future this robot can be used in nursing homes, hospitals, and private homes.

---

* http://rtc.nagoya.riken.jp/RIBA/index-e.html.

for successful innovation of health care systems people should not be regarded as passive users but as active co-creators (Meulendijk et al., 2011).

Although robots still take the shape of clumsy behemoths with metallic voices, the dream of developers is to build multifunctional robots that are able to respond to emotions and will show emotions themselves. Developers wish to create a robot that can help people and keep them company. Both at universities and in industry, specialists are working on the realization of the robot as the caregiver of the future. Despite all of this, during the next 10 years care support robots may not yet widely enter the field of care, due to the complexity of the care tasks. For a mobile robot, even the most fundamental abilities, such as localization (the identification of its current position in the environment) and safe navigation (the planning of trajectories which involve smooth curves and maximize the distance from obstacles), appear very difficult to deal with (Cesta & Pecora, 2005).

BOX 3.3   FOUR CARE ROBOT PROJECTS

Mobiserv (An Integrated Intelligent Home Environment for the Provision of Health, Nutrition and Mobility Services to the Elderly) (FP7 Project, 2009–2012)*
The social purpose of this project is to develop a personal robot that serves to support senior citizens during their normal daily activities within their homes. Here, the emphasis is on supporting personal health care and increasing independence and quality of life.

CompanionAble (FP7 Project, 2008–2011)†
This project supports senior citizens suffering from mild memory problems and aims to help them to continue living independently at home for as long as possible through the integration of robotics and automation. A companion robot is being developed that will integrate with a smart home environment. Actions might include assistance in the fields of medication intake, memory support training, management of the appointment calendar, and social interaction and safety.

KSERA (Knowledgeable Service Robots for Aging) (FP7 Project, 2010–2013)‡
This project investigates the use of robots as "wise family friends" for senior citizens. It is largely focused on COPD patients, who are people with chronic obstructive pulmonary disease. The idea is that special housing is provided that includes a house robot and other smart home facilities. The robot must be able to follow the patient through the house to learn his or her habits, offer advice and, if necessary, alert a doctor immediately. Entertainment is also included in the form of the Internet and video communication.

HOBBIT (The Mutual Care Robot) (FP7 Project, 2011–2014)§
This project is developing a socially assistive robot that helps senior citizens at home. The focus of HOBBIT is the

---

* www.mobiserv.eu/.
† www.companionable.net/.
‡ http://ksera.ieis.tue.nl/.
§ http://hobbit.acin.tuwien.ac.at/.

development of the mutual care concept: building a relationship between the human and the robot in which both take care of each other. This is similar to how a person learns what an animal understands and can do, so it is like building a bond with a pet. The main task of the robot is fall prevention and fall detection. To achieve this, the robot will clear the floor of all objects and thus reduce the risk of falling. It will detect emergency situations so that help can be called in time. The purpose of the mutual care approach is to increase the acceptance of the home robot.

3.3.2 General Ethical Issues Relating to Care Robots

The ethical issues we already have discussed in relation to domotics (see Section 3.2.2) are also applicable to care robots: privacy, human contact, quality of life, and the competence of caregivers and care recipients. For example, most care robots will collect data, since they will monitor the care recipient, which might infringe the right to privacy, and working with a lifting robot, such as RIBA II, requires specific skills of caregivers: knowing how to steer the robot and predicting potential failures. Here, we discuss three other ethical issues that are at stake for care robots in general: their safe use, design, and physical appearance.

3.3.2.1 Safety

We are not far from the time when people will live and interact with care robots and, thus, safety will become fundamental. Robot designers should produce safe products for humans no matter what failure, malfunction, or mishandling may occur. Because care robots will need to work around people and to touch them, errors or faults in these robots could result in serious or even fatal accidents. Conventional safety strategies for industrial robots cannot be applied to social robots, and especially not to care robots. A lot of research has been done in this area (see, e.g., Ikuta & Nokata, 2003; Mukai, Hirano, Nakashima, Sakaida, & Guo, 2011; CompanionAble Project). Up to 2014, there were no internationally recognized safety regulations or guidelines. As a consequence, companies have been reluctant to take the risk of investing in and launching a new robot product in case something goes wrong and they are taken to court. In this context, it is not surprising that there are still so few commercial care robots about today, although there is clearly a market for them. It is difficult for an individual company, or even a single country, to define acceptable safety regulations. In order to allow the industry to move forward, international discussion and consensus are needed; in February 2014, the International Organisation of Standards released ISO 13482 regarding safety requirements for personal care robots based on the standardization efforts on robot safety of the European Union funded project euRobotics.[*] The eagerly awaited standard gives researchers, manufacturers, and regulators a basis against which to measure and monitor products. ISO 13482:2014 specifies requirements and guidelines for the inherently safe design of, protective measures for, and information about the use of care robots.[†] The standard describes hazards associated with the use of these robots and sets down requirements to eliminate the risks associated with these hazards or reduce them to an acceptable level. The scope of the standard is limited primarily to human care–related hazards but, where appropriate, it includes domestic animals and property (defined as safety-related objects) when the personal care robot is properly installed and maintained and used for its intended purpose or under conditions that can reasonably be foreseen.[‡]

There are, however, some concerns with regard to this standard. For example, Brian Scassellati, Associate Professor of Computer Science at Yale and an expert on socially assistive robotics and human–robot interaction (HRI) who is involved with the *Socially Assistive Robotics* project[§] (a National Science Foundation–funded initiative designed to improve the performance of child-centered care robots), states that the ISO 13482 standard may give the false impression that care robots are ready for the mainstream and for widespread industry development and that this may cause more harm than good. According to Scassellati, we do not have a clear understanding of the basic science behind HRI: "Roboticists and other stakeholders don't have a clear

---

[*] http://www.eurobotics-project.eu.
[†] https://www.iso.org/obp/ui/#iso:std:iso:13482:ed-1:v1:en.
[‡] For a more detailed analysis of ISO 13482, we refer to Jacobs and Gurvinder (2014).
[§] http://www.robotshelpingkids.com/index.php.

understanding of the roles that care robots should play, the kind of support the robots should or should not provide, and the impact that robot care will have on their users."*

3.3.2.2 Designing Care Robots   Despite the need for care robots for the rapidly aging population and the potential success of some care robots, other care robots have had a poor response. Broadbent, Stafford, and MacDonald (2009) attribute this poor response to the fact that designers do not properly assess the needs of the human user—the caregiver as well as the care recipient—and then match the robot's role, appearance, and behavior to these needs. The perceptual worlds of caregiver, care recipients, and designers vary due to differences in background and experiences. For designers the effects will involve technical goals, and caregivers are mainly interested in effects on workload and quality of care, while care recipients are influenced by usability effects. In general, designers neglect the views of the elderly people who are the users of the care robots. Frennert and Östlund (2014) found that in most health care research, older people are described as objects and that designers do not involve older people as subjects in their research, and therefore not in the development of robotic technologies either. In contrast to the stereotypical view of older people, evidence indicates that they are far from passive consumers; instead, they are technogenarians: older individuals who creatively adapt and utilize technological artifacts to suit their needs (Joyce & Loe, 2010). Designers should take into account the older user's capabilities and limitations. As people age, motor behaviors, such as disrupted coordination, change; sensory abilities, such as vision and hearing, reduce; aspects of memory decline; and so on. According to Rogers and Mynatt (2003), designers must recognize and accommodate those abilities that do decline while at the same time capitalize on the abilities that remain intact. A perfect single design method for every care robot is unlikely to exist, but in any case, designers should take into consideration the wishes and needs of caregivers as well as those of care recipients in their design process. Both of these user groups should be involved as early as

---

* http://www.roboticsbusinessreview.com/article/new_international_standards_boon_to_personal_care_robotics.

possible in the design process in such a way that the technological knowledge of the designers and the contextual knowledge of the users are married in a design (Van der Plas, Smits, & Wehrmann, 2010). The European projects Mobiserv, CompanionAble, KSERA, and HOBBIT have shown how this can be done. In the KSERA project, the designers adopted a user-centered design framework to link the design with the needs and the context of the lives of the people to which it is addressed (Johnson et al., 2014). This framework is explicitly intended to be a dynamic process in which the end users are involved from the beginning of the project, not as subjects but as active agents that influence decisions, development, and implementation.

To include the ethical aspects into a design process relating to care robots, one must first identify the significant moral values and then describe how to operationalize these values (Van Wynsberghe, 2013). This ethical evaluation ensures that the design and introduction of a care robot do not impede the promotion of moral values and the dignity of caregivers. For this ethical evaluation of care robots, Van Wynsberghe developed a framework that incorporates the recognition of the specific context of use, the unique needs of users and the tasks for which the robot will be used, as well as the technical capabilities of the robot, based on a value-sensitive design approach. Such ethical evaluation provides guidance for robotic design and development on how to proceed and what to strive for in current and future work (Nylander, Ljungblad, & Villareal, 2012).

3.3.2.3 Physical Appearance   Another aspect of designing care robots is physical appearance, since the robot's appearance influences how people appraise the abilities of the robot and has profound effects on its acceptance (Wu, Fassert, & Rigaud, 2011). For example, participants in the study of Wu et al. (2011) were reluctant to interact with some humanoid robots that have inauthentic expressions and offer ersatz interactions and companionships. Acceptability further depends on the acceptance of and attitudes of others toward the robot (Salvani, Laschi, & Dario, 2010), facilitating conditions, perceived usefulness, perceived ease of use, and perceived enjoyment and trust (Heerink, Kröse, Evers, & Wielinga, 2010).

3.4 Specific Ethical Issues with Regard to the Role of Care Robots

In this section, we will deepen our ethical reflection on the use of health care robots for the elderly by distinguishing between three different roles: (1) the robot as companion for the care recipient; (2) the robot as cognitive assistant for the care recipient; and (3) the robot as (supporter of the) caregiver. Distinguishing these three types of care robots enables us to introduce various relevant ethical concerns in a contextualized manner, that is, in relation to the specific role a care robot is envisioned as playing in certain social practice. In this way, the five following relevant ethical concerns will be discussed (cf. Vallor, 2011):

1. *Deception*: The potential for the "relationship" with the care robots to be inherently deceptive or infantilizing.
2. *Autonomy*: The potential of robots to enlarge or reduce the opportunities, freedom, autonomy, and/or dignity of the care recipients.
3. *Dehumanization*: The ethical issue of senior citizens being reduced to objects who are basically problematic and whose care problems can be solved with the use of robot technology.
4. *Quality of care*: The ethical discussion point about the quality of care that can be provided by care robots.
5. *Human contact*: The potential of robots to increase or reduce the care recipient's human contact with family and caregivers.

Before we start discussing these ethical issues, it is important to realize that the actual deployment of these types of robots is not to be expected in the short term. Some companion robots are already being commercially produced, but the social interaction of these robots is very limited. This is because even the best robots are no match for the social and thinking abilities used in the interaction of the average toddler (Bringsjord, 2008). Breakthroughs in AI are needed in order to make the companion robot a success story (see also Chapter 2). In contrast to the robot as cognitive assistant for the care recipient, the final type of robot we will discuss is the one that is able to provide physical assistance by supporting basic activities such as eating, bathing, going to the toilet and getting dressed, helping with basic household tasks, and providing assistance with walking. It will take a while before these robots enter the field of care. As we have seen in Chapter 2, these tasks require very complex decisions, and designers also have to deal with the frame problem, which will probably not be solved within the next 10 years. The robot as cognitive assistant has been developed as part of an intelligent assistive environment for elderly people. It monitors and supervises the activities of daily living and monitors health and safety. A lot of research is being done in relation to these robots, such as the four robot projects mentioned in Box 3.3 in the previous section. Many of the cognitive assistant robots are still in the development and testing phase. Some have been commercially produced, but they turned out not to be successful (Broadbent et al., 2009). The lack of uptake of these care robots is due to a number of factors, such as users losing interest after a certain period of time, users not perceiving any benefits, and the commercial market proving to be a difficult area to break into. Not all care robots can be put in just one of the aforementioned three categories. A robot can, for example, be programmed to perform monitoring tasks and at the same time can provide companionship.

3.4.1 The Robot as a Companion

3.4.1.1 Deception The robot as companion technology raises controversial images: lonely elderly people who only have contact with robot animals or humanoid robots. The ethical concerns about the pet robot focus on the degree of human contact that such technology brings about or, instead, on depriving and deceiving patients with dementia, for example (Borenstein & Pearson, 2010; Coeckelbergh, 2010; Sharkey & Sharkey, 2012; Sparrow & Sparrow, 2006; Turkle, 2006). Sparrow and Sparrow (2006) describe care robots for the elderly as "simulacra" replacing real social interaction (see also Section 2.3). They think that robots, which can be neglected, paused or disabled, probably cannot build a meaningful relationship with the user. Characteristic of our relationship with another human being is that the other party has its own needs and wishes, regardless of our own needs and desires. "The demands that our friends … make on us are therefore unpredictable, sometimes unexpected and often inconvenient. This is an essential part of what makes relationships with other people, or animals, interesting, involving and rewarding," according to Sparrow and Sparrow (2006, p. 149). In addition, they wonder to what degree a robot can provide entertainment for longer periods; their expectation is that once the novelty of the robot has worn off, it ends up idle, having been chucked in a corner. Borenstein and Pearson (2010) are more positive about the deployment of robots; they believe that although robots cannot provide real friendship, the deployment of a companion robot, such as the seal robot Paro (see Box 3.4), relieves feelings of loneliness and isolation and may increase conversational opportunities with the robot itself as well as with other humans. According to Sorell and Draper (2014), a companion robot may even have a role in assisting elderly people who are isolated to keep up their skills of social interaction.

3.4.2 The Robot as a Cognitive Assistant for the Care Recipient

Cognitive assistance care robots can meet the need for senior citizens to live independently for an increasingly long time. These robots are deployed as "cognitive prosthesis": a robot can assist someone to remember to take medication or to eat on time, for example. *Kompai* is an example of such an assistance robot (see Box 3.5). This robot is at the service of the elderly: Kompai helps people to remember appointments, can provide information when prompted, and can also ask the care recipient questions, such as do you have pain in your leg and then log this. Robots that support the elderly may also be used to keep them active by encouraging them to move about or go out. This allows senior citizens to come out of their isolation. Many initiatives have been taken to develop all kinds of cognitive assistance care robots with specific tasks, such as the meal assistant robot *Brian 2.1*.

Brian 2.1 provides cognitive assistance, targeted engagement, and motivation for elderly individuals in order to encourage them to consume a meal via meal-related cues and motivating statements using natural verbal and nonverbal communication means (McColl & Nejat, 2013).

3.4.2.1 Autonomy The use of care assistant robots also raises questions. How pushy may a robot become, for example, in reminding someone to take medication? What if someone refuses to take

BOX 3.4   PARO: A "SEAL-TYPE THERAPEUTIC ROBOT"

Paro, a soft seal robot, was developed in Japan by the *National Institute of Advanced Industrial Science and Technology* (AIST) (Lau et al., 2009). Paro has a white plush coat, is able to move his head, eyelids and limbs, and produces sound (see Figure 3.2). Sensors under Paro's fur register touch, and he responds by moving or emitting sounds. Paro is also equipped with a sound sensor to record sounds and will eventually be able to recognize people's voices. Paro weighs 2.8 kilograms (or 6.2 pounds), is charged via the electric mains, and can be on active duty for over an hour before requiring the next "feeding" (Inoue, Wada, & Ito, 2008). According to AIST, Paro is able to show emotions. The institute also claims that Paro can exhibit different kinds of behaviors: proactive, reactive, and psychological behavior (Wada & Shibata, 2007).

Paro is already deployed in health care. The Danish government, for instance, placed an order—at the end of 2008—for 1000 Paros, which were delivered in 2011. Almost all Danish institutions for the elderly now own a Paro.*

Figure 3.2   Paro. (Photo courtesy of Bart van Overbeeke.)

---

* www.breitbart.com/article.php?id=D94IJ0V00&show_article=1.

Makers of this robotic creature claim that interaction with it has therapeutic value: senior citizens will become encouraged by Paro and children will relax. There are several studies on Paro's effect in inpatient elderly care.* It is concluded in these studies that the mood of the elderly improves and that depression levels decrease; in addition, their mental condition is improved, lowering the risk of burnout for caregivers and significantly improving communication between the senior citizens and strengthening their social bonds. Personal interaction with Paro also resulted in other psychological improvements (relaxation, motivation). In addition, by means of urine tests, physiological improvements were measured. However, the scientific value of the studies examining the effectiveness of social robots like Paro appears to be unreliable. Besides the fact that the number of field studies is very limited, they are based on exploratory studies in Japan with only a small number of robots (Bemelmans, Gelderblom, Jonker, & de Witte, 2012) and poor research designs (Broekens et al., 2009).

* See among others Hansen, Andersen, and Bak (2010), Inoue et al. (2008), Wada, Shibata, Saito, and Tanie (2006), and Wada and Shibata (2007).

the medication? The danger of paternalism comes into play. In this case, the robot technology will force users to take a particular course of action on the basis that the developers know what is best for these users (Van de Poel & Royakkers, 2011). Sharkey and Sharkey (2012) also mention this point and argue that the degree to which a robot intervenes will affect the freedom of a care recipient. The risk is that we are on a slippery slope and may create authoritarian care robots (Sharkey & Sharkey, 2012), which is in contrast to the meaning of "care" (also see the next section). Care is more than the achievement of a material goal. It is about entering into a relationship, that is, a caregiver having contact with a patient. Ultimately, caring for someone is not only about achieving a goal (a cure, taking medication) but is in essence about a relationship with the person in need of care.

BOX 3.5   KOMPAI: "ROBOTIC-ASSISTED DAILY LIVING"

The French company Robosoft has developed *Kompai*, a robot that gives support to dependent people in their private home. This robot is directly connected to the Internet, thereby enabling contact via videoconferences between a resident and a doctor. This robot on wheels can talk, recognize human speech, has a touchscreen, and can provide access to the Internet (see Figure 3.3). The robot understands commands and can perform actions, such as leaving the room on command, playing music, or creating a shopping list. Kompai is also able to determine its location within the house and will return independently to its station when its batteries need recharging. The robot also has no arms, by the way, and is unable to express human emotions, but Robosoft expects this functionality will be added to Kompai at a later date. Robosoft has launched Kompai as a basic robot platform, enabling

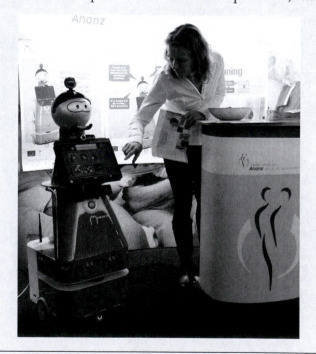

Figure 3.3   Kompai. (Photo courtesy of Ananz.)

technicians to continue tinkering so that in time new applications can be developed and added.*

Kompai is now deployed and will be further developed within several European projects. In the aforementioned Seventh Framework project, Mobiserv examines how innovative technologies help to support the elderly in a user-friendly way so that they can live independently for as long as possible.† To this end, Kompai is part of a "personal robotic system" in which sensors in clothing, as well as tele-alarm and nutrition support systems, are operative elements.

The *Ambient Assistive Living* Project DOMEO,‡ an open robotic platform, is also being developed for personalized home services and for cognitive and physical support for senior citizens, with the aim of enabling them to live at home longer and more safely.§ The *robuMATE*, developed from the Kompai robot, serves in this project as cognitive support for the elderly. The robot is designed to enable visual and verbal interaction with residents. It also serves as a memory assistant for the elderly. If there is an emergency, a video link will help analyze the situation. This robot is also equipped with the ability to encourage older people to move and to monitor the behavior of senior citizens.

---

* www.robosoft.com/img/data/2010-03-Kompai-Robosoft_Press_Release_English_v2.pdf.
† www.mobiserv.eu/index.php?option=com_content&view=article&id=46&Itemid=27&lang=en.
‡ http://www.aal-europe.eu/projects/domeo/.
§ www.aal-domeo.eu/.

3.4.3 The Robot as a (Supporter of the) Caregiver

3.4.3.1 Dehumanization   When robots are used to replace the caregiver, there is a risk that care becomes dehumanized (Veruggio & Operto, 2008). Sharkey and Sharkey (2012) also point to the ethical objections concerning the dehumanization of care for senior citizens, which is caused by using care robots. When robots take over tasks such as feeding and lifting, the care recipients may consider themselves to be objects. On top of that, they foresee the possibility that senior citizens may get the idea that they have less control over their lives with care robots compared to just receiving care from human care-givers. It has, however, been suggested that one of the advantages of using care robots is that they do not have any "social baggage" and they do not judge. Studies indicate that in some situations the risk of dehumanization for the older user may decrease if they receive robotic care instead of human help (Breazeal, 2011). Decker (2008), however, foresees dehumanization of the care recipient with the use of domotics, since the present technology is invisible and actions are decentralized; this is in contrast to care robots, which directly interact with the client.

With regard to the "dehumanization of the patient," some authors propose "mutual care" (see, e.g., Lammer, Huber, Weiss, & Vincze, 2014), which means giving the user simple opportunities to care for the robot, thereby increasing the feeling of being part of a team and, consequently, the acceptance of the robot's help. This requires that the robot is capable of asking the user for help in reciprocal dialogues. For example, the robot could remind the user that the flowers need water and can apologize because it cannot do this task properly, but says it would love to show the user which flowers need water if the user would like to accompany the robot. Research has suggested that mutual care makes the care recipient feel needed and important, which creates a sense of power and control, and that the care recipient does not con-sider herself to be an object.

3.4.3.2 Quality of Care   The ethical objection of "dehumanization of the patient" is consistent with the idea that robots cannot provide *actual care*. Decker (2008) also shows that the way people inter-act with robots is important. He thinks robots should only be used instrumentally. Coeckelbergh (2010) agrees with this and would only deploy care robots for "routine care jobs." These are tasks in which no emotional, intimate, or personal involvement is required. The act of "care giving" itself is reserved for people. In this, ethi-cists touch on an important basic question regarding care robots: are robots able to provide *care*? Care is concern about the welfare of people, entering into a relationship with them, dealing with their discomforts, and finding a balance between what is good for that person and whatever it is that they are asking for. Robots seem to be the epitome of effective and efficient care: the ultimate rationalization of a concept that perhaps cannot be captured in sensors, figures, and data. The use of care robots requires a vision of care practice, and the discussion should be about what exactly we mean by "care," taking into consideration aspects such as reciprocity, empathy, and warmth. The substantiation is that robots are devices that cannot mimic empathic abilities and the reciprocity of humans in care relationships. Human contact is often seen as crucial for the provision of good care (Coeckelbergh, 2010). According to Sparrow and Sparrow (2006), robots are unable to meet the emotional and social needs that senior citizens have in relation to almost all aspects of caring.

According to Borenstein and Pearson (2010), there is a clear and expressly made choice to be made as to when robots should support caregivers in their tasks: "robots have the potential to make caregiving a genuine choice" (p. 283). Caregivers can, for example, deploy care robots for "routine care jobs" so that the caregiver can put more effort into entering into a relationship with the care recipient. They see robots as technologies that facilitate the provision of care. Care robots may relieve the caregivers' workload and may expand the freedom of caregivers to provide care, and thereby shed a different light on the discussion of the care of senior citizens as well as end-of-life care. Borenstein and Pearson give the following example: Hardwig (1997) argues in defense of a duty to die, primarily on the grounds that no individual has the right to impose significant burdens (e.g., emotional, social, or financial) on his/her family members. The duty to die concern might diminish significantly if robots fill deficits in the caregiving, and caregivers can focus on care tasks in which personal involvement is required. Borenstein and Pearson also show us that we should be prepared for unexpected, opposite shifts that will not lead to any time savings. This takes us back to the point of Van Oost and Reed (2011): when reflecting on deploying care robots, one must not focus only on those persons directly involved, but the entire socio-technical context must also be examined. In addition, we should pay attention to a possible shift in responsibilities and we must reflect critically on the actual time saved by the use of care robots.

3.4.3.3 Human Contact   Sharkey and Sharkey (2012) find that the deployment of care robots will deprive elderly people of social interaction with their fellow humans and that as a consequence their welfare will suffer. The absence of human contact affects the physical as well as the psychological well-being of the elderly. Some researchers believe that contact with care robots cannot compensate for the lack of human contact. An objection to this could be that, rather, the use of robots gives people the chance to live longer independently without relying on the care of others. Take, for instance, smart robot technologies that help to put on compression stockings or that assist in showering or using the toilet. In this way, robots provide an opportunity for a more dignified existence. Borenstein and Pearson (2010) also see these benefits: under certain conditions, care robots may ensure that people with disabilities become more independent. It is important, however, that the person in question exerts control over the robot; this is in contrast to the control that human caregivers often have over their patients. Parks (2010), however, notes that there is already isolation among older people living both at home and in nursing homes and thinks that robots would make things worse. The existing isolation is one reason for older people to resist the use of robots. Visits by human caregivers are so important that many older people prefer their help with showering and other hygiene-related needs even though this is something many people would prefer not having another person do for them. However, this is, according to Parks, often the only social contact available to them.

3.5 Observational Conclusions: The Long Term

Robot developers have high expectations: in the long term, care robots will take the workload away from caregivers. The scenario looms irrevocably of a nursing home or our own home full of smart devices and machines that take care of us. Will this image of the future become reality as aging reaches its peak in 2040? We are very doubtful that this will happen.

The argument that robots can solve staff shortages in health care has no hard evidence basis. Instead of replacing labor, the deployment of care robots rather leads to a shift and redistribution of responsibilities and tasks and forms new kinds of care. Therefore, both in practice and in the policy context, all innovation and deployment-of-care technologies should be reexamined. In most policy documents on domotics (including robotics), cognitively able but physically not very capable elderly people are the central client group (see, e.g., Department of Health, 2010). Domotics might enable these people to continue to live in their own home rather than in hospitals or residential care homes, publicly funded ones in particular.

Care that depends significantly on domotics is already a reality, but the technology typically used is not robotic. In long-term care practice, innovation proves to be a difficult matter (Berwick, 2003). Moving from the laboratory to practice and from prototype to actual production is a complicated process. Butter et al. (2008) note that a lot of time is needed to implement important innovations in health care. Also, in the field of robotics, there are many skeptical voices that can be heard coming from care practitioners, developers, and users. In recent years, many initiatives have been started in Europe to integrate robots into domotics, such as Mobiserv and KSERA, in which the development of care robots for remote care plays a major role. Without a commercial follow-up, however, budgets will diminish and tested robot applications will fail to take root in practice, especially because the stage of development in the field of care robotics is still too far from practical application with "real patients" (Butter et al., 2008), and because of the complexity of caregiving tasks related to the frame problem (see Section 2.2). Because of the high costs of care robotics, this might also lead, in practical application in the future, to unfair access to the care robots (Datteri & Tamburrini, 2009).

Robots can, however, add to the potential benefits of domotics. A common benefit is that a robot—because of its embodiment—has "presence": it is there with the older person (Sorell & Draper, 2014). This presence enables a robot to provide companionship to the elderly and to move around with them, to prompt them to undertake beneficial behaviors, to communicate through a touchscreen, to react to the elderly person's commands, and so on.

The question is, however, what role might robots have in the future care of the elderly at home? Robots can play different roles: companion, cognitive assistant, and caregiver. These roles, nowadays reserved for human beings, are rationalized so that they can be performed by robots. This requires a vision of care, and the role is taken up by robotics for this purpose. As we have seen, the rationalization of care by the deployment of care robots gives rise to several ethical issues depending on the kind of role a robot has. A robot as companion can lead to misleading relationships, and the risk of paternalism comes into play with a robot as cognitive assistant in terms of the extent to which a robot may enforce actions. The use of robots as caregivers raises social issues relating to the human dignity of the care recipient (Sharkey, 2014). Through the use of these robots, care recipients could be dehumanized.

Another important drawback, related to dehumanization, put forward by ethicists is the expected reduction in human contact caused by the use of domotics, and especially caused by the use of care robots. Care recipients will no longer have direct contact with human caregivers, but will instead have contact via devices or remotely, mediated by technology. The question underlying all of this is: how much right has a care recipient to receive real human contact? Or, to put it more bluntly, how many minutes of real human contact is a care recipient entitled to receive each day? It is important to observe the choice of the care recipient. Some people might prefer a human caregiver, while others may prefer support robots, depending on which one gives them a greater sense of self-worth. Robots can thus be used to make people more independent or to motivate them to go out more often. Thus, the elderly may, for example, keep up their social contacts as they can go outside independently with the help of robots; robots here are used as technology to combat loneliness. Or, when deploying robots that assist people when showering or going to the toilet, the robots are the key to independence. Again, the manner in which robots are deployed and the tasks they carry out are both of crucial importance. The more control the care recipient has over the robot, the less likely he or she is to feel objectified by the care robot.

The use of tele-technologies and robotics should therefore be tailor-made but should not lose sight of the needs of care recipients. Too little attention is currently paid to these needs and the expectations that exist in relation to long-term care. One of the main conclusions for designers of care robots is that they should consider elderly people not only as recipients of care but also as users whose needs have to be carefully identified and who should actively participate in the design process. This will increase the acceptance of robots by elderly people (Frennert & Östlund, 2014).

Sorell and Draper (2014) propose an ethical framework for the design of care robots. The framework defines the values that should be promoted or at least respected by care robot designers. These values are as follows:

- *Autonomy* (being able to set goals in life and choose the means of achieving them)
- *Independence* (being able to implement one's goals without the permission, assistance, or material resources of others)
- *Enablement* (having or having access to means of realizing goals and choices)
- *Safety* (being able to readily avoid pain or harm)
- *Privacy* (being able to pursue and realize one's goals and implement one's choices unobserved)
- *Social connectedness* (having regular contact with friends and loved ones and safe access to strangers one can choose to meet)

It is nearly impossible to avoid conflicts between these values in a design, and trade-offs are necessary. In such trade-offs, a balance must always be sought between increasing the quality of life—by allowing older people to remain at home longer—and protecting the individual rights of people and their physical and mental well-being (Sharkey & Sharkey, 2012).

Interview with Hans Rietman (Professor of Physical Medicine and Rehabilitation, University of Twente, the Netherlands)

"The work will change, but personal contact is irreplaceable."
Hans Rietman is not an engineer working in health care, but a doctor turned technology expert—which makes a difference. He admires sophisticated exoskeletons that allow "the infirmed to walk," but wonders if the designers will think of preventing bedsores.

"New robotic devices are best designed in a collaborative effort of patients who know what they want, medical experts who know what's needed, and engineers who know what's feasible," Hans Rietman says. "That's my favorite take-home message."

His triple professional position makes him see perhaps more clearly than others how important all three parties are. At the University of Twente (UT) in the Netherlands, he teaches future engineers about rehabilitation technology. As a medical practitioner, he sees patients at Roessingh, a specialized rehabilitation center in Enschede. Finally, he is the director of Roessingh's separate R&D division. They cooperate not only with the rehabilitation center but also with UT, which is located within easy cycling distance.

Coaching Role

Among the projects that Roessingh R&D participates in is the development of therapeutic robots. These are feats of engineering and will potentially benefit both patients and doctors. "After a stroke, people need intensive exercise to recover their motor skills, especially those of the limbs, so that after some time they can return home. Ideally, they should exercise several times a day, preferably doing useful and motivational activities and getting direct feedback. But there is a personnel problem: walking exercises on a treadmill, for instance, require the presence of two therapists, and it's such a strain on them that they can't do this for more than fifteen minutes at a time—no more than thirty minutes a day, if we follow workplace legislation to the letter. Therefore, together with UT and other partners, we have considered whether robots might take over this job, and perhaps even be better at it. The result has been LOPES, an exoskeleton full of sensors, which supports the patient's weight and will correct their walking movements where needed. Something similar has been developed for arms. Hands are too complex at this stage, but there is an international project working on them too.

My vision for the future is that in say ten years we'll have exercise areas with some two or three leg-exercise robots and five arm-exercise robots. They will offer the users mostly game-like activities, some in 2D with touchscreens, others in 3D with a console similar to a Wii. People will exercise there under the supervision of a therapist, who will divide their attention among a number of patients. Theirs will be more of a coaching role than at present. So the day-to-day work will change, but it will remain vital: only the therapist can assess how well the recovery is going. Also, they will still be the most accurate observer of a patient's movements, and can tell best what corrections are needed."

Subjective Perception

"We are evaluating the quality of this innovation with conventional therapy. The results so far indicate that they're pretty close when it comes to walking speed, endurance, quality of gait, and so on. The robot is not yet superior, but it is just as good.

We also set great store by the patients' perception of the therapy. It is particularly this subjective, emotional aspect that Roessingh want to bring to the collaboration with the more technology-oriented parties such as UT and several companies. We pay attention to apparent details that matter greatly to patients. For instance, how long does it take to strap on the device, sensors and all? It shouldn't be more than three minutes—we have it down to five now. On the positive side, it turns out that patients positively enjoy the arm-exercise robot. Today's patients, including the elderly, are comfortable with computer games. Their main complaint is that they would like more variety.

Those with misgivings about the devices are not so much the patients, but rather the therapists and nurses, probably for two reasons. First, as professionals, they often have to work with prototypes, which tend to have teething troubles. This frustrates them no end, and I understand that completely. Second, it is often claimed that robots will replace therapists and nurses. Personally, I think that these professionals will remain of vital importance. The emotional value of personal contact is irreplaceable. I do expect that contacts with therapists will become less frequent, as patients will be able to exercise with a robot in between. Importantly, patients tend to be happy about that."

Preferring the Wheelchair

Other robotic devices that Roessingh R&D are developing include smart prostheses and orthoses—artificial body parts and external body supports, in everyday language. Thanks to their sensors, they can predict the user's next movement and adjust their action accordingly. But yet again, there are some snags.

"We now have leg prostheses with very advanced computers in the knees. The need for them is greatest among the elderly, but the benefits will mostly go to young people, because insurance companies consider them too expensive for elderly patients. In the future, these prostheses will be motorized, so as to support the legs even more. But many elderly patients will have trouble controlling such a powerful device. Their motor intelligence is no longer up to it, so accidents are bound to happen. I believe they will benefit more from nonrobotic improvements in prostheses."

Orthoses pose their own problems. Some of them are technical: how to enable the user to control them if their nerve signals are either absent, as in paraplegics, or inadequate, as in people with spasticity. Others result from "these developments being technology-driven," Rietman says. "Designers realize insufficiently that the patient will be within a device that grates against the skin, irritating it. You get pressure sores, comparable to bedsores and wheelchair sores. Perhaps these can be avoided by using innovative materials, but this is not the priority I think it ought to be, in line with the patients' preferences. Actually, I expect that 10 years from now, most patients will still find a wheelchair preferable."

Betrayed by a Bot

Will health care see an invasion of service robots over the next decade? Rietman strongly doubts it. "We are doing work on a patient lift that is so smart you might well call it a robot. But it will be mostly for professional settings, as it is way too expensive for home use. Besides, the legal requirements for home use are even more stringent, as there is no professional around to intervene when things go pear-shaped.

Yet another development in this field, though not here at Roessingh, are social robots to keep people company and chat with them. Personally, they make me a bit uneasy. Of course, they will get increasingly sophisticated, responding correctly to emotions and questions. It is conceivable that at some point it will be hard to tell their conversations apart from talking with humans. Interestingly, it doesn't even matter when the robot doesn't look human. I remember seeing a research presentation from Canada about a robot of only 1.30 meters (or 4.27 feet) tall. Its head was the classical tin affair, but it was endowed with human speech. When they let a patient talk with it, he started treating it as if it were a sentient being within 5 minutes. He even tried to pull the robot's leg! Afterward, when the robot told the therapists that the patient had failed to do some of his exercises, the patient felt let down. Betrayed.

Obviously, this gets you into all sorts of ethical discussions. Therefore, I've been very happy to notice that engineering students, when I teach them about the relationship between health care and rehabilitation technology, often spontaneously bring up the ethical questions. What will these technologies mean for individual patients and for society? I think it is very good and important that they think about these questions right from the start."

References

Aarts, E., & Marzano, S. (2003). *The new everyday: Views on ambient intelligence.* Rotterdam, the Netherlands: 010 Publishers.

Akrich, M. (1992). The description of technical objects. In W. Bijker, & J. Law (Eds.), *Shaping technology/building society: Studies in sociotechnical change* (pp. 205–224). Cambridge, MA: MIT Press.

Aldrich, F. K. (2003). Smarthomes: Past, present and future. In R. Harper (Ed.), *Inside the smart home* (pp. 17–39). London, UK: Springer-Verlag.

Bemelmans, R., Gelderblom, G., Jonker, P., & de Witte, L. (2012). Socially assistive robots in elderly care: A systematic review into effects and effectiveness. *Journal of American Medical Directors Association, 13*(2), 114–120.

Berwick, D. M. (2003). Disseminating innovations in health care. *Journal of American Medical Association, 289*(15), 1969–1975.

Bharucha, A. J., Anand, V., Forlizzi, J., Dew, M. A., Reynolds, C. F., Stevens, S., & Wactlar, H. (2009). Intelligent assistive technology applications to dementia care: Current capabilities, limitations, and future challenges. *The American Journal of Geriatric Psychiatry, 17*(2), 88–104.

Boissy, P., Corriveau, H., Michaud, F., Labonte, D., & Royer, M. P. (2007). A qualitative study of in-home robotic telepresence for home care of community-living elderly subjects. *Journal of Telemedicine and Telecare, 13*(2), 79–84.

Borenstein, J., & Pearson, Y. (2010). Robot caregivers: Harbingers of expanded freedom for all? *Ethics and Information Technology, 12*(3), 277–288.

Breazeal, C. (2011). Social robots for health applications. *Proceedings of the annual international conference of the IEEE engineering in medicine and biology society* (pp. 5368–5371), Boston MA, August 30–September 3. http://ieeexplore.ieee.org/xpl/articleDetails.jsp?arnumber=6091328 (accessed January 23, 2015).

Bringsjord, S. (2008). Ethical robots: The future can heed us. *AI & Society, 22*(4), 539–550.

Broadbent, E., Stafford, R., & MacDonald, B. (2009). Acceptance of healthcare robots for the older population: Review and future directions. *International Journal of Social Robotics, 1*(4), 319–330.

Broekens, J., Heerink, M., & Rosendal, H. (2009). Assistive social robots in elderly care: A review. *Gerontechnology, 8*(2), 94–103.

Butter, M., Rensma, A., van Boxsel, J., Kalisingh, S., Schoone, M., & Leis, M. (2008). *Robotics for healthcare* (final report). Brussels, Belgium: European Commission, DG Information Society.

Center for Practice Improvement and Innovation. (2008). *Communicating with patients electronically*. Washington, DC: ACP. http://www.acpon-line.org/running_practice/technology/comm_electronic.pdf (accessed May 24, 2014).

Cesta, A., & Pecora, F. (2005). The RoboCare project: Intelligent systems for elder care. *Proceedings of the AAAI fall symposium on caring machines: AI in elder care*, Arlington, VA, November 4–6 (pp. 25–28). Menlo Park, California: The AAAI Press. http://aaaipress.org/Papers/Symposia/Fall/2005/FS-05-02/FS05-02-005.pdf (accessed January 23, 2014); http://www.academia.edu/2809421/The_robocare_project_Intelligent_systems_for_elder_care (accessed November 16, 2014).

Coeckelbergh, M. (2010). Health care, capabilities, and AI assistive technologies. *Ethical Theory and Moral Practice*, *13*(2), 181–190.

Courtney, K. L. (2008). Privacy and senior willingness to adopt smart home information technology in residential care facilities. *Methods of Information in Medicine*, *47*(1), 76–81.

Czaja, S. J., & Sharit, J. (2012). *Designing training and instructional programs for older adults*. Boca Raton, FL: CRC Press.

Datteri, E., & Tamburrini, G. (2009). Ethical reflections on health care robotics. In R. Capurro, & M. Nagenborg (Eds.), *Ethics and robotics* (pp. 35–48). Amsterdam, the Netherlands: IOS Press.

Decker, M. (2008). Caregiving robots and ethical reflection: The perspective of interdisciplinary technology assessment. *AI & Society*, *22*(3), 315–330.

Department of Health. (2010). *Building the national care service*. London, UK: Crown.

European Commission. (2012). *The 2012 Ageing Report: Economic and budgetary projections for the 27 EU Member States (2010–2060)*. Brussels, Belgium: European Commission.

Frennert, S., & Östlund, B. (2014). Review: Seven matters of concern of social robots and older people. *International Journal of Social Robotics*, *6*(2), 299–310.

Hansen, S. T., Andersen, H. J., & Bak, T. (2010). Practical evaluation of robots for elderly in Denmark—An overview. *Proceedings of the fifth ACM/IEEE international conference on human–robot interaction* (pp. 149–150), Osaka, Japan, March 2–5. Piscataway, NJ. U.S.: IEEE Press. http://dl.acm.org/citation.cfm?id=1734517 (accessed October 26, 2014).

Hardwig, J. (1997). Is there a duty to die? *Hastings Center Report*, *27*(2), 34–42.

Heerink, M., Kröse, B., Evers, V., & Wielinga, B. (2010). Assessing acceptance of assistive social agent technology by older people: The Almere model. *International Journal of Social Robotics*, *2*(4), 361–375.

Ikuta, K., & Nokata, M. (2003). Safety evaluation method of design and control for human-care robots. *The International Journal of Robotics Research*, *22*(5), 281–297.

Inoue, K., Wada, K., & Ito, Y. (2008). Effective application of Paro: Seal type robots for disabled people in according to ideas of occupational therapists. In K. Miesenberger, J. Klaus, W. Zagler, & A. Karshmer (Eds.), *LNCS:* Vol. 5105. *Computers helping people with special needs* (pp. 1321–1324). Berlin, Germany: Springer Verlag.

Jacobs, T., & Gurvinder, V. (2014). ISO 13482—The new safety standard for personal care robots. *ISR/Robotik 2014, joint conference of 45th international symposium on robotics and eighth German conference on robotics* (pp. 698–703). Berlin, Germany: VDE-Verlag.

Johnson, D. O., Cuijpers, R. H., Juola, J. F., Torta, E., Simonov, M., Frisiello, A., ..., Beck, C. (2014). Socially assistive robots: A comprehensive approach to extending independent living. *International Journal of Social Robotics*, *6*(2), 195–211.

Joyce, K., & Loe, M. (2010). A sociological approach to ageing, technology and health. *Sociology of Health and Illness*, *32*(2), 171–180.

Kort, S. M. (2005). Ethics in domotics (reply). *Gerontechnology*, *3*(3), 54.

Lammer, L., Huber, A., Weiss, A., & Vincze, M. (2014). Mutual Care: How older adults react when they should help their care robot. *Proceedings of the third international symposium on new frontiers in human–robot interaction— AISB 2014*, London, UK, April 1–4. http://hobbit.acin.tuwien.ac.at/publications/AISB2014-HRIpaper.pdf (accessed January 23, 2015).

Lau, Y. Y., van't Hof, C., & van Est, R. (2009). *Beyond the surface. An exploration in healthcare robotics in Japan.* The Hague, the Netherlands: Rathenau Institute.

M&ICT. (2010). Factsheet project KOALA. http://www.m-ict.nl/wp-content/uploads/2011/09/Factsheet-Koala-def.pdf (accessed November 16, 2014).

McColl, D., & Nejat, G. (2013). Meal-time with a socially assistive robot and older adults at a long-term care facility. *Journal of Human–Robot Interaction*, *2*(1), 152–171.

Meulendijk, M., van de Wijngaert, L., Brinkkemper, S., & Leenstra, H. (2011). AmI in good care? Developing design principles for ambient domotics for elderly. *Informatics for Health and Social Care*, *36*(2), 75–88.

Mihailidis, A., Cockburn, A., Longley, C., & Boger, J. (2008). The acceptability of home monitoring technology among community-dwelling older adults and baby boomers. *Assistive Technology*, *20*(1), 1–12.

Mukai, T., Hirano, S., Nakashima, H., Sakaida, Y., & Guo, S. (2011). Realization and safety measures of patient transfer by nursing-care assistant robot RIBA with tactile sensors. *Journal of Robotics and Mechatronics*, *23*(3), 360–369.

Mykityshyn, A. L., Fisk, A. D., & Rogers, W. A. (2002). Learning to use a home medical device: Mediating age-related differences with training. *Human Factors*, *44*(3), 354–364.

Nagai, Y., Tanioka, T., Fuji, S., Yasuhara, Y., Sakamaki, S., Taoka, N., ..., Matsumoto, K. (2010). Needs and challenges of care robots in nursing care setting: A literature review. *Proceedings of the sixth international conference on natural language processing and knowledge engineering* (NLP-KE), Beijing, China, August 21–23. http://ieeexplore.ieee.org/stamp/stamp.jsp?tp=&arnumber=5587815 (accessed September 9, 2014).

Nylander, S., Ljungblad, S., & Villareal, J. J. (2012). A complementing approach for identifying ethical issues in care robotics—Grounding ethics in practical use. *Proceedings of the 21st IEEE international symposium on robot and human interactive communication, RO-MAN* (pp. 797–802), Paris, France, September 9–13. http://ieeexplore.ieee.org/xpl/articleDetails.jsp?arnumber=6343849 (accessed November 11, 2014).

Oudshoorn, N. (2008). Diagnosis at a distance: The invisible work of patients and healthcare professionals in cardiac telemonitoring technology. *Sociology of Health & Illness, 30*(2), 272–288.

Parks, J. A. (2010). Lifting the burden of women's care work: Should robots replace the "human touch"? *Hypatia, 25*(1), 100–120.

Pearson, Y., & Borenstein, J. (2013). The intervention of robot caregivers and the cultivation of children's capability to play. *Science and Engineering Ethics, 19*(1), 123–137.

Pols, J. (2010). The heart of the matter. About good nursing and telecare. *Health Care Analysis, 18*(4), 374–388.

Pols, J. (2011). Wonderful webcams. About active gazing and invisible technologies. *Science, Technology & Human Values, 36*(4), 451–473.

Riva, G. (2003). Ambient intelligence in health care. *CyberPsychology & Behavior, 6*(3), 295–300.

Rodriguez, M. D., Favula, J., Preciado, A., & Aurora, V. (2005). Agent-based ambient intelligence for healthcare. *AI Communications, 18*(3), 201–216.

Rogers, W. A., & Mynatt, E. D. (2003). How can technology contribute to the quality of life of older adults? In M. E. Mitchell (Ed.), *The technology of humanity: Can technology contribute in the quality of life?* (pp. 22–30). Chicago, IL: Illinois Institute of Technology.

Salvani, P., Laschi, C., & Dario, P. (2010). Design for acceptability: Improving robots' coexistence in human society. *International Journal of Social Robotics, 2*(4), 451–460.

Sävenstedt, S., Sandman, P. O., & Zingmark, K. (2006). The duality in using information and communication technology in elder care. *Journal of Advanced Nursing, 56*(1), 17–25.

Sharkey, A. (2014). Robots and human dignity: A consideration of the effects of robot care on the dignity of older people. *Ethics and Information Technology, 16*(1), 63–75.

Sharkey, A., & Sharkey, N. (2012). Granny and the robots: Ethical issues in robot care for the elderly. *Ethics and Information Technology, 14*(1), 27–40.

Shaw-Garlock, G. (2011). Loving machines: Theorizing human and sociable-technology interaction. In M. H. Lamers & F. J. Verbeek (Eds.), *LNICTS*: Vol. 59. *Human–robot personal relationships* (pp. 1–10). Heidelberg, Germany: Springer.

Sorell, T., & Draper, H. (2014). Robot carers, ethics, and older people. *Ethics and Information Technology, 16*(3), 183–195.

Sparrow, R., & Sparrow, L. (2006). In the hands of machines? The future of aged care. *Mind and Machines, 16*(2), 141–161.

Turkle, S. (2006). *A nascent robotics culture: New complicities for companionship* (AAAI Technical Report Series). http://mit.edu/sturkle/www/nascentroboticsculture.pdf (accessed October 14, 2014).

Vallor, S. (2011). Carebots and caregivers: Sustaining the ethical ideal of care in the twenty-first century. *Philosophy* and *Technology, 24*(3), 251–268.

Van der Plas, A., Smits, M., & Wehrman, C. (2010). Beyond speculative robot ethics: A vision assessment study on the future of the robotic caregiver. *Accountability in Research Policies and Quality Assurance, 17*(6), 299–315.

Van de Poel, I. R., & Royakkers, L. M. M. (2011). *Ethics, technology, and engineering. An introduction.* Oxford, UK: Wiley-Blackwell.

Van Hoof, J., Kort, H. S. M., Markopoulos, P., & Soede, M. (2007). *Gerontechnology, 6*(3), 155–163.

Van Oost, E., & Reed, D. (2011). Towards a sociological understanding of robots as companions. In M. H. Lamers, & F. J. Verbeek (Eds.), *LNICTS*: Vol. 59. *Human–robot personal relationships* (pp. 11–18). Heidelberg, Germany: Springer.

Van Wynsberghe, A. (2013). Designing robots for care: Care centered value-sensitive design. *Science and Engineering Ethics, 19*(2), 407–433.

Veruggio, G., & Operto, F. (2008). Roboethics: Social and ethical implications of robotics. In B. Siciliano, & O. Khatib (Eds.), *Springer handbook of robotics* (pp. 1499–1524). Berlin, Germany: Springer Verlag.

Wada, K., & Shibata, T. (2007). Living with seal robots: Its sociopsychological and physiological influences on the elderly at a care house. *IEEE Transaction on Robotics, 23*(5), 972–980.

Wada, K., Shibata, T., Saito, T., & Tanie, K. (2006). Robot assisted activity at a health service facility; an interim report of a long-term experiment. *Journal of Systems and Control Engineering, 220*(6), 709–715.

Wu, Y.-H., Fassert, C., & Rigaud, A. S. (2011). Designing robots for the elderly: Appearance issue and beyond. *Archives of Gerontology and Geriatrics, 54*(1), 121–126.

Zwijsen, S. A., Niemeijer, A. R., & Hertogh, C. M. P. M. (2011). Ethics of using assistive technology in the care for community-dwelling elderly people: An overview of the literature. *Aging & Mental Health, 15*(4), 419–427.

4

DRONES IN THE CITY

Toward a Floating Robotic Panopticon?

4.1 Introduction: Amazon Prime Air

A few weeks before Christmas 2013, Amazon, the world's largest e-commerce company, revealed plans about Prime Air, a delivery system that uses drones (unmanned aerial vehicles) to deliver products to customers in 30 minutes or less (see Figure 4.1). Its CEO, Jeff Bezos, said on television that the drones could be ready to take flight in 4 to 5 years.* Amazon's drones are small, unmanned octocopters or eight-rotor helicopters (helicopters with eight tiny rotors) that use global positioning system (GPS) navigation and have electric motors. The current models have a range of 16 kilometers (or 9 miles), travel over 80 km/h (or 50 mph), and can carry products weighing less than 2.5 kilograms (or 5.5 pounds) (a criterion that covers 86% of the products offered on Amazon). Prime Air could mean quick and efficient deliveries for consumers. Amazon's goal is to make this drone delivery service available to customers worldwide as soon as it is permitted by legislation.

When Amazon first announced the project in a video† released in December 2013, a convenient time, just when the holiday shopping season was heating up, many regarded it as a publicity stunt. The company's petition of July 2014, asking that outdoor research and development (R&D) testing for Prime Air be allowed in the United States, sent to the U.S. air safety regulatory agency the Federal Aviation Administration (FAA), however, reveals how

---

* http://www.cbsnews.com/news/amazon-unveils-futuristic-plan-delivery-by-drone/.
† http://www.amazon.com/b?node=8037720011.

Figure 4.1   Amazon's Prime Air. (Photo courtesy of Amazon.)

serious this project is. In the petition, Amazon claims that it has made advancements toward the development of highly automated aerial vehicles for Prime Air, including:

> Testing a range of capabilities for our eighth- and ninth-generation aerial vehicles, including agility, flight duration, redundancy, and sense-and-avoid sensors and algorithms; developing aerial vehicles that travel over 50 miles per hour, and will carry 5-pound payloads, which cover 86% of products sold on Amazon; and attracting a growing team of world-renowned roboticists, scientists, aeronautical engineers, remote sensing experts, and a former NASA astronaut.[*]

Although Amazon has clearly invested a lot of capital and time into this initiative, its technical completion and legislative approval might take a long period. As indicated in the petition, existing U.S. laws hinder Amazon's R&D projects. According to the *Economic Times* (based on anonymous sources), Amazon plans to test unmanned aerial vehicles for deliveries in Mumbai and Bangalore, where it maintains warehouses.[†] India has been chosen because it lacks the more strident regulations on commercial drone usage found in other countries. Regulatory clearances aside, Amazon will have to face a number of challenges to get Prime Air off the ground. Losing communication with the drones or other technical mishaps that cause a drone to crash, possibly injuring people and/or damaging property

---

[*] http://www.amazon.com/b?node=8037720011.

[†] http://articles.economictimes.indiatimes.com/2014-08-20/news/53028827_1_prime-air-drones-outdoors-amazon.

in the process, would cost Amazon dearly. The fallout from such an incident would not be limited to India but would in fact ruin Amazon's ambitions to launch Prime Air in general. Amazon itself refused to confirm or deny that it has chosen India as the testing ground for Prime Air. BGR News said its sources inside Amazon in India claim that Prime Air will not be launched in India anytime soon and that they had in fact indicated that a 2015 start may represent an overly ambitious timeline.*

Jeff Bezos's announcement on TV not only kick-started a good bit of media attention for Amazon, but also made the idea of drones for popular commercial application suddenly seem like a viable proposition. The drone market nowadays is dominated by military applications, with barely any significant spending on commercial drones (see Chapter 6). Commercial drone space, however, is projected to become a multibillion dollar industry over the next 10 years. *Business Insider Intelligence* published a report in January 2014 that predicts that there will be U.S. $98.2 billion in cumulative global spending on aerial unmanned vehicles over the next 10 years, with U.S. $11.8 billion of that spending on commercial applications (Rubin, 2014).

Amazon may not be the only company to eventually use drones for carrying everyday items. Drones could be deployed to deliver items such as prescription medication from pharmacies, meals from restaurants, and food from supermarkets. While not able to fly yet due to the FAA restriction, the *TacoCopter*, designed in Silicon Valley, is already able to deliver tacos direct to people's doorsteps in San Francisco via unmanned helicopters.† It is expected that commercial drones will quickly become a reality. The FAA currently estimates that as many as 7500 small commercial unmanned aircraft systems (UASs) may be in use by 2018, assuming the necessary regulations are in place, which they expect by September 30, 2015 (see also Section 4.5). In Europe, Deutsche Post AG, Europe's largest postal service, has already begun, since October 2014, to deliver medication and other urgent goods to the North Sea island of Juist Island (Germany) using unmanned

---

* http://www.bgr.in/news/amazon-prime-air-drone-delivery-service-to-start-from-india-not-us-report/.

† http://www.huffingtonpost.com/2012/03/23/tacocopter-startup-delivers-tacos-by-unmanned-drone-helicopter_n_1375842.html.

helicopters, which are called *parcelcopters*, after securing approval from state and federal transportation ministries and air traffic control authorities to operate in a restricted flight area.*

Besides delivery, a wide range of current and potential applications of drones exist. They can be outfitted with sophisticated sensors depending on their particular needs. For example, a thermal sensor is an obvious choice for a fire-lookout drone or a drone patrolling a power plant. Another example of one of the latest developments is the ambulance drone (see Figure 4.2), developed by Delft University of Technology in the Netherlands.† When the emergency services receive a cardiac arrest call, this drone can quickly deliver a defibrillator to the emergency scene. Via a live-stream video and audio connection, the drone can also provide direct feedback to the emergency services and the people on-site can be instructed how to treat the patient. The drone autonomously finds the patient's location via the caller's mobile phone signal and makes its way there using GPS. The drone can fly at around 100 km/h (or 62 mph), weighs 4 kilograms (or 8.8 pounds), and has a payload of another 4 kilograms (or 8.8 pounds). Regulation is needed

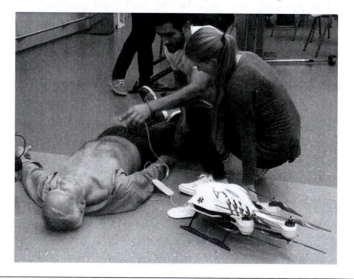

Figure 4.2   The ambulance drone. (Photo courtesy of Delft University of Technology.)

* http://www.bloomberg.com/news/2014–09–25/dhl-beats-amazon-google-to-first-scheduled-drone-delivery.html.
† http://www.tudelft.nl/en/current/latest-news/article/detail/ambulance-drone-tu-delft-vergroot-overlevingskans-bij-hartstilstand-drastisch/.

to reap the societal benefits of these drones and to deal with the social and ethical issues that these drones evoke.

It is common to make a distinction between public and civil drones. Civil drones are drones for private or commercial use. In Section 4.2, we will discuss some civil applications of drones that seem likely in the near future. A public drone is one that is used by the national, state, or local agencies. In Section 4.3, we discuss the public drones that are used by law enforcement agencies, in particular some applications of police drones. Before these drones can be successfully introduced into society, there is a lot of ground to cover. Although drones are already being used under fragmented regulatory frameworks, there are still major safety, technological, and legislative hurdles to the deployment of drones for general commercial purposes. Next, we will look at various safety (Section 4.4) and privacy (Section 4.5) issues that the deployment of civil and public drones raises. In particular, camera-equipped domestic drones for surveillance have elicited many privacy concerns. Regulation concerning safety and privacy issues will be discussed in Section 4.6, and we will end with some observational conclusions in Section 4.7.

4.2 Civil Applications of Drones

Primary civil applications are monitoring and security, exploration, aid efforts, disaster recovery, entertainment, farming, journalism and photography, and so on. This wide diversity suggests that there are few industries and service sectors that could not potentially be touched by drones. This section considers three civil application areas—recreational use, drone journalism, and precision farming—in more detail to illustrate the potential for drones and related issues that may need attention before deploying drones.

4.2.1 Recreational Use

Drones are popular because it does not require many skills to fly them, so they can easily be used to take pictures and create videos. Significant technological advances and associated cost reduction have made drones appealing to an ever-wider public. For example, the leading community of personal drones, DIYDrones (DIY stands for do-it-yourself), has a robust community of almost 60,000 hobbyists. Recreational

drone flying is already a fairly established category in the toy industry. One can, for example, buy more than 100 different drones with
a camera in Amazon's Drone Store section, costing from U.S. $50 to
$2000.* The biggest thing holding drones back today is limited flight
endurance. Most drones cannot fly for more than a few minutes. The
more expensive drones have a flight time of about 15 minutes.

Probably the best-known drone is the quadcopter helicopter *Parrot
AR.Drone 2.0* (see Figure 4.3) developed by the French company
Parrot.† The drone can be controlled by a smartphone or tablet and
comes equipped with a high-definition camera. The drone was a trail-
blazer in developing technology that allows toys to be remotely controlled by an iPhone. Its flight time is approximately 12 minutes, it has
a maximum speed of 18 km/h (or 11 mph) with a range of 50 meters
(or 165 feet), and weighs less than 500 grams (or 1.1 pounds). The
Parrot Company refuses to reveal sales figures, but the leading

Figure 4.3   Parrot AR.Drone 2.0. (Photo courtesy of Rinie van Est.)

---

* http://www.amazon.com/.
† http://ardrone2.parrot.com/.

industry website sUAS News claims that 500,000 have been sold worldwide since it was first launched in 2010.* The price of this drone is between U.S. $300 and $400.

For U.S. $1000 you can get more advanced drones, such as the *DJI Phantom*,† also popular, which has a flight control system that will automatically control the aircraft so that it returns home and lands without causing injury or damage, a range of 1 kilometer (or 0.6 mile), and a flight endurance of almost half an hour.

Due to the latest technology, drones are becoming smaller and smaller. One of the smallest drones for sale that has a camera is the *Hubsan X4*, which is 6 × 6 cm (or 2.5 × 2.5 inches).‡ Delft University of Technology (TU Delft) claims that it has developed the world's smallest autopilot.§ Its goal is to eventually create drones *without* a controller that are small enough to fit into your pocket. Hence, TU Delft aims to create a tiny autopilot system of 2 × 2 cm (or 0.8 × 0.8 inches) that weighs 2 grams (or 0.04 ounce). Bart Remes, project manager at the Micro Aerial Vehicle Laboratory at the TU Delft, predicts that within 5 years a lot of people will carry a drone in their pocket, for example, to film themselves while they are skiing.¶

Recreational use of drones is considered to be model aircraft use, and this does not need permission from federal governments in almost all countries. There are some restrictions on its use, though. Most national governments forbid flying a drone higher than 150 meters (or about 500 feet), or within populated areas or airports, or one that weighs more than 25 kilograms (or 55 pounds). Drones must remain in view, and can only be used for recreational rather than commercial purposes. In contrast with the relatively easy use of drones, it takes a lot of time to learn how to fly a model aircraft, which is actually part of the challenge of having such an aircraft. As a consequence, people flying a model aircraft may be more careful than people flying a drone. Since flying a drone is relatively easy, the bar to buying such

---

* http://www.suasnews.com/2013/01/20637/qgroundcontrol-and-flight-recorder-bring-autonomous-flight-to-ar-drone/.
† http://www.dji.com/product/phantom-2.
‡ http://www.hubsan.com/products/helicopter/h107.htm.
§ http://www.gizmag.com/tu-delft-worlds-smallest-autopilot/28845/.
¶ http://www.tedxamsterdam.com/2013/11/bart-remes-a-drone-inside-everybodys-pocket/.

a drone has become considerably lower. A problem, therefore, is that drones could be a magnet for reckless pilots.*

The use of drones to invade someone's private home or business, etc., could potentially be viewed as criminal harassment or voyeurism. The problem is that it would be difficult to find out who is flying the drone, since the operator may well be out of sight and the drone does not have identifying features. Citizens of Vancouver, for example, have already lodged more than 10 drone complaints in the first months of 2014 with the police.†

While national governments are investing time and money developing strict standards for commercial drone use, recreational use has largely escaped the regulatory spotlight. Recreational drone users, by definition, are just having fun and may not care about regulations. The activities of recreational drone users are also difficult to monitor. This aspect of use, combined with the increasing availability and affordability of drones, means that these users are likely to pose an even greater threat to safety and privacy than the commercial drone users.

4.2.2 Drone Journalism

On January 1, 2014, spokesman photographer Jesse Tinsley reportedly used his personal quadcopter to capture an aerial view of the annual Polar Bear Plunge which takes place on Sanders Beach on Lake Coeur d'Alene in Idaho (United States) as a New Year's Day tradition. Because publication of the video constitutes commercial use, the FAA responded, saying that this activity is decidedly illegal, though they do not intend to pursue action against the paper or the filmmaker (and the video is still available).‡ In 2013, the University of Missouri's Drone Journalism Program and the University of Nebraska–Lincoln's drone program have been suspended by the FAA until they obtain a Certificate of Authorization (COA) that allows them to use drones.§

---

* http://motherboard.vice.com/read/the-worlds-most-popular-drone-is-a-magnet-for-reckless-pilots.

† http://arstechnica.com/tech-policy/2014/08/vancouver-man-creeped-out-by-drone-buzzing-near-his-36th-story-condo/.

‡ http://www.inlander.com/Bloglander/archives/2014/01/06/faa-takes-issue-with-recent-spokesman-drone-video.

§ http://www.columbiatribune.com/news/education/faa-hampers-mu-j-school-drone-classes/article_162d9e8c-0b51-11e3-83dd-10604b9ffe60.html.

Obtaining a COA for a certain location can take several months, and if one wants to fly in another location, one has to follow the same lengthy process, which is not well suited to journalism, since news is often momentary. Through these actions of the FAA, it has de facto put a ban on the use of drones for news-gathering purposes.

In 2014, 16 major U.S. news organizations, therefore, came together to accuse the FAA of curtailing freedom of the press by restricting the use of drones.* The United States prohibits the use of drones for commercial purposes, although the FAA grants rare exceptions for government and law enforcement use by a COA. According to these news organizations, the FAA's position is untenable as it rests on a fundamental misunderstanding about journalism, since news gathering is not a commercial purpose but a First Amendment right. An advantage of using drones for journalism is that viewers can be shown the scope of the news from perspectives that were previously unprocurable, and it also increases the scope for having an observation platform from which to hold a position and monitor the situation instead of just flying over it (Clarke, 2014a), which would not otherwise be possible. Similarly, drones could aid reporting of fires and other natural disasters. Furthermore, drones have the potential for providing data in investigative journalism as well.

Drone journalism can raise specific ethical issues when covering people. For example, University of Nebraska journalism professor Matt Waite states that drones can be very intrusive, for example, when a number of journalists with small drones cover an event in which a mother is grieving over the gruesome murder of her children, or when paparazzi use drones to take pictures or videos of the private retreats of celebrities.† Clarke (2014a) calls this "voyeurnalism," that is, voyeurism by journalists: a form of corrupted journalism in which information regarding events and issues is gathered and presented that is not in the public interest, but rather is what the public is or may become interested in. Voyeurnalism with drones will create a paparazzi aloft. Continuous monitoring of celebrities can be undertaken at locations such as the target's front door,

---

* http://www.ndtv.com/article/world/us-drone-ban-infringes-press-freedom-519566.
† http://www.dailymail.co.uk/news/article-2746231/Attack-drones-Hollywood-celebrities-besieged-paparazzi-spies-sky-Worried-You-ll-soon-regular-fixture-YOUR-home.html.

and tracking becomes much easier. A growing fleet of drones is forcing celebrities to run for cover even inside their own homes, as cameras
swoop in above swimming pools, tennis courts, and balconies. Paparazzi
drones have terrorized stars including singer Rihanna and actress Jennifer
Garner.* Another example of voyeurnalism could be the incident at the
World Cup in 2014, when France asked the world soccer organization
FIFA to investigate its suspicions that a drone was used to spy on the
French national team's preparations for its World Cup opening match
against Honduras. It was probably a fan who used the drone, but if a
journalist had operated the drone, it would have been voyeurnalism.†

4.2.3 Precision Farming

Drones may provide a big lift in precision agriculture, which allows
farmers, through the use of information and communication technology
(ICT), to micromanage their land and make detailed plans for the work
to be done. Precision agriculture refers to two segments of the farm
market: remote sensing and precision application (Jenkins & Vasigh,
2013). A variety of remote sensors are used to scan plants for health
problems, record growth rates and hydration, and locate disease out-
breaks. Precision application, a practice especially useful for crop farmers and horticulturists, utilizes effective and efficient spray techniques to
more selectively cover plants and fields. Solutions have existed since the
1980s but have not resulted in widespread adoption (Bramley, 2009).
This could change with the rapid development of low-cost, open-source
drones and advances in camera technology over the last couple of years.

Drones can be programmed to fly low over fields and stream photos and videos to a ground station, where the images can be stitched
together into maps or analyzed to gauge crop health. Compared to
satellite imagery, drone images are much cheaper and offer higher
resolution (Barrientos et al., 2011). Drones can provide farmers
with three types of detailed views (Anderson, 2014). First, seeing
a crop from the air can reveal patterns that expose everything from

---

* http://www.dailymail.co.uk/news/article-2746231/Attack-drones-Hollywood-
  celebrities-besieged-paparazzi-spies-sky-Worried-You-ll-soon-regular-fixture-
  YOUR-home.html.
† http://www.businessinsider.com/someone-used-a-drone-to-spy-on-frances-world-
  cup-team-2014-6.

irrigation problems to soil variation and even pest and fungal infestations that are not apparent at eye level. Second, airborne cameras can take multispectral images, capturing data from the visual as well as the infrared spectrum, which can be combined to create a view of the crop that highlights differences between healthy and distressed plants in a way that cannot be seen with the naked eye. Finally, a drone can survey a crop every week or every day, or even every hour. A time series animation such as this can show changes in the crop, revealing trouble spots or opportunities for better crop management. It is part of a trend toward increasingly data-driven agriculture. This will also allow for the creation of a historical database, which farmers might use to predict future crop yields and soil health.

The drone developed by InventWorks, Inc. and Boulder Labs, Inc. presents a nice example.* This drone, which weighs 2 kilograms (or 4.5 pounds) and has a 1.8 meters (or about 6 feet) wide wingspan, carries multispectral cameras that take high-resolution, geo-tagged photos every few seconds. InventWorks and Boulder Labs claim that the deployment of their drone could potentially save 80% per acre on herbicide costs, which would translate to nearly U.S. $10,000 in cost savings to the average farmer per crop cycle. David Mulla, director of the Precision Agriculture Center of the University of Minnesota, estimates that the use of drones could save U.S. $10–$30 an acre in fertilizer and in related costs by examining the progress of crops while they are still in the ground.†

Agriculture could provide the ground for commercial drone applications, partly because operating in rural areas far from crowds, buildings, and airports alleviates privacy and safety concerns. A 2013 study by Jenkins and Vasigh (2013) estimates that future commercial drone markets would be largely in agriculture. Huang, Thomson, Hoffmann, Lan, and Fritz (2013) state that it will be some years before the farming drones are successful, mainly because of the limitations in payload and flight endurance. They therefore think that for the next decade at least, human-piloted aircraft and ground equipment (tractor mounted) will still dominate and drones will only be used to inspect and treat small sections of fields, especially those that large equipment cannot reach (see Box 4.1).

---

* http://www.inventworksinc.com/project-gallery/single-gallery/18407868.
† http://www.startribune.com/business/269913801.html.

BOX 4.1   THE JAPANESE FARMING DRONE RMAX

Unmanned helicopters have been in use in Japan for the last 20 years. In 1983, the Ministry of Agriculture, Forestry and Fishery asked Yamaha to start developing an unmanned aircraft for farming, because of the aging workforce and lack of a younger generation of successors. In 1990, Yamaha delivered the first unmanned helicopter for crop dusting. A massive breakthrough for the drone was the development of the Yamaha Attitude Control System (YACS), characterized by its vastly improved ability to hover in a stationary position and the fact that complete novices could fly it (Sato, 2003). Today, the most popular farming drone is the Yamaha remote-controlled helicopter *RMAX* (see Figure 4.4), and it costs about U.S. $100,000. It weighs 100 kilograms (or 220 pounds) and has a total length of 3.63 meters (or 11.9 feet) and a height of 1.08 meters (or 3.54 feet). It has a 28 kilogram (or 62 pounds) payload and two liquid sprayer tanks that can hold 8 liters each. Liquids and granules can be dispersed across a 400 meter (or 0.25 mile) range from the location of the operator, covering nearly a hectare in 6 minutes. The drone allows operators to spray weeds or crops, or to spread seed in any terrain in a more cost-effective and accurate manner. In Japan, where rice fields are on average about 5 acres and are often surrounded by residential or commercial development, the drone provides a safe, efficient, and accurate method for applying agricultural sprays (Sato, 2003).

Today, there are around 2400 RMAX drones flying in Japan, representing a 77% market share. The number of people capable

Figure 4.4   The RMAX. (Photo courtesy of Rich Pedroncelli/Hollandse Hoogte.)

of operating them has grown to about 7500 nationwide.* In 2013, the total area of farmland in Japan being sprayed by these drones reached 2.4 million acres.[†]

The drones have received much attention from other countries. For example, the University of California, Davis (UC Davis) studies where and how the mini helicopter might play a role in U.S. agriculture.[‡] The FAA has approved Yamaha's partnership with UC Davis for conducting experimental flights for data collection and demonstration for agricultural uses. A test field of this study is the Napa Valley's vineyards, with their relatively small plantings, adjacent development, and often hilly terrain. The drone can go where a standard-sized helicopter or fixed-wing aircraft could not go and, in some situations, represents less risk to the operator than a tractor-drawn spray rig. Yamaha has exported 100 RMAXs to South Korea, and in 2013 Yamaha Sky Division Australia introduced the RMAX by enabling franchisees and contractors to maintain land and crops remotely. Rather than selling such a drone, the manufacturer leases the drone to an operator with a trained and certified pilot who flies it for farmers.[§]

The RMAX has already been deployed for purposes other than agriculture. At the end of March 2000, the 732 meter (or 2400 foot) high, active volcano Mount Usu erupted. The Japanese government asked Yamaha to perform surveillance of the volcano area at Mount Usu in April 2000. It was able to gather valuable and precise data that it would not have been possible to gather using manned aircraft. It was the first case in the world of successful drone operation out of the range of sight by means of a GPS autonomous flight system (Sato, 2003).

Japan can be seen as a model for the successful use of unmanned aircraft in agriculture. It worked well, because the Japanese Ministry of Agriculture, Forestry and Fishery commissioned the technology rather than inhibiting the commercialization of drones by imposing specific regulations and an operator licensing system to operate the drones safely.

---

* http://rmax.yamaha-motor.com.au/history.

[†] http://www.yamahaprecisionagriculture.com/rmax.aspx.

[‡] http://news.ucdavis.edu/search/news_detail.lasso?id=10623.

[§] http://www.theland.com.au/news/agriculture/machinery/general-news/franchisees-required-for-unmanned-helicopters/2668468.aspx.

4.3 Drones for Law Enforcement

4.3.1 Robocops

In 1924, in the U.S. magazine *Science and Invention*, a police robot is discussed: the *Radio Police Automaton* (see Figures 4.5 and 4.6). This robot has spotlights, loudspeakers, tear gas, and rotating discs that use bullets to disperse a crowd. Radio technology allows the robot to be steered from a police car. According to the author, this police robot is "matchless" since it is resistant to bullets and has a strong gasoline engine of 20–40 horsepower (HP). This futuristic 1920s view

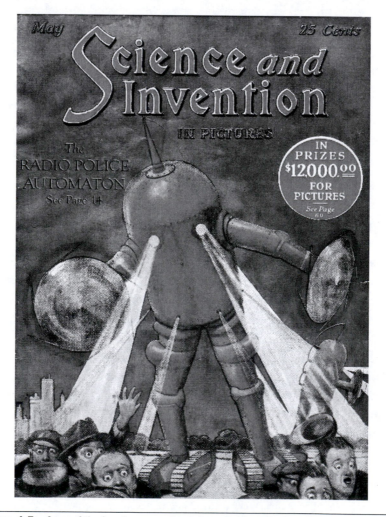

Figure 4.5 Cover of the American magazine *Science and Invention* from 1924 with an image of the Radio Police Automaton.

Figure 4.6 Explanation of Radio Police Automaton by Gernsback (1924).

on police work conjures up—especially given the front cover of the magazine—Orwellian images: inhumanly strong robots that can be deployed to maintain public order and security.

In today's era, police robots inevitably evoke associations with films such as *Robocop* and *The Terminator*. The android robots of these movies play the lead roles in epics in which evil forces are overcome. These spectres present Orwellian scenes in which society comes under control from omnipotent and omniscient robots. These futuristic images stand in stark contrast with reality. According to critical British robot professor Noel Sharkey (2008), these specters go too far, but there are certainly caveats. Using a road map for the future, Sharkey is trying to visualize the implications of current developments in the field of police robots. The road map outlined by Sharkey begins in the present and ends in the year 2084—a clear reference to Orwell's *1984*. Sharkey shows that currently, robots may be used for mobile surveillance and detection of explosives. In 2084, there may be a humanoid police robot walking down the street that is connected to a network of cameras, databases, and other robots, which assists in road safety, riot control, crowd control, locating missing persons, or interrogating dangerous suspects. According to Sharkey, in the future robots will take over all kinds of police tasks and they will increasingly perform these tasks more independently.

Police robots today resemble remote-controlled toy vehicles rather than police officers, as even opening a door can already present an insurmountable obstacle.

While the "Robocop" will remain to be science fiction for a long time, robots are already being used by the police for dangerous, dull, and dirty police work. Robots can perform boring work or dangerous jobs without loss of concentration and can reach places that are otherwise difficult to access by a human or another technical tool. In Great Britain, robots have been used since the Irish Republican Army (IRA) era for removing bombs, and since the late 1980s the remotely controlled so-called V-A1 robots have been used to inspect possible hazardous situations in the State of Virginia (United States).* These V-A1 robots are equipped with cameras, chemical detection equipment, and a mechanical arm to grab objects. They enable agent-operators to assess dangerous situations from a distance without running risks themselves.

Although there are some land-based police robots, we see an increasing use of drones by law enforcement agencies. This is mainly due to the huge investments and developments in military drone technology (see Chapter 6), from which law enforcement agencies benefit. Furthermore, law enforcement agencies expect a great deal of good to come from robots for surveillance and reconnaissance purposes, and drones may ideally serve these purposes. In this section, we will discuss law enforcement drones, mainly those used by the police.

---

* http://ww2.roanoke.com/news/roanoke/wb/230806.

4.3.2 Tasks of Police Drones

Police drones have mainly been developed for surveillance and reconnaissance purposes, tasks using cameras and sensors for observation and communication with the environment. For example, in 2009 a mini helicopter drone was purchased for the Dutch police's task force for organized hemp growers. The drone has odor-detection equipment and the ability to use video so it can trace hemp fields from the air. The police have already uncovered a cannabis factory during the first test flight of the mini helicopter. Seven suspects were arrested and several kilos of hemp were intercepted.*

In 2014, Belize, a country on the eastern coast of Central America, deployed drones in the fight against illegal fishing.† Illegal fishing, with worldwide industry losses worth U.S. $23 billion, represents a major global problem because it can also damage fragile maritime ecosystems. Belize is one of the countries hardest hit by illegal fishing. It was so rampant that in March 2014, that the European Union suspended all seafood imports from Belize (as well as from Cambodia and Guinea), stating that Belize had not acted forcefully enough to prevent illegal fishing. The fixed-wing drones, the so-called *Skywalkers*, used against illegal fishing can fly for over an hour, have a range of 50 kilometers (or 30 miles), and are capable of taking high-definition photos and videos. They will be used to patrol areas that are difficult to reach, such as coastal mangrove forests, at a fraction of the cost of a conventional, manned aircraft. Once the illegal activity is located, authorities can dispatch a vessel and perform a seagoing search much more efficiently.

In 2009, the National Aerospace Laboratory (NLR)—the independent knowledge enterprise on aerospace in the Netherlands—wrote a report for the National Police Agency on their assistance in the operational testing and evaluation of a number of market-available, remote-controlled reconnaissance aircraft (Ladiges & Van Stijl, 2009). On the basis of 11 scenarios, the report provides a picture of the type of tasks for which law enforcement agencies may use drones and what results may be achieved. The wide range of scenarios includes the search for

* http://www.dutchdailynews.com/dutch-police-hunt-cannabis-hemp-growers-with-canna-chopper/.
† http://news.nationalgeographic.com/news/2014/07/140718-drones-illegal-fishing-pirate-belize-ocean/.

a missing person or a kidnapped daughter of a wealthy family in a nature reserve, monitoring a burning ship in a harbor when there is a threat of a massive explosion, recognizing marijuana containers at an industrial site and ensuring security at a gathering of international VIPs (see Box 4.1 for an example). Based on these scenarios, it can be determined which requirements the drone must meet to adequately fulfill these types of operations (see Box 4.2).

4.3.3 Examples of Police Drones

Numerous types of drones are developed for law enforcement. Most of these drones fit easily into a police car and can be assembled and ready for flight in less than 5 minutes to provide a rapidly deployable eye in the sky, transmitting live video directly to the operator at a fraction of the cost of a manned aircraft. We will give three examples to provide an idea of the breadth and adaptability of police drone technology.

The *Qube* is a drone with four rotor blades, is 90 centimeters (or 3 feet) long, and weighs 2.5 kilograms (or 5.5 pounds).* This drone is at the forefront of the use of drone technology by police forces in the United States. It can swoop back and forth, has hover-and-stare capability, and uses high-resolution color and thermal cameras. The Qube's flight endurance time is 40 minutes with payload. The range (1 kilometer, or 0.6 mile, line of sight) and operational altitude (30–150 meters, or 100–500 feet) are restricted by the FAA.

The Grand Forks Police Department in North Dakota, for example, uses this drone. Since 2013, the drone has flown 11 active missions.[†] These missions involve taking digital photos over a crime or traffic accident that might provide detectives with useful intelligence from an aerial perspective, monitoring the annual threat of flooding from the Red River and searching for people. In one mission, a suspected drunk driver ran away into a field covered in corn stalks rising more than 2 meters (or 7 feet) high that made visibility on the ground impossible. With the help of the Qube's thermal-imaging device, which can detect any live presence through body heat, the suspect was found in 3 minutes.[‡]

---

[*] http://www.avinc.com/uas/small_uas/qube.

[†] http://www.theguardian.com/world/2014/oct/01/drones-police-force-crime-uavs-north-dakota.

[‡] http://motherboard.vice.com/read/police-used-a-drone-to-chase-down-and-arrest-four-dui-suspects-in-a-cornfield.

BOX 4.2   SCENARIO 9: EXPLORATION IN URBAN AREAS

Case Scenario

An illegal arms factory is probably located at industrial estate X. The building has a courtyard and many people walk in and out of the building. Police want a daylight raid, but first want to know what the exact situation is.

Scenario Setting

Industrial estate, remote, many major roads
Assignment scenario
Is this plant in fact an illegal weapons factory?
What happens in the courtyard?
How many people are in the building?
Where are all the entrances and unmanned aerial vehicle (UAV) requirements

The following operational UAV requirements are derived from the previous scenario description:

a. It can observe a permanent building for a long time (>1 hour) and all its access points, continuously monitoring all objects (people, vehicles) approaching from any direction and those present in the courtyard can also be seen.
b. The ability by observation to determine how many people are hiding in and around the area observed, using a UAV.
c. Recognizing and recording certain objects (weapons) in the area observed.
d. The ability to observe the area unnoticed by means of a UAV.

Source: Ladiges, R., & Van Stijl, M.C., OT&E scenario's voor KLPD UAS, NLR-CR-2009-598, Nationaal Lucht-en Ruimtevaartlaboratorium, Amsterdam, the Netherlands, 2009.

Figure 4.7   The ShadowHawk. (Photo courtesy of Vanguard Defense Industries.)

A more advanced drone is the drone helicopter *ShadowHawk* (see Figure 4.7), which has a rotor span of almost 2 meters (or 7 feet) and weighs 16 kilograms (or 35 pounds).* The drone has a range of 24 kilometers and can fly at speeds of up to 89 km/h (or 55 mph). The ShadowHawk can maintain aerial surveillance of an area (i.e., a house, vehicle, person, etc.) at more than 200 meters (or 650 feet) without being heard or seen, unlike a full-sized aircraft. The Montgomery County Sheriff's Office (MCSO) in Texas, for example, uses this drone for emergency management, missing person recovery, and watching over operations, for example, filming above Special Weapons and Tactics (SWAT) team activities.† In April 2014, the drone of the MCSO, which cost U.S. $250,000, suffered a malfunction and crashed during an exercise over Lake Conroe.

The *Nano Hummingbird* (see Figure 4.8) is a pocket-sized drone with a wingspan of 16.5 centimeters (or 6.5 inches) and a weight of 19 grams (or 0.67 ounce) that is equipped with a video camera for surveillance and reconnaissance purposes.‡ The drone has been developed by AeroVironment under a Defense Advanced Research Projects Agency (DARPA)-sponsored research contract to develop a new class of air vehicle systems capable of indoor and outdoor operation. The drone is able to fly at speeds of up to 18 km/h (or 11 mph),

---

* http://vanguarddefense.com/productsservices/uavs/.

† http://www.chron.com/neighborhood/woodlands/article/250K-police-drone-crashes-into-Lake-Conroe-5435343.php.

‡ http://www.avinc.com/nano.

Figure 4.8 The Nano Hummingbird. (Photo courtesy of AeroVironment, Inc.)

hover, and fly sideways, backward, and forward, and has a flight time of 11 minutes. The Nano Hummingbird has not yet been deployed by the police, but in the near future it could be deployed to perform reconnaissance and surveillance in urban environments, and might perch on windowsills or power lines, or enter buildings to observe its surroundings, relaying camera views back to its operator.*

These drones open up a new field of possibilities for surveillance purposes. In order to realize this potential, there are still some hurdles to be overcome. The flight times of the air robots, for example, are still limited. In addition to the restricted energy supply, wind force can constrain their use. At a given wind force the air robot loses its stability, which keeps it from producing usable pictures or video images. Furthermore, programming the flight is a fragile and delicate process, and a small programming mistake can easily cause the drone to crash.

Police drones can also be equipped with "intelligent" camera systems. Using particular video analysis software, a camera can remotely register incidents, including burglaries and fires, for example, on a

---

* http://articles.latimes.com/2011/feb/17/business/la-fi-hummingbird-drone-20110217.

shopping street or on a construction site, and pass information on to an emergency room. Integrating intelligent camera drones can provide very useful police applications. By equipping a drone with intelligent eyes, it can recognize suspect behavior in a public space and prevent escalation. It would be of great importance if the smart cameras could sort out certain movements that are perhaps beyond the norm and where emotional outbursts of one individual may influence bystanders. Escalation can be avoided if the drone can make timely and accurately predictions about group dynamics and call for help, or perhaps take a supervisory and monitoring role in a panic-induced situation.

4.3.4 Legal and Ethical Issues

Hambling (2010) asks other important questions in relation to tele-operations: Will the ability to operate at a distance make the police officer less able to intervene when serious trouble is about to erupt? Will the police, as a result of becoming habituated to the use of drones, lose the skills acquired through extensive training, which are crucial when things get tense? This is referred to as the risk of de-skilling: the loss of essential skills as a result of habituation to new technologies. This can also appear in using police drones and deserves attention from those designing task innovation and replacement. The increasing deployment of police drones requires new skills from police officers. First, they need to be able to operate these drones, and second, they need to be able to perform police actions using robotic technology; both put different operational and strategic demands on police personnel. The downside is that a loss of essential police skills—skills acquired through extensive training and experience—may occur as a result of getting used to the deployment of police drones, after which police officers would be less able to intervene in serious problems that cannot be solved using drones.

A legal complication regarding the deploying of drones for police purposes is that it is not yet clear how these robots can be deployed in accordance with existing laws and regulations. For example, which specific restrictions apply on the basis of the current legal framework regarding the deployment of drones over a festival crowd, or a fire in the dunes, or at night, or out of sight of the operator? The political challenge is to adapt Aviation Acts and privacy regulations in order to increase the use of drones for law enforcement purposes without risking the safety in the air or on the ground and without compromising the privacy of citizens too much. The use of drones for law enforcement could prove to be one of the most far-reaching and potentially controversial uses of drones as a new, relatively cheap surveillance tool. Drones raise the prospect of much more pervasive surveillance. Using them would allow police to record the activities of the public below with high-resolution, infrared, and thermal-imaging cameras. The deployment of these drones has driven debates about the boundaries of privacy. The main problem is that drones will be used to watch citizens.

From the perspective of the police, the Aviation Acts and regulations can be seen as an obstacle to the use of drones for specific purposes. Law enforcement agencies in the United States have been clamoring for the FAA to allow the rapid deployment of drones. They tout drones as a tactical game changer in scenarios such as hostage situations and high-speed chases. As we will see, the FAA is reluctant to simply open up airspace for law enforcement, even for small drones, since they have concerns about safety. Also, in Europe, different countries are currently considering the adaption of existing regulatory frameworks that hamper the commercial use of drones. In July 2014, the Spanish government approved a provisional regulatory framework for commercial operations of drones, which sets out requirements according to the weight of the unmanned remotely controlled vehicles and the obligations of pilots and operating companies.* The regulation will enable the use of drones in carrying out aerial works such as investigation and development activities, aerial agriculture relating to treatments that require the spreading out of substances over the surface or atmosphere, including products for extinguishing fires, aerial surveys, aerial observation and surveillance, including filming and forest fire surveillance activities, aerial advertising, radio and TV emissions, emergency operations, search and rescue, and other types of special work.

---

* http://www.twobirds.com/en/news/articles/2014/spain/spain-temporary-regulations-on-commercial-use-of-drones-approved.

In 2007, former Amsterdam police chief Eric Nordholt worried about the encroaching cameras in an interview:

> With a system of cameras and computers, the police want to control access to different parts of the city. I do not think we want to live in such a society. But the measures proposed do work towards that. Therefore, we should think about how we do wish to have it. It's about whether you believe that you are shaped by the city or you help shape the city. It is about the question what the police contribution can be. As chief of police one is not there to further doomsday predictions.

De Jong and Schuilenburg (2007)

Drones will add to the strong emphasis on monitoring and law enforcement instead of finding a balance between what one does in the technical sphere and in the sphere of controlling and monitoring—with, for example, drones—and what one does in the social field: "to give help to those in need." According to Nordholt, the police should be more intelligent in thinking about the question: How do we connect technologies that enable us to control with a strong socially oriented police force?

The next two sections discuss the two main ethical issues raised by the use of drones—safety and privacy—in more detail.

4.4 Safety

While the applications of public and civil drones are numerous, they are not without risks. One of the most important risks is safety. The FAA reports, obtained by the news station ABC15, show that police departments are trying to crack down on illegal drone use, but are struggling with the task.* The reports show that at least 23 investigations have been launched in recent years, but most of the illegal drone operators have never been found. The 10 that were tracked down only received official warnings, although one operator, Raphael Pirker, was fined U.S. $10,000 for taking commercial photographs above the University of Virginia (see Box 4.3). Civilian drones, however, flown with the FAA's permission and under its scrutiny, are also susceptible

---

* http://www.abc15.com/news/local-news/investigations/commercial-drones-a-serious-safety-concern.

BOX 4.3   THE CASE OF RAPHAEL PIRKER

Pirker first caught the media's attention in 2010 when he uploaded footage he took of New York City and the Statue of Liberty using his first-person-view model plane. Authorities were notified, but no action was taken because Pirker was not paid for the flight. Therefore, his model aircraft was subject to the 1981 Advisory Circular and not the 2007 notice about commercial applications for drones. Then, in October 2011, Pirker was contacted by the advertisement agency Lewis Communications and was asked to film aerial footages of the University of Virginia campus. For the shoot, Pirker flew his drone under trees, through a tunnel and near a person—a spotter working for Pirker. The FAA caught wind of Pirker's commercial flight. Two years later, Pirker received a fine for U.S. $10,000, on the grounds that he flew his UAV for commercial purposes and that this was in violation of an FAA regulation, stating that "no person may operate an aircraft in a careless or reckless manner so as to endanger the life or property of another."

Pirker fought the enforcement action before the National Transportation and Safety Board (NTSB), arguing that the FAA did not have any authority to fine someone operating a drone because it had not issued any formal rules governing their use. In March 2014, an NTSB administrative law judge agreed with Pirker. There was no enforceable FAA rule in place, and if the scope of the FAA's existing regulations were correct, its position "would then result in the risible argument that a flight in the air of, e.g., a paper aircraft, or a toy balsa wood glider, could subject the 'operator' to" these regulations. The FAA appealed, and the appeal is pending. In a press release, it states that "[t]he FAA is appealing the decision of an NTSB Administrative Law Judge to the full National Transportation Safety Board, which has the effect of staying the decision until the Board rules. The agency is concerned that this decision could impact the safe operation of the national airspace system and the safety of people and property on the ground."*

---

* http://www.faa.gov/news/press_releases/news_story.cfm?newsId=15894.

to crashes. According to the *Washington Post*, 23 accidents and 236 incidents in which registered civilian drones were involved have been deemed "unsafe" by the FAA since November 2009.[*]

4.4.1 Aerial Safety

With the increasing use of civil drones and without proper oversight, the unregulated use of drones will become a recipe for disaster. In 2013, that terrifying vision nearly came true, when a pilot of an Alitalia jetliner came within 60 meters (or 200 feet) of a drone while on its approach to JFK Airport.[†] According to the FBI, the drone, described as a small helicopter, was flying at an altitude of about 530 meters (or 1700 feet) over a densely populated neighborhood. The drone's operator has never been found. A midair collision could be calamitous. The potential results of a collision can be compared with the 2009 "Miracle on the Hudson," in which a flock of geese flew into the engine of a U.S. Airways' Airbus A320 and forced the pilot to conduct an emergency landing on the water.[‡]

The problem with commercial and private drones is that these small objects do not appear on radar and thus are often undetected by traffic collision avoidance systems. Although there are basic flight regulations, such as that these drones may not fly any higher than about 120 meters (or 400 feet) and must operate at a distance of about 5 kilometers (or 3 miles) from commercial airports, many operators do not obey the regulations. According to a yearlong investigation by the *Washington Post*, drones flew dangerously close to airports of passenger aircrafts in 15 cases during 2013 and 2014. They also found that a NASA database of confidential complaints filed by pilots and air traffic controllers has recorded 50 other reports of risky midair encounters.

4.4.2 Improper Operations

Drones can also cause serious injuries if not operated properly. There are several reasons for incorrect operation; for example, the remote control

---

[*] http://www.washingtonpost.com/sf/investigative/2014/06/23/close-encounters-with-small-drones-on-rise/.

[†] http://www.abc15.com/news/local-news/investigations/commercial-drones-a-serious-safety-concern.

[‡] http://rt.com/usa/157972-drone-american-airlines-near-miss/.

device sends bad signals, the drone has a defect, or operators try to carry out a maneuver above their skill level. Several incidents have already happened. In 2013, a 19-year-old man was killed when a remote-controlled model helicopter that he was piloting in a Brooklyn Park struck him on the head.* Also in 2013, a small helicopter drone flying high above buildings on the east side of Manhattan crash-landed just feet away from a businessman during the Monday evening rush hour.† And a wedding photographer almost ruined a happy couple's big day during a photo shoot while trying to get an aerial shot of the pair when his quadcopter smashed at full speed into the groom's head and caused the groom to suffer a cut eye and cheek.‡ In 2014, a triathlete, competing in the Endure Batavia Triathlon in Geraldton, Western Australia, suffered head injuries requiring three stitches when a drone fell some 10 meters (or 30 feet) from above the course.§ The drone was operated by a local video company filming the race. The drone's pilot claims that the drone had been hijacked or hacked, possibly by someone nearby who used a mobile phone to take control away from the operator.

4.4.3 Hacking of Drones

Regardless of the claim of the drone operator, this illustrates that it need not be the drone's operator that intends the harm to be done. A drone may be hijacked and its behavior controlled by someone other than the original operator. Drones rely on GPS signals to navigate and are controlled by operators via a two-way radio transmission link. The military protects the communications and navigation links it uses to control drones with highly advanced encryption technology. Researchers from the University of Texas, however, have proven that in spite of this highly advanced encryption technology, military drones can be hacked. The technique they use is called "spoofing": a technique that leads the drone to mistake the signal from hackers for the one sent from GPS satellites. The team transmitted false signals

---

* http://www.nytimes.com/2013/09/06/nyregion/remote-controlled-copter-fatally-strikes-pilot-at-park.html?_r=1&.
† http://7online.com/archive/9270668/.
‡ http://www.dailymail.co.uk/news/article-2395933/Fail-Photographers-drone-smacks-groom-head-looked-perfect-shot.html.
§ http://www.cnet.com/news/drone-falls-out-of-the-sky-and-injures-athlete/.

that fooled the drone onto "thinking" that it was flying high up when in fact it was plummeting toward the ground. The team changed its course at the last minute to avoid a crash. The danger of hacking could be immense. Computer viruses can cause drones to become uncontrollable. Moreover, hacking could hijack unmanned combat aircraft in order to make them crash into a building or to turn them on civilians. Civilian drones generally rely on unencrypted satellite links and radio transmissions that can easily be hacked, jammed, or spoofed. Todd Humphreys, who led the research team from the University of Texas, states that it would be neither easy nor cheap to build defenses against hackers. He predicts that if effective solutions are not in place before commercial drones are on the market "hackers will come out of the woodwork."[*] Clarke and Bennett Moses (2014), however, believe that the increasingly low prices of drones make it unlikely that hacking will become common. The only motivation left for hijackers of drones would be to obscure the hijacker's identity.

4.4.4 Drone Hunting

A drone's behavior may be affected not only by hacking but also by direct physical attack using a projectile, including attack by another drone. Reports about hunting drones are on the rise. In 2012, an animal rights group in Pennsylvania attempting to document the cruelty of pigeon shoots at a shooting club had their camera-equipped drone blown out of the air more than once.[†] According to the police, in October 2014, a New Jersey man was arrested after he shot down a neighbor's remote-controlled drone.[‡] Under a proposed ordinance, the town Deer Trail in Colorado would grant hunting permits to shoot drones. The permits would cost U.S. $25 each. The town would also encourage drone hunting by awarding U.S. $100 to anyone who presents a valid hunting license and identifiable pieces of a drone that has been shot down. The proposed ordinance, a clear publicity stunt, was a

---

[*] http://www.washingtonpost.com/sf/investigative/2014/06/23/close-encounters-with-small-drones-on-rise/.

[†] http://www.suasnews.com/2012/11/19719/activists-drone-shot-out-of-the-sky-for-fourth-time/.

[‡] http://philadelphia.cbslocal.com/2014/09/30/new-jersey-man-accused-of-shooting-down-neighbors-remote-control-drone/.

symbolic protest against small, civilian drones flying over their land and violating their privacy, and was set against a background of an emerging and increasingly sinister U.S. surveillance state. In response to the proposed ordinance, the FAA issued a stern warning that it is against federal law to shoot down unmanned drones by releasing a statement in which it states that a drone "hit by gunfire could crash, causing damage to persons or property on the ground, or it could collide with other objects in the air. Shooting at an unmanned aircraft could result in criminal or civil liability, just as would firing at a manned airplane."[*] Since local officials split on whether to approve the ordinance, the issue went to the voters, who finally decided against the idea in 2014.[†]

Apparently, people feel uncomfortable with a neighbor's drone buzzing overhead and spying on them. Several states and local governments in the United States have, therefore, enacted laws regulating private drone use. They have taken varying approaches; they have articulated permissible uses and created new causes of action and established new crimes (Syed & Berry, 2014). The state of Idaho has passed the most sweeping legislation on private drone use. Its law prohibits people from using drones "to photograph or otherwise record an individual, without such individual's written consent, for the purpose of publishing or otherwise publicly disseminating such photograph or recording."[‡] In 2013, the city of Charlottesville in Virginia became the first U.S. municipality to pass an anti-drone resolution.[§] The resolution means that the use of drones for surveillance and other uses would only be allowed from 2016. Today, many states have also legislation against drone use in an effort to safeguard privacy.

4.5 Privacy

Any review of potential drone risks is not complete without mentioning privacy. Privacy concerns have dominated the discussion about drones for domestic use, since drones significantly change

---

[*] http://www.huffingtonpost.com/2013/07/19/faa-guns-drones_n_3624940.html.
[†] http://time.com/46327/drone-hunting-deer-trail/#46327/drone-hunting-deer-trail/.
[‡] http://www.legislature.idaho.gov/idstat/Title21/T21CH2SECT21–213PrinterFriendly.htm.
[§] http://www.usnews.com/news/articles/2013/02/05/city-in-virginia-becomes-first-to-pass-anti-drone-legislation-.

visual surveillance in at least three ways (Clarke, 2014b): drones offer views from new angles and enable ground-level obstructions to be overcome; they avoid ground-level congestion and greatly increase the feasibility of pursuit; and drones dramatically reduce the cost profile by combining inexpensive devices with low costs and limited labor content. In addition to camera equipment, drones can be equipped with microphones, thermal-imaging devices, and the capacity to intercept wireless communications. It appears that it is not only "Big Brother," but also "Little Brother," in possession of a drone, who threatens our privacy (see Section 4.7). So, surveillance is carried out not only by national law enforcement agencies but also by citizens or companies, which is called *sousveillance* (see Mann, Nolan, & Wellman, 2003).

California's Democrat Senator Dianne Feinstein claims that a drone was flown outside the window of her house while a demonstration was taking place there. During a Senate Commerce Committee hearing on drone policy, she implored lawmakers to consider this kind of violation of privacy as they look at allowing commercial drone use.* Recently, Supreme Court Justice Sonia Sotomayor spoke about the privacy concerns drones raise, and that U.S. citizens should be more concerned about their privacy being invaded by the spread of drones. She encourages citizens to take a more active role in the privacy debate:

> There are drones flying over the air randomly that are recording everything that's happening on what we consider our private property. That type of technology has to stimulate us to think about what is it that we cherish in privacy and how far we want to protect it and from whom. Because people think that it should be protected just against government intrusion, but I don't like the fact that someone I don't know ... can pick up, if they're a private citizen, one of these drones and fly it over my property.†

Democratically elected politicians are not the only ones who have expressed privacy concerns about the increased use of drones.

---

* http://www.politico.com/story/2014/01/senator-dianne-feinstein-encounter-with-drone-technology-privacy-surveillance-102233.html.

† http://blogs.wsj.com/law/2014/09/12/justice-sotomayor-americans-should-be-alarmed-by-spread-of-drones/?mod=e2tw.

More than 30 nongovernmental U.S. organizations, including the American Civil Liberties Union (ACLU), the Bill of Rights Defense Committee, and the Electronic Privacy Information Center (EPIC), have submitted a petition to the FAA that states that drones pose "substantial threats to privacy."* Already, drones are causing civilians problems concerning their privacy: in 2014, a woman reportedly assaulted a 17-year-old boy who was flying his drone with an attached camera over a beach in Connecticut. The boy was allegedly confronted and attacked by the woman, who accused him of taking perverted pictures of her.†

Fixed surveillance cameras or other observation methods have been in place for some time, but there are significant differences between these technologies and drones. Because drones can move freely in the air, they can follow moving objects and can hover over a location, that is, they can lie in wait. Since drones are not fixed, it is difficult to identify who is the owner of the drone. Furthermore, because drone cameras can reach greater heights than cameras that are attached to lampposts or buildings, larger areas can be observed. This leads to an increase in the probability that civilians are observed, among them also people who are on the beach or in their own garden. At a certain altitude, drones are not always visible to the public, since drones are quite small. This is in contrast to helicopters and manned aircraft, since they are usually visible or audible. Moreover, drones are more agile than fixed cameras, so vertical and angled shots can be taken. Finally, drones are becoming more popular and affordable, and it is expected that the domestic use of drones will be widespread in the future, which could lead to more voyeurism by civilians.

These unique features of camera-equipped drones for domestic use are capable of taking surveillance to a whole new level. According to Calo (2011), "[d]rones represent the cold, technological embodiment of observation." Drones make it possible (and, in many cases, cheap and easy) to gather detailed information about individuals to an

---

* http://www.thenewamerican.com/usnews/constitution/item/8197-privacy-rights-groups-fight-faa-on-use-of-drones-in-us.
† http://betabeat.com/2014/06/woman-violently-assaults-teen-boy-flying-drone-over-beach/.

extent never possible before. Using domestic drones for surveillance therefore raises some serious privacy issues.

4.5.1 Reasonable Expectations of Privacy

"Reasonable expectation of privacy" is the extent to which someone in a specific situation can reasonably expect that his or her privacy is respected. In the U.S. case *Kyllo v. United States*, the Supreme Court determined, based on the privacy protection set out in the Fourth Amendment, that the use of radars and sense-enhancing technology to gather information about activities occurring within the home was an invasion of the individual's reasonable expectation of privacy.[*] The Court reasoned that since sense-enhancing technology was not available for general public use, individuals could not reasonably expect that they would need to protect their privacy interests against this type of technology. The Achilles heel of this decision is that the Court qualified its decision by saying that it only related to technology that was "not in general public use" (Molko, 2013). Applying this reasoning that drones will become ever present in our society and will become "normalized," an important question surfaces: "has the availability of drones for general use, combined with public knowledge of drone operation destroyed society's privacy expectations to the degree that individuals have no reasonable expectation of privacy from drone surveillance?" (Schlag, 2013, p. 15; see also Thompson II, 2013). Another open privacy issue is that both in the United States and in the European Union, the protection of reasonable expectation of privacy mainly protects individuals against actions by law enforcement agencies rather than against actions by private citizens. Furthermore, drones will pose a greater invasion of privacy in the nonpublic space, since, in contrast to fixed surveillance cameras, drones can easily monitor and follow citizens in their private environment. With respect to public space, there is a reduced or "zero" expectation of privacy in the United States. As a consequence, pervasive tracking or extensive data collection with drones could be allowed, which, according to Cavoukian (2013), calls for regulations providing strong privacy protection.

---

[*] *Kyllo v. United States*, 533 U.S. 27, 37 (2001).

4.5.2 Voyeurism

A drone can hover over a particular person in order to track that person. Drones provide considerably greater empowerment to the voyeur than installed cameras. In June 2014, Seattle police investigated a woman's report of a drone peeping into her apartment window.* The woman was concerned that the drone was looking into her apartment while she was not fully dressed. The incident raises the following question: can other people use drones to spy on us, potentially for nefarious reasons? Policy makers and the public, until recently, have paid less attention to the use of privately owned domestic drones. As the Seattle incident illustrates, hobbyists can now buy a drone and use it to engage in privacy-intrusive behavior such as voyeurism. This type of behavior can border on harassment, trespassing, and privacy invasion. With the increasing use of drones by civilians, we may also see increasing instances of drones being used to spy on people. As Wright (2014) puts it: "Is it so hard to imagine a testosterone-packed teenager directing his drone to watch the object of his affection (or lust) sunbathing in the supposed privacy of her back yard?" (p. 226).

Drone journalism is going to be a common news-gathering method, which will increase "voyeurnalism" (see Section 4.2.2). In their code of ethics for drone journalists, the Professional Society of Drone Journalists (PSDJ), the first international organization dedicated to establishing the ethical, educational, and technological framework for the emerging field of drone journalism, established in 2011, has added the following requirement with regard to privacy:

> PRIVACY. The drone must be operated in a fashion that does not needlessly compromise the privacy of non-public figures. If at all possible, record only images of activities in public spaces, and censor or redact images of private individuals in private spaces that occur beyond the scope of the investigation.†

This requirement does not really discourage the paparazzi, since their scope of investigation is precisely the private space of public figures.

---

* http://verdict.justia.com/2014/06/26/drones-new-peeping-toms.
† http://www.dronejournalism.org/code-of-ethics.

4.5.3 Big Brother Drone Is Watching You

Since 2001, influenced by terrorist attacks, observation and monitoring technology has been implemented at a pace that was unthinkable until recently, a development that from the perspective of privacy laws is at least questionable (Phillips, 2004). With a combination of increased powers and new technology, such as drones, the opportunities for obtaining information about the public have grown. Drones could complement the many security cameras operating throughout the cities. By way of the increasing use of drones, the possibility exists that the daily activities of citizens will be monitored 24 hours a day by drones and fixed cameras and that this information will be stored.

Former Mayor Michael Bloomberg of New York predicts in an interview on Christian talk radio station WOR 710 that New Yorkers of the future will be more watched and have less privacy than they do today:

> We're going into a different world, unchartered. And, like it or not, what people can do or governments can do is different, and you can to some extent control, but you can't keep the tides from coming in. We're going to have more visibility and less privacy. I don't see how you stop that. And it's not a question of whether I think it's good or bad. I just don't see how you could stop that because we're going to have them [drones].*

The "Big Brother is watching you" reality is thus progressing in that these systems keep track of what we are doing and with whom, with the aim of subtly influencing our lives. Sharkey (2008) fears that in the current political climate the all-seeing closed-circuit television (CCTV) cameras will advance quickly and refers to the ease with which in recent years a massive surveillance network of CCTV cameras has invaded society, "despite all of the 1984 Orwellian rumbling in our stomach" (Sharkey, 2008, p. 4). According to a survey held in 2009 by *Big Brother Watch*, a British civil rights movement that defends the privacy of citizens, around 60,000 cameras are

---

* http://www.wor710.com/articles/local-news-465659/bloomberg-tells-john-gambling-drones-are-11097567/.

monitoring British citizens, that is, 1 camera per 1000 inhabitants.* Sharkey considers the steep rise in the number of surveillance cameras as an important precursor for the way police drones are going to be accepted, if we do not dwell on the social costs in time. He wonders whether we really want such a society: "There may be no hiding place in the city of the future" (Sharkey, 2008, p. 2). Our private identity may, therefore, be challenged by the intrusive drone surveillance.

4.5.4 The Chilling Effect

Since drones are not always visible and their purpose is often unknown, it is difficult for people to estimate whether or not they are being filmed by drones. This pervasive drone surveillance could have a chilling effect on the public's behavior. When people get the feeling that they can be observed everywhere and all the time, a large part of their autonomy is lost (cf. Solove, 2004). So, privacy relating to personal behavior is concerned with the freedom of the individual to behave as they wish, without undue observation or interference from others (Clarke, 2014c). With the increasing use of drones, this chilling effect will become a bigger issue, since the chance will increase that one may be monitored unnoticed by national security agencies (Big Brother) as well as by citizens or companies (Little Brother).

Volovelsky (2014) concludes in his research that the use of drones will infringe the fundamental right to privacy and that given the advantages inherent in the use of drones, and the desire to avoid a chilling effect, it is essential to affect a solution that would enable the use of drones while concurrently protecting the right to privacy. In the next section, we will discuss whether regulation can contribute to such a solution.

4.6 The Regulation of Drones

In this section, we discuss the regulation of drones in the United States and the European Union. The FAA and the European Commission have responsibility for regulating the use of drones for the United States and the European Union, respectively. We will see that the

---

* http://www.bigbrotherwatch.org.uk/cctvreport.pdf.

current requirements of the FAA and the European Commission for drone operations are minimal and largely perfunctory. As a consequence, many states and countries have enacted or proposed their own laws, since they are afraid of the threat that unregulated drone use poses to safety and privacy.

4.6.1 Regulations in the United States

4.6.1.1 The U.S. Federal Aviation Administration   In 2007, the FAA issued a policy notice declaring that "no person may operate a UAS [unmanned aircraft system] in the National Airspace System without authority." This policy still holds and applies to both public and private drones. The FAA has two methods for granting authority to operate a drone, depending on whether it is a civil or a public drone. Public drones, operated by the military or other federal, state or local agencies, must obtain a Certificate of Waiver or Authorization (COA) from the FAA. Since 2006, the FAA has issued COAs to more than 100 unique public entities throughout the United States for, for example, law enforcement, firefighting, border patrol, and search and rescue. In 2009, the FAA issued 146 COAs. In 2013 (as of October 31), the number grew to 373.

Obtaining an experimental airworthiness certificate for a particular drone is currently the only way to operate civil drones. The certificates have been issued on a limited basis for flight tests, demonstrations, and training. Commercial drone operations are limited and require the operator to have certified aircraft and pilots, as well as operating approval. To date, only two drone models (the *ScanEagle* and *AeroVironment's Puma*) have been certified, and they can only fly in the Arctic. The ScanEagle will help to survey ocean ice floes and migrating whales in Arctic oil exploration areas, and the Puma is expected to support emergency response crews that are on call for those carrying out monitoring of oil spills and wildlife surveillance in the Beaufort Sea. For recreational use of airspace by model aircraft, no certificate is needed. The use of these drones is limited by the FAA to operations below 400 feet (about 120 meters) above ground level and away from airports and air traffic. Recreational use of airspace only applies to those flying model aircraft purely as a hobby, and specifically excludes individuals or companies flying model aircraft for business purposes.

The policy statement mentioned earlier from 2007 also indicates that the FAA would undertake a safety review of drones and would then provide new rules. However, new rules have still not been proposed. Frustrated by the FAA's delay in promulgating comprehensive regulations, and recognizing the growing demand to use this technology, in 2012, the U.S. Congress enacted the FAA Modernization and Reform Act (FMRA). The Act requires the FAA to devise a "comprehensive plan to safely accelerate the integration of civil unmanned aircraft systems into the national airspace" by September 2015. The Act mandates a series of deadlines to be met by the FAA, such as a requirement to develop and implement operational and certification requirements for public drones by December 13, 2015, a deadline which will probably not be met.

In addition to allowing law enforcement to use small drones by loosening up the processes for doing this via the issue of COAs, establishing the six test sites and releasing its first annual Integration of Civil Unmanned Aircraft Systems in National Airspace System Roadmap (Syed & Berry, 2014), very little has been achieved with respect to safety (Clarke & Bennett Moses, 2014). The Roadmap is merely aspirational; it offers no schedule for actual changes to regulations and publication of standards and provides only a conceptual timeline.

With regard to privacy, the FAA pointedly refuses to regulate privacy in a broad fashion. By way of the *FAA Modernization and Reform Act*, Congress instructed the FAA to devise rules for safe and wider use of drones inside the United States. Nowhere did the Act mention privacy. The FAA (2013), however, has thought about some privacy considerations and has incorporated them into the drone test sites:

> The FAA's mission is to provide the safest, most efficient aerospace system in the world and does not include regulating privacy. At the same time, the FAA recognises that there is substantial debate and difference of opinion among policy makers, industry, advocacy groups, and members of the public as to whether UAS [unmanned aircraft system] operations at the test sites will raise novel privacy issues that are not adequately addressed by existing legal frameworks.

In particular, the FAA had developed privacy requirements which mean that test-site operators have to maintain a record of all drones operating at the test site and that each drone operator has to develop a written plan for the use and retention of data collected by the drone. A disadvantage of this is that the FAA has not abdicated its important role in developing specific privacy requirements for drones (EPIC, 2014). Furthermore, the FAA has missed the chance to make the test-site operators conduct specific tests related to privacy and surveillance. For example, they could have required drone operators to test how accurate their surveillance systems are and how much data those systems collect.

The FAA is primarily concerned with balancing safety concerns with its obligation to facilitate access to airspace by drones. They disregard matters pertaining to the regulation of specific drone use based upon their perceived intrusiveness or the scope of data that can be collected by drones. Unlike some state and local governments, the U.S. Congress has not seriously considered or passed a bill setting general privacy standards or regulating drones and privacy specifically (Bennett, 2014), with the result that law enforcement agencies are deploying drones without any established privacy guidelines being in place (Farber, 2013).

4.6.1.2 The U.S. Federal Government   Some U.S. senators have pushed the FAA to develop and release privacy rules and guidelines for their use as quickly as possible. For example, Senator Charles E. Schumer has warned that New York will become a wild, wild west for commercial and private drones until clear, smart regulations are put in place: "[m]ore and more, small drones are being used by private investigators to spy on unaware New Yorkers or for illegal purposes like drug deliveries, and the lack of clear rules from the FAA holds a great deal of the blame for confusion as to what is legal, and the blatant abuses of this great technology."*

Some federal legislators have been unable to wait until the FAA establishes clear rules and have proposed bills. Possibly, the two most comprehensive proposals for legislation in Congress are the

---

* http://www.schumer.senate.gov/Newsroom/record.cfm?id=355151.

Drone Aircraft Privacy and Transparency Act,* introduced in March 2013 by Senator Edward Markey, and the *Preserving American Privacy Act of 2013,*† introduced in February 2013 by Senators Ted Poe, Trey Gowdy, and Zoe Lofgren. The *Drone Aircraft Privacy and Transparency Act* requires law enforcement agencies to file a data minimization statement that explains how the agency will minimize the collection and retention of data unrelated to a criminal investigation. The bill also requires private operators to submit a "data collection statement" to the FAA prior to receiving a license to operate a drone. The statement must include the name of the operator of the drone; where it will be operated; what data will be collected and how the data will be used and retained; and whether data will be sold to third parties. The *Preserving American Privacy Act of 2013* would significantly restrict the use of private drones. The bill prohibits private drone operators from capturing "highly offensive" data involving personal or familial activity in relation to which the person has a reasonable expectation of privacy. Law enforcement agencies require a court authorization for their activities with drones, with exceptions for exigent circumstances.

None of these proposed bills has been passed; however, they reflect the privacy concerns of the federal government. According to Schlag (2013), the proposed bills do not fully address privacy concerns. Drones could still be used for surveillance purposes when there is an open and visible area, even if monitored individuals have a certain degree of an expectation of privacy in that area, and none of these bills addresses the private use of drones to surreptitiously collect information about individuals.

4.6.1.3 Local and State Governments It is not a surprise that local and state governments have outpaced federal lawmakers when it comes to regulating the use of drones. As we have already seen in Section 4.4, the local government of the city of Charlottesville in Virginia was the first to pass an anti-drone resolution. Conoy Township in

---

* H.R. 2868, 113th Congress (2013–2014) (http://thomas.loc.gov/cgi-bin/bdquery/z?d113:h.r.2868:).

† H.R. 637, 113th Congress (2013–2014) (http://thomas.loc.gov/cgi-bin/bdquery/D?d113:26:./temp/~bdx2aJ::).

Pennsylvania has passed an ordinance that prohibits the operation of a drone over property not owned by the operator and without permission of the property owner.*

Thirteen states have already enacted laws regulating drone use, and about thirty states are considering legislation.† North Carolina and Virginia, for example, have a 2-year moratorium on drone use until 2015, with limited exceptions, to give legislators time to observe the progression of drone technology and draft sensible regulations. Idaho has prohibited any person, entity, or state agency from using surveillance by drones without a warrant to conduct surveillance on, gather evidence or information about, or photograph or record specifically targeted persons or property without consent. Most states that have addressed the use of drones for law enforcement have stipulated that a warrant is required or have enumerated exceptions for law enforcement surveillance. The states' drone legislation varies greatly in scope and focus. A question is whether these local and state regulations are legally valid at all, since formally the federal government has exclusive sovereignty of domestic airspace (Elias, 2012).

4.6.2 Regulations in the European Union

4.6.2.1 The European Commission In April 2014, the European Commission proposed a set of actions to regulate the operations of public and civil drones from 2016 onward. Public drones are increasingly being used in Europe, in countries such as Sweden, France, the Netherlands, and the United Kingdom, in different sectors, but under a fragmented regulatory framework. According to the European Commission, the lack of harmonized regulations across Europe and of validating technologies form the main obstacle to opening the drone market and to integrating drones in European nonsegregated airspace, and the industry is urging that rapid steps are taken toward the establishment of an enabling European regulatory framework

---

* http://lancasteronline.com/news/conoy-township-passes-ordinance-that-limits-drones-air-space/article_c975a727–24b0–5b35-bc60-cd14c31341b4.html.

† https://www.aclu.org/blog/technology-and-liberty/status-2014-domestic-drone-legislation-states.

for the operation of drones. The core of the European Commission's (2014) drone strategy is as follows:

> The European strategy aims at establishing a single RPAS [remotely piloted aircraft systems] market to reap the societal benefits of this innovative technology and at dealing with citizens' concerns through public debate and protective action wherever needed. It should also set the conditions for creating a strong and competitive manufacturing and services industry able to compete in the global market.

The new actions proposed by the Commission address issues such as safety, security, privacy, data protection, insurance, and liability. Just like the FAA, the European Commission sees safety as the paramount objective of aviation policy. EU standards will be based on the principle that civil drones must provide an equivalent level of safety to manned aviation operations. The European Aviation Safety Agency (EASA) is starting to develop specific EU-wide standards for remotely piloted aircraft. EASA will also develop the necessary security requirements, particularly those that will protect information streams, since drones can be subject to potential unlawful actions and security threats. Furthermore, the European Commission will stimulate research and development efforts with respect to technologies for the safe integration of drones.

In contrast to the FAA, the European Commission is concerned about privacy. The Commission states that drones must not lead to fundamental rights being infringed, including respect for the right to private and family life and the protection of personal data. According to the European RPAS Steering Group (2013), the European Commission should ensure that drone operations fully comply with existing privacy and data protection legislation, and states that new legislation may be needed if the utilization of drones could result in new issues that are not adequately addressed by the current regulation.

Just like the FAA, however, the European Commission is not progressing well. No concrete schedule for actual changes to regulations or publication of standards has yet been published by the European Commission. The only result so far is an aspirational road map produced by the European RPAS Steering Group (2013) that merely provides a conceptual timeline.

Furthermore, the European Commission has recently been accused of having too cosy a relationship with the drone industry and of being more preoccupied with public acceptance than with exploring and addressing public attitudes around concerns such as privacy: "Vague commitments to 'privacy by design' and 'data protection' are no substitute for proper regulation. The idea that such commitments will somehow allay deep-seated public fears is laughable" (Hayes, Jones, & Töpfer, 2014, p. 79).

4.6.2.2 European Countries   In the current situation, it is up to individual countries to regulate the use of drones. Some EU member states, such as Sweden, France, Denmark, Italy, Germany, the Czech Republic, Lithuania, the Netherlands, and the United Kingdom, have, therefore, adopted legislation for public drone operations to make these operations possible. In the United Kingdom, for example, the Civil Aviation Authority (CAA) has banned drones weighing more than 20 kilograms (or 44 pounds) from flying in British civilian airspace. Drones that weigh less than 20 kilograms (or 44 pounds) can be flown normally, as long as data or images from the flight are not used and the drone remains at least 150 meters (or 500 feet) away from people. Since 2010, the CAA has required operators of small drones used for aerial work and those equipped for data acquisition and/or surveillance to obtain permission. Organizations currently with such a permission include three police forces, the Ministry of Defence's Defence Science and Technology Laboratory (DSTL), the BBC, universities, and a large number of film production and photography companies.* In the Netherlands, an amendment has been approved that allows Dutch municipalities to use mobile cameras, including drones, to monitor residents. According to the amendment, existing legislation relating to camera surveillance needed to be extended to allow law enforcement agencies to intervene in the event of persistent disturbances that move between areas—for example, a riot spreading between

---

* http://www.theguardian.com/world/2014/oct/26/drones-permit-uk-british-airline-pilots-association-unmanned-aircraft-house-of-lords.

neighborhoods. Under the new legislation, it is now up to the mayor of a city to decide what form of camera surveillance should be used: fixed, vehicle mounted, or airborne.*

According to the European Commission, the national procedures do not provide a coherent framework with the necessary legal safeguards in relation to concerns about safety, security, privacy, and liability built in. For example, in 2012, the Dutch police used drones for civilian surveillance on average once every other day, but only 50% of the flights were made public in advance, which may be in violation of privacy regulations.†

4.6.3 Proliferation of Drone Regulations

Pan-European or Pan-American coordination of drone regulation is fragmented. During the last few years, we have seen drone regulation at state level (in the case of the United States), at national level, at supranational level (as in the case of the European Union), and even at international level (from the International Civil Aviation Organization). Basic national or state safety and privacy rules apply to all drone operators, but the rules differ across the European Union and the United States, and a number of key issues with regard to privacy are not addressed in a coherent way.

A proposal for dealing with the privacy issues is offered by Schlag (2013). He proposes developing a baseline consumer protection law that details permissible uses of drones in domestic airspace by both law enforcement agencies and private parties:

> A baseline consumer protection law would need to address drone surveillance, data collection, and the various drone technological capabilities. ... This would ensure both governmental and private parties were not using drones in a manner that would violate an individual's privacy. Similarly, a federally enacted baseline law would ensure baseline privacy expectations are consistent between states while also creating a way for private parties to comply more easily with privacy laws. Absent some

---

* http://www.zdnet.com/dutch-authorities-now-allowed-to-film-citizens-using-drones-7000028019/.
† http://www.dutchnews.nl/news/archives/2013/03/use_of_drone_aircraft_by_polic.php.

baseline mechanism for control, drone use may become so common-place that it dissolves current privacy expectations to the degree that individuals will have no reasonable expectation to privacy. A baseline drone use consumer protection law would be the best and most proactive way to establish strong privacy protection prior to drone implementation and privacy invasions.

Schlag (2013, pp. 21–22)

The lack of harmonized drone regulation across Europe and the United States forms the main obstacle to opening the drone market and integrating drones into European and U.S. nonsegregated airspace. The industry is urging rapid steps toward the establishment of an enabling regulatory framework for the use of drones (European Commission, 2014). Such a well-constructed regulatory framework that supports the commercial interests of business will be critical to states and nations winning the race to regulate drones (Magriña, 2014).

4.7 Concluding Observations: Drones Create a Floating Robotic Panopticon

Drones are already beginning to appear in our skies but there are no clear general rules that put in place the necessary safeguards to protect people's safety, security, and privacy. There are serious worries about the slow progress of the FAA, the European Commission, and European countries in regulating the use of drones, since they are not adapting "existing aviation rules sufficiently rapidly to cater for the drone explosion, particular in relation to the various categories of small drones" (Clarke & Bennett Moses, 2014). This holds especially for the commercial and private drones.

Despite the myriad of legal and ethical issues, the overriding concern for the use of drones is safety. For example, for the European Commission (2014), safety is "the paramount objective of EU aviation policy." Almost all existing regulations or proposals that have limited (or even banned) the use of drones have been driven by safety concerns and prevent drones from operating in certain circumstances. Current safety legislation is based on fragmented rules for ad hoc operational authorization. A regulatory framework should reflect the wide variety of drones and the applications of drones and should indicate the requirements for each kind and use of drones to allow for a safe integration of drones into the aviation system. These requirements could be technical, and maybe the technologies that are needed to guarantee safety are not yet available. Detect and avoid technologies, for example, will need further development and validation before certain applications of commercial drones, such as delivery drones flying over populated areas, can be approved.

Drones with an infrared surveillance and thermal-imaging security camera offer many opportunities and develop a new field of possibilities when deployed by the law enforcement agencies, such as detecting violations of law, searching for missing people, and crime prevention. Police drones are already being used regularly, for purposes such as detecting cannabis plantations, ensuring security at festivals, clearing buildings, and exploring urban areas to prevent burglaries. Drones thus offer many opportunities and develop a new field of opportunities for gathering information and also constitute a potential violation of the privacy of citizens. As a result, mass control can be exercised in a manner that was previously impossible. In itself, the breach of privacy is allowed if and when good reasons can be given, in particular with regard to the safety of citizens, because the government has responsibility for the safety of its citizens.

It is possible to monitor the daily activities of citizens 24 hours a day and to store this information. Regardless of how or whether the information will actually be used, people would feel observed. The deployment of drones creates the so-called panopticon (which means "all-seeing"). The panopticon, designed by ethicist Jeremy Bentham (1748–1832), is a dome-shaped, circular prison in which a prison warder can see all prisoners (see Figure 4.9). They are kept in cells that together form a ring and whose windows face inward. The warder can observe all prisoners, but the prisoners cannot see the warder. The idea behind this is simple: if individuals are checked by an all-seeing eye (without the eye being seen), they will allow themselves to be disciplined and will be controllable.

The very essence of a constitutional state will be in danger if there is no balance between protection of the public *by* the government and

Figure 4.9 The panopticon. (Photo courtesy of Magnum Photos/Hollandse Hoogte.)

protection *against* the government. Without privacy protection, the government is a potential threat to the rights of its own citizens. On the other hand, the government loses its legitimacy when it cannot guarantee safety for its citizens from outside threats by other parties. It is important that it is clear what data may be collected and stored, when data may be collected and for what purpose drones may gather data, and that a clear distinction continues to exist between the monitored public space and private life. Moreover, the risk of sound and image recordings being manipulated should also be factored in, as should the risk that data might end up in the hands of the wrong people.

Perhaps even more disturbing is the threat to our privacy from "Little Brother." Drones are now available on the civilian market. In addition, the drones deployed by companies for various legitimate purposes, such as maintaining tall buildings (e.g., to detect heat loss or leaks via infrared cameras), measuring cadastral boundaries, detecting sources of a fire, monitoring radiation, dyke monitoring, and so on, drones can be used by everyone for all kinds of less legitimate purposes. Anyone can buy, for less than U.S. $100, a drone at, for example, Amazon that has built-in cameras. Tens of thousands of these drones have already been sold. Although they are still limited in their ability (they only have a range of 50 meters (or 165 feet) and have a flight time of 15 minutes), these commercial drones carry a great risk. They are a threat to privacy, because they are ideal for spying on your neighbors without their noticing or without their knowing who is spying on them. Privacy legislation will have little or no effect on dissuading malicious drone owners from using their drones, as it is almost impossible to trace them.

Given the rapid developments arising from generally legitimate uses, it is certainly not inconceivable that drones can be bought that, almost silently, can penetrate private properties to take pictures or record conversations, or even to grab objects and may drop. For the paparazzi, stalkers, criminals, and terrorists, this development provides unprecedented opportunities and provides a virtual panopticon for the innocent citizen in which he may also be a constant target of invisible, malicious drone owners.

As we have seen, the FAA and the European Commission see safety as the paramount objective of aviation policy, and both—so far—have failed to take seriously concerns about privacy. From a fundamental rights perspective, all that matters is ensuring that drones process all the data they capture in accordance with data protection laws. This completely misunderstands the issues mentioned earlier and many of the reasons why people are concerned about drones in the first place.

Currently, little or no political reflection or debate exists on the issues discussed in this chapter in the political arena or in society in general, let alone a societal agreement or political endorsement. For various reasons, there is an urgent need for a debate in which the various issues at stake in the use of public as well as civil drones should be weighed to protect our dignity, autonomy, and privacy.

It is conceivable that the use of drones for surveillance activities may seriously erode our expectations of privacy. However, maybe the opposite will happen, as Calo (2011) points out: "Drones may help restore our mental model of a privacy violation. They could be just the visceral jolt society needs to drag privacy laws into the twenty-first century."

Interview with Mark Wiebes (Innovation Manager with
the Dutch National Police, the Netherlands)

"Aggression from police robots, that's definitely a line not to cross."
An eye in the sky, a beast of burden, or a fearless handler of risky
objects: robots can help the police in many ways, keeping officers out
of harm's way and operating in places that are normally inaccessible.
But how will society feel about these high-tech cops? Or had we better worry about the force's moderate pace of innovation?

As the interview with Police Commissioner Mark Wiebes, responsible for innovation at the Central Unit of the Dutch National Police,
draws to a close, he says, "I know that my story is low key rather
than sensational. Popular expectations of robotics are often based on
developments in health care. But there, the working environment is
controlled and friendly, much unlike ours. Granted, neither are we up
against the chaotic and hostile conditions that the armed forces often
face, but then, they have a license to do damage, which makes the
deployment of military robots much easier. The police face a certain
amount of hostility, but we must strictly avoid collateral damage. As a
result, innovation of police technologies, including robotics, is always
less sophisticated by several degrees than Hollywood would have you
expect."

*Sophisticated or not, robots are making their way into the station. To
do what?*

"We're using some types already. Robot helicopters give us a view from
the sky in situations where a manned helicopter can't fly, for instance
when there's dense smoke or toxic fumes. They were used around the
Chemie-Pack disaster in 2011, when a chemical plant some way south
of Rotterdam exploded and burned down. Robot helicopters are also
the better option when the aircraft may come under fire, and they're
the only option when airspace is closed, as in 2010 after the volcanic
eruption in Iceland. On other occasions, they're just cheaper.

Second, robots can perform useful tasks in hazardous conditions.
They can handle objects, take photographs, carry a message and so
on. I expect that before long it will be standard practice for robots
to defuse explosives. The current devices cannot handle all situations yet.

Load-carrying robots are yet another type. One of these was also used after the Chemie-Pack accident. There was some piece of evidence, a vessel or chest I believe, that was too heavy for a person in cumbersome protective clothing to carry. It was hauled off the factory grounds by a mobile robot. However, the drag-rope had to be attached by human hands.

Finally, robots are good at very specific tasks in places that people just can't enter, but where a small vehicle with a camera will fit. Low crawling spaces are good examples. Sewerage companies have found these devices very helpful."

Will robots have an impact on surveillance as well?

"Quite possibly. We could use all sorts of sensors: visual, acoustic, and so on. If you integrate these into an autonomous device, you may call it a robot. But it may also be useful to place them in a fixed location or attach them to a police car. There is software available that can tell from the sound of a loud group of people whether they are celebrating or aggressive. Camera footage too can be automatically analyzed. A person seeking their own car in a parking lot moves differently from someone on the prowl for a car to break into. In Britain, software has been developed to detect left luggage in public places. But false positives are still a problem: on one occasion, the computer mistook an elderly lady sitting still for a suitcase.

False positives are also a problem with reasoning and decision support engines, which can draw police attention to suspicious digital activities. In New York State, the police searched a house after the words 'Boston bombing,' 'pressure cooker,' and 'backpack' had been Googled from one IP address. It turned out the queries were from different people and the investigation was dropped. But imagine the engine finds not just three, but a thousand digital clues pointing to one address and yet there is no case. Will human officers then have the open mind to notice this and the guts to admit it, even when the system is *nearly* always right? That requires the user to understand how the engine works, so they can judiciously assess any alarm signal and avoid tunnel vision. Doctors set a good example that we ought to follow: they use decision support engines, but not unquestioningly."

Are operators of security cameras liable to replacement by robots?

"Robots won't replace them, but they will reduce the information overload. Operators are typically looking for a needle in a haystack. Robots can remove lots of the hay and point out potential needles; that is, atypical behavior. With their help, the cameras can become much more effective."

The robot applications you've mentioned so far strike me as quite peaceable.

"That's right. I do expect robots to be somewhat proactive in the future. They may do things like open a door behind which there's something nasty. But the use of aggression, that's definitely a line not to cross. The police only ever uses firearms in self-defense, and a robot by definition can never plead self-defense. But there's still a gray area. Suppose we know the whereabouts of this really dangerous suspect and we send a robot in to handcuff him. I would expect a lot of public discussion there."

Will robotics affect the work of the ordinary officer on the beat?

"Perhaps they will have a miniature aircraft in the boot of their patrol car. Even now, toy helicopters with four propellers cost less than U.S. $300. It can be equipped with a camera and remote controlled with a smartphone app."

That raises a lot of privacy issues. Will people accept these flying cameras?

"I'm sure it will not help acceptance if we start watching people from all sorts of angles. On the other hand, when I take a simple stepladder, I can snoop over fences and in people's gardens. The crux is that we may only do these things for a good reason. In practice, our actions are circumscribed more by rules than by technological limitations. We are able to enter any house, but only under certain conditions are we entitled. The means must be proportionate to the ends."

We've discussed police use of robots so far. Is the police also likely to be confronted with criminals using robots?

"Most definitely. Any technology we can use for our own purposes, will also be used against us. Indeed, more so, because we are bound by rules and scruples. Criminals may use a remote-controlled car to transport an explosive. Or they may smuggle substances with autonomous or remote-controlled vehicles."

Which makes it all the more important for the police to be as well-equipped as possible. Does the force appreciate the importance of robotics sufficiently?

"There's always a certain degree of conflict between innovative developments and the wish to get quick results. Unfortunately, the early stage brings costs only, while results are some way off. You're finding out what works, you're taking small steps in important developments. These investments are not without risk, and setbacks occur. Which is why, to my mind, we should only spend resources on things of considerable potential benefit. The successful innovations have to make up for the setbacks.

What may be harder here than elsewhere is to develop a long-term vision. The police have a culture of action, but it's reactive rather than proactive. After all, our main job is to respond to a disruption of some sort. But perhaps the problem is not specific to us after all. It's probably just part of the human condition. The short term always trumps the long term."

References

Anderson, C. (2014). Agricultural drones. *MIT Technology Review, 117*(3), 58–61.

Barrientos, A., Colorado, J., del-Cerro, J., Martinez, A., Rossi, C., Sanz, D., & Valente, J. (2011). Aerial remote sensing in agriculture: A practical approach to area coverage and path planning for fleets of mini aerial robots. *Journal of Field Robotics, 28*(5), 667–689.

Bennett, W. C. (2014). *Civilian drones, privacy, and the federal-state balance.* Washington, DC: Brookings, Center for Technology Innovation.

Bramley, R. (2009). Lessons from nearly 20 years of precision agriculture research, development, and adoption as a guide to its appropriate application. *Crop and Pasture Science, 60*(3), 197–217.

Calo, M. R. (2011). The drone as privacy catalyst. *Stanford Law Review Online, 64*, 29–33.

Cavoukian, A. (2013). *Surveillance, then and now: Securing privacy in public spaces.* Toronto, Ontario, Canada: Information and Privacy Commissioner. http://www.ipc.on.ca/images/Resources/pbd-surveillance.pdf (accessed May 24, 2014).

Clarke, R. (2014a). Understanding the drone epidemic. *Computer Law & Security Review, 30*(3), 230–246.

Clarke, R. (2014b). What drones inherit from their ancestors. *Computer Law & Security Review, 30*(3), 247–262.

Clarke, R. (2014c). The regulation of civilian drones' impact on behavioural privacy. *Computer Law & Security Review, 30*(3), 286–305.

Clarke, R., & Bennett Moses, L. (2014). The regulation of civilian drones' impacts on public safety. *Computer Law & Security Review, 30*(3), 263–285.

De Jong, A., & Schuilenburg, M. (2007). Een cultuur van controle. Interview met Eric Nordholt. *Gonzo (circus), 79*, 12–15.

Elias, B. (2012). *Pilotless drones: Background and considerations for congress regarding unmanned aircraft operations in the national airspace system* (CRS Report for Congress, R42718). http://fas.org/sgp/crs/natsec/R42718.pdf.

EPIC (2014). *Drones: Eyes in the sky.* https://www.epic.org/privacy/surveillance/ spotlight/1014/drones.html (accessed January 23, 2015).

European Commission (2014). *A new era for aviation. Opening the aviation market to the civil use of remotely piloted aircraft systems in a safe and sustainable manner.* Brussels, Belgium: European Commission. http://eur-lex. europa.eu/legal-content/EN/TXT/?uri=CELEX%3A52014DC0207 (accessed September 4, 2014).

European RPAS Steering Group (2013). *Roadmap for the integration of civil remotely-piloted aircraft systems into the European aviation system.* http:// ec.europa.eu/enterprise/sectors/aerospace/uas/ (accessed September 4, 2014).

FAA (2013). *Unmanned aircraft system test site program.* Washington, DC: Department of Transportation.

Farber, H. B. (2013). Eyes in the sky: Constitutional and regulatory approaches to domestic drone deployment. *Syracuse Law Review, 64*(1), 1–48. http:// papers.ssrn.com/sol3/papers.cfm?abstract_id=2350421 (accessed October 14, 2014).

Gernsback, H. (1924). Radio police automaton. *Science and Invention*, May 14.

Hambling, D. (2010, February 10). Future police: Meet the UK's armed robot drones. *Wired.* www.wired.co.uk/news/archive/2010-02/10/future-police-meet-the-uk's-armed-robot-drones#comments (accessed March 17, 2014).

Hayes, B., Jones, C., & Töpfer, E. (2014). *Eurodrones Inc.* London, UK: Statewatch.

Huang, Y., Thomson, S. J., Hoffmann, W. C., Lan, Y., & Fritz, B. K. (2013). Development and prospect of unmanned aerial vehicle technologies for agricultural production management. *International Journal of Agricultural and Biological Engineering, 6*(3), 1–10. http://www.ijabe.org/index.php/ ijabe/article/view/900 (accessed October 5, 2014).

Jenkins, D., & Vasigh, B. (2013). *The economic impact of unmanned aircraft systems integration in the United States.* Arlington, VA: Association for Unmanned Vehicle Systems International.

Ladiges, R., & Van Stijl, M. C. (2009). *OT&E scenario's voor KLPD UAS* (NLR-CR-2009-598). Amsterdam, the Netherlands: Nationaal Lucht- en Ruimtevaartlaboratorium.

Mann, S., Nolan, J., &Wellman, B. (2003). Sousveillance: Inventing and using wearable computing devices for data collection in surveillance environments. *Surveillance & Society, 1*(3), 331–355.

Magriña, E. (2014, June 27). The global race for drone regulation. *Inline*. http://inlinepolicy.com/2014/the-global-race-for-drone-regulation/ (accessed August 25, 2014).

Molko, R. (2013). The drones are coming! Will the fourth amendment stop their threat to our privacy? *Brooklyn Law Review*, *78*(4), 1279–1333.

Phillips, D. J. (2004). Privacy policy and PETs: The influence of policy regimes on the development and social implications of privacy enhancing technologies. *New Media & Society*, *6*(6), 691–706.

Rubin, R. (2014). *Drones: Quickly navigating toward commercial application, starting with e-commerce and retail.* New York: BI Intelligence.

Sato, A. (2003). *The RMAX helicopter UAV* (Public report). Shizuoka, Japan: Aeronautic Operations Yamaha Motor Co., Ltd.

Schlag, C. (2013). The new privacy battle: How the expanding use of drones continues to erode our concept of privacy and privacy rights. *Pittsburgh Journal of Technology Law & Policy*, *13*(2), 1–22. http://tlp.law.pitt.edu/ojs/index.php/tlp/article/view/123 (accessed October 26, 2014).

Sharkey, N. (2008). 2084: *Big robot is watching you. Report on the future for policing, surveillance and security.* http://www.dcs.shef.ac.uk/~noel/Future%20robot%20policing%20report%20Final.doc (accessed November 16, 2014).

Solove, D. J. (2004). *The digital person: Technology and privacy in the information age.* New York: New York University Press.

Syed, N., & Berry, M. (2014). Journo-drones: A flight over the legal landscape. *Communications Lawyer*, *30*(3), 1–10. http://www.lskslaw.com/documents/CL_Jun14_v30n4_SyedBerry.pdf (accessed October 26, 2014).

Thompson II, R. M. (2013). *Drones in domestic surveillance operations: Fourth amendment implications and legislative responses.* Washington, DC: Congressional Research Service. http://fas.org/sgp/crs/natsec/R42701.pdf (accessed October 26, 2014).

Volovelsky, U. (2014). Civilian uses of unmanned aerial vehicles and the threat to the right to privacy—An Israeli case study. *Computer Law & Security Review*, *30*(3), 306–320.

Wright, D. (2014). Drones: Regulatory challenges to an incipient industry. *Computer Law & Security Review*, *30*(3), 226–229.

5

WHO DRIVES THE CAR?

5.1 Introduction

Can a car read a traffic sign?
Can it shut its grill vents to save fuel?
Can it stop itself to help prevent an accident or park itself in a tight spot?
With a host of technologically advanced features, this one can.
The all-new Ford Focus: Start more than a car.*

Making the car more intelligent—in other words, car robotization—has taken off in recent years. The development of smart driver assistance systems in cars is currently rapidly progressing. In part, this is caused by the decrease in the price of the required components, such as video cameras, microprocessors, and sensors. Therefore, such systems are no longer just built into expensive car models but are also increasingly fitted in middle-class models. In addition, car manufacturers especially compete with each other in terms of comfort and safety, because there is actually not much more to improve in the quality of cars (Gusikhin, Filev, & Rychtyckyj, 2008). Intelligence, therefore, becomes the unique selling point for a new car, as we can see from the advertising text of the Ford Focus. Future systems will evolve from "driver assistance" to "fully automated (autonomous) driving," completely piloting a car along highways and through urban environments. Although the idea of driverless cars on the road may seem futuristic, industry leaders anticipate that autonomous cars will hit the road within the next decade. This projection is due to the fact that the majority of the technologies necessary to build a fully autonomous car already exist (Pawsey & Nath, 2013). This vision of the path toward fully autonomous cars assumes that car robotization is a continuum between conventional, fully human-driven vehicles and vehicles that

---

* https://www.youtube.com/watch?v=LXDTgmm2iak.

require no driver at all. Actually, it is an ongoing automation and interconnection of single vehicles and traffic's infrastructure that aims at fully self-driving or autonomous cars. The National Highway Traffic Safety Administration (2013) developed a five-level hierarchy to conceptualize this continuum as guidance "to help states implement this technology safely so that its full benefits can be realized":

- *Level 0 (no automation)*: The human driver is in complete control of all functions of the vehicle.
- *Level 1 (function-specific automation)*: One control function is automated.
- *Level 2 (combined-function automation)*: More than one control function is automated at the same time (e.g., steering and acceleration), but the driver must remain constantly attentive.
- *Level 3 (limited self-driving automation)*: The driver functions are sufficiently automated so that the driver can safely engage in other activities (e.g., reading a paper).
- *Level 4 (fully self-driving automation)*: The car can drive itself without a human driver.

The automation of control functions started with the features of convenience and safety, such as cruise control in 1958[*] and anti-lock braking system (ABS) in 1972[†] (level 1). Automation of multiple and integrated control functions, such as adaptive cruise control with lane centering, has become more common recently. They are classified as level 2, because they still allow the driver to override them: the driver is responsible for monitoring the roadway and is expected to be available for control all the time. The next step in this continuum is taking over driving tasks through cooperative systems—in conjunction with traffic management (level 3). Ultimately, it is expected that this will lead to autonomous vehicles (level 4).

In this chapter, we define car robotization as a combination of developments in the following technologies: driver assistance systems, traffic management, and cooperative systems. In fact, these technologies are the building blocks for the fully autonomous car, which we

---

[*] http://www.conceptcarz.com/vehicle/z5628/Chrysler-Imperial-D-Elegance.aspx.
[†] https://history.gmheritagecenter.com/wiki/index.php/1972,_First_Automotive_Anti-lock_Brake_System_%28ABS%29.

will call the "connected autonomous car," since this car is connected to numerous information and communication networks through navigation systems, roadside systems, and traffic services.

There is also a second path toward fully autonomous cars, which does not adhere to the five-level hierarchy proposed by the National Highway Traffic Safety Administration. Instead of trying to gradually integrate the technology, these cars sense their surroundings with techniques such as radar, lidar,* global positioning system (GPS), and computer vision, and do not depend on cooperative systems. The Google car is an example of following this path (see Section 5.5.1).

The Dutch Institute TNO (the Netherlands Organisation for Applied Scientific Research) describes car robotization in the twenty-first century as a gradual but revolutionary process, but as one that is making irreversible changes to the role of the driver and the impact of information and communication technology (ICT).† TNO expects that the autonomous car will appear on the market in about 25 years from now. Big automobile manufacturers expect the autonomous car already by 2020: Nissan plans to have commercially viable autonomous-drive vehicles on the road by 2020‡; Daimler, the maker of Mercedes-Benz and Smart cars, has announced that it will start selling a self-driving car by 2020.§ The first changes are already visible. Increasingly, the driver is supported in driving tasks, and it will not be long before the driving tasks of motorists will be taken over. In this chapter, we explore the social and ethical issues that await us with the advance of the intelligent car. To this end, we outline in Section 5.2 problems relating to modern road traffic; it is expected that the robotization of the car will present a major contribution to solving these problems. We then present car robotization in terms of the levels of autonomous vehicles: driver assistance systems (levels 1 and 2) in Section 5.3; limited self-driving automation (level 3) in Section 5.4; and autonomous cars (level 4) in Section 5.5. In Section 5.4, the

---

* Lidar is a remote sensing technology that measures distance by illuminating a target with a laser and analyzing the reflected light.

† www.tno.nl/content.cfm?context=thema&content=inno_case&laag1=894&laag2=914&item_id=852.

‡ http://www.nissanusa.com/innovations/autonomous-car.article.html.

§ http://www.dailymail.co.uk/sciencetech/article-2418526/Self-driving-Mercedes-Benz-sale-2020-unveiled.html.

autonomous car is central, being the most far-reaching form of car robot control. Social and ethical issues of car robotization are discussed in Section 5.5. We will end with some concluding observations.

As we have pointed out in the introduction, car robotization is developing along three lines: driver assistance systems, traffic management, and cooperative systems. In this section, we will discuss developments and gain an insight into some of these applications and how the car is gradually developing into a more "intelligent" vehicle.

5.2 Problems for Modern Road Traffic and the Costs

Road traffic leads to negative effects in terms of traffic accidents, congestion, and environmental pollution. The costs of these negative effects during the last few years for the United States are roughly estimated to be $800 billion/year (Morgan Stanley, 2013, see Section 5.5). Therefore, it pays to reduce these costs. Salvation is largely expected from car robotization. The European Commission (2010) mainly encourages the robotization of the car in terms of safety. The Commission aims to reduce the number of fatalities by half by 2020 by ensuring the safety of vehicles and also by promoting intelligent transportation systems, which communicate between vehicles and between a vehicle and the infrastructure.* In this section, we briefly map these three negative effects of road traffic.

5.2.1 Traffic Victims

In 2012, 27,700 people were killed in the European Union (EU) as a consequence of road collisions, and around 313,000 people were recorded by the police as seriously injured (Jost, Allsop, & Steriu, 2013). In the United States, 32,000 people were killed in traffic accidents in 2011 (Anderson et al., 2014). Approximately, 1.24 million people globally die every year as a result of car accidents (World Health Organization, 2013). For every death on the roads, there are

---

* In addition, the European Commission is funding numerous research projects, in which both car manufacturers and research institutes are involved, with respect to driver support systems, traffic management, and cooperative systems. For an overview of European projects related to road safety see http://ec.europa.eu/transport/road_safety/specialist/projects/index_en.htm.

an estimated 4 permanently disabled, 10 seriously injured, and 40 slightly injured people (European Commission, 2013; Mackay, 2005). Speed and alcohol use are the primary factors determining the crash fatalities.

After an increase in the 1950s and 1960s, the number of traffic accidents in Europe and the United States has shown a gradual, continuing downward curve ever since 1991. This downward trend is attributed to a number of policies and measures in the fields of construction and redevelopment and the management and maintenance of infrastructure, such as the design of roundabouts and lower speed limits on some roads. It is, furthermore, also due to vehicle developments such as Electronic Stability Control (ESC), airbags, safety certification, and EuroNCAP—a standard crash tests for cars—improved training being required to gain car and moped licenses, law enforcement, and communication and behavioral changes. Proportionally, the number of fatalities among car occupants has decreased more than that among pedestrians, moped drivers, cyclists, and motorcyclists. Furthermore, the aging world is mirrored in the road safety statistics of the European Commission (2014), with the elderly making up an increasing share of the total number of road safety fatalities.

More than half of all serious injuries occur in urban areas, especially affecting pedestrians and vulnerable road users: children, the elderly, and young drivers. Road traffic accidents are a major cause of serious head and brain injuries. Accidents involving cars and powered two-wheelers are the most dangerous in this respect (World Health Organization, 2013).

5.2.2 Traffic Congestion

Traffic congestion has long been at the center of attention, both in politics and in society. According to the European Commission, congestion is one of the worst transportation problems. Congestion costs Europe about 1% of its gross domestic product (GDP) every year and also produces large amounts of carbon dioxide (CO_2) and other unwelcome emissions (European Commission, 2013). A study of the Texas A&M Transportation Institute shows that congestion costs are increasing every year in the United States: the cost of extra time and fuel in 498 urban areas was U.S. $121 billion in 2011, $94 billion in 2000, and \$24 billion in 1982 (Schrank, Eisele, & Lomax, 2012). More specifically, the average commuter spent an extra 38 hours/year traveling in 2011, up from 16 hours in 1982, and wasted 72 liters of fuel per year in 2011—a week's worth of fuel for the average U.S. driver—up from 30 liters in 1982. The main reasons for this increase are population growth, an increase in the number of people who are employed, and increased car ownership.

5.2.3 Pollution

Climate change is one of the most pressing challenges our society faces today and for the foreseeable future. This has led to the introduction of regulations and innovative pollution-control approaches throughout the world that have resulted in a reduction of exhaust emissions, particularly in developed countries (HEI Panel on the Health Effects of Traffic-Related Air Pollution, 2010). The EU has also acknowledged the impact of a rising global temperature and is committed to reducing CO_2 emissions, which make up the largest part of greenhouse gas emissions that cause climate change.[*] Road transportation contributes substantially to greenhouse gas emissions—especially CO_2—air pollution (including acidifying emissions and particulates floating in the air) and noise. In the EU, road transportation accounts for almost 25% and 50% of man-made CO_2 and NO_X emissions, respectively.[†] A reduction is needed to meet the EU's targets included in the Europe 2020 strategy. The European Commission sent a clear signal about the role transportation will have to play by setting the sector an objective of reducing its CO_2 emissions by 60% by 2050 compared to 1990 levels (European Environment Agency, 2012). Although the EU is generally moving in line with the "target path," this does not mean that transportation-related impacts are on a continued downward trend every year. For example, transportation energy consumption actually rose slightly in 2011 compared with 2010. So, substantial opportunities for further efficiency improvements remain.

---

[*] http://www.etrera2020.eu/library/22-smart-green-and-integrated-transport/51-towards-low-carbon-transport-in-europe.html.

[†] http://ec.europa.eu/research/rtdinfo/en/25/05.html.

5.3 Driver Assistance Systems (Levels 1 and 2)

Advanced Driver Assistance Systems (ADAS) support the driver and have been developed in order to improve comfort and safety in driving. These systems do not allow for automated driving in traffic, but are a necessary step toward cooperative and autonomous driving. We can distinguish five types of driver assistance systems with respect to their intended action: (1) controlling the vehicle, (2) preventing traffic violations, (3) supporting detecting and/or interpreting traffic situations, (4) limiting injury in collisions, and (5) intervening in temporarily impaired driver capacity.

The application of driver assistance systems is developing rapidly, especially because the expectations of these systems are running high regarding safety effects. The driver assistance systems now available are probably only harbingers of major developments that will lead to a progressive automation of the driving task, perhaps even going so far as to allow the driver to take their hands off the steering wheel entirely: driving in an autonomous car (see Section 5.5). To get a picture of the applications and developments of driver assistance systems, we will briefly discuss some of these systems. The other driver assistance systems mentioned in Table 5.1 are listed in Box 5.1 along with a short explanation.

5.3.1 ABS and ESC

The ABS has become standard, and from 2008 it has been made compulsory by the European Commission. ABS is one of the first

Table 5.1   Driver Assistance Systems

FUNCTION	DRIVER ASSISTANCE SYSTEMS
Controlling the vehicle	ABS, ESC, ACC, stop-and-go
Preventing traffic violations	Drive alert, traffic sign recognition, intelligent speed assistant
Supporting detecting and/or interpreting traffic situations	Lane departing warning, park assist, blind spot information system
Limiting injury in collisions	Pedestrian and cyclist airbags, airbags
Intervening in temporarily impaired driver capacity	Pre-crash, hill start assist, lane keeping aid, intelligent speed authority

BOX 5.1   SOME COMMERCIAL DRIVER ASSISTANCE SYSTEMS

DRIVE ALERT

This system continuously analyzes the driver's vigilance level by monitoring direction and behavior. When the system detects that the driver is losing attention, the driver is alerted by a warning signal and an additional visual warning sign.

TRAFFIC SIGN RECOGNITION

This system recognizes traffic signs on the road, for example, "no U-turn," "speed limit," "do not enter," "wild animal warning," or "no overtaking." The system shows these signs on the instrument panel.

INTELLIGENT SPEED ADAPTION

A collective of any system that constantly monitors vehicle speed and the local speed limit on a road and implements an action when the vehicle is detected to be exceeding the speed limit. This can be done through an advisory system, where the driver is warned (intelligent speed assistant) often via built-in car navigation systems, or through an intervention system where the driving systems of the vehicle are controlled automatically to reduce the vehicle's speed (intelligent speed authority).

LANE DEPARTURE WARNING AND LANE KEEPING AID

The lane departure warning alerts the driver via short vibration pulses in the steering wheel and a warning light in the instrument panel when the driver is about to veer off the road inadvertently. The lane keeping aid does not only warn the driver, but, if no action is taken by the driver, automatically takes steps to ensure the vehicle stays in its lane. In order to ignore the system, the driver can steer quite sharply in the desired direction.

BLIND SPOT INFORMATION SYSTEM

This system displays, through a light in the mirror, that a car is situated in the driver's blind spot.

PARK ASSIST

Via a sensor, this system detects an available parking space on the left or right side of the road. The driver then only needs to press the throttle and brake. The car steers itself into the available parking space.

HILL START ASSIST (HSA)

This system prevents the vehicle from rolling backward when it is stationary on a slope and drives it off from a standstill.

autonomous subsystems of commercial vehicles,* as it prevents the wheels from locking during forceful braking, functioning independently of the driver. A sensor in the wheels detects whether a wheel is about to lock while braking. If that is about to happen, ABS takes over and diminishes the braking force. When the actions of the ABS end, the driver's brake action is restored until the moment that the wheels are about to lock again. Because of this technique, the wheels maintain a continuous grip on the road surface and the vehicle can be steered. The successor of ABS is electronic stability control, ESC. In many cases, this security system will prevent slipping. When an incipient slip movement starts, there are differences between the actual movement of the vehicle and the driver's intention. In such a situation, ESC will brake separate targeted wheels with the same technique that is used by ABS. In addition to the wheel action, ESC also often intervenes directly in the throttle. The gas supply—and thus the power—is reduced, causing the vehicle to move more slowly, and therefore, it becomes more manageable (Ferguson, 2007).

In 2009, the European Commission decided that as of November 2011, all new models of vehicles, both cars and lorries, must be equipped with ESC, and from November 2014 this applies to

---

* Automatic gear switching is probably the first autonomous subsystem introduced in the car.

all new vehicles sold, including existing models (Schwab, 2009). The positive effect of ESC on the road is estimated to be high in various studies. In particular, a large proportion of single vehicle crashes with passenger cars, with only one vehicle moving, could be prevented by ESC; according to some studies, this amounts to between 30% and 62% of fatal single vehicle crashes (Erke, 2008; Ferguson, 2007).

5.3.2 Adaptive Cruise Control and Stop-and-Go Systems

In 1997, Toyota introduced Adaptive Cruise Control or ACC. ACC is an extension of the conventional cruise control systems. ACC not only maintains the speed set by the driver, but also adjusts the speed of the vehicle to that of the vehicle in front and helps to maintain a preset travel time interval between one's own car and the vehicle just in front. ACC uses a radar or laser sensor on the front of the vehicle to detect the vehicle ahead, monitors the vehicle's speed and distance, and then controls the throttle or brakes lightly. ACC cannot detect stationary objects and is only active in a speed range of 30–200 km/h (or 20–120 mph). Therefore, the system is not suitable, for example, for driving in traffic congestion. If a greater speed reduction is required, the driver is alerted by an audible signal. If the slower-moving vehicle in front is no longer located in the same lane, the speed of the vehicle will be returned to the preset speed. An ACC field trial by the Institute for Road Safety Research (SWOV) in the Netherlands shows that if all vehicles were equipped with ACC, the number of traffic accidents on highways could decrease by about 13% and on secondary roads by 3.4%.[*] In the same study, a reduction of fuel consumption by 3% and an exhaust emission reduction of 5%–10% on highways were found with an average use of ACC. Nowadays, ACC is already present in many car models.

ACC's successor is already on the market: *stop-and-go*. In addition to having the functions of the ordinary ACC, this system has the ability to bring the vehicle to a complete stop and also works in traffic congestion. In traffic congestion, the system may control

---

[*] www.swov.nl/rapport/Factsheets/UK/FS_ACC_UK.pdf.

the speed, for example, by adapting the speed to the speed of the vehicle ahead. Model calculations for traffic congestion on narrow roads reveal that the total loss time decreases by 30% when just 10% of the vehicles are equipped with stop-and-go, and decreases to 60% when 50% of the vehicles are equipped with stop-and-go, therefore having a positive effect on the environment (Van Driel & Van Arem, 2010).

5.3.3 Pedestrian and Cyclist Airbags

In order to increase the safety of cyclists, TNO is working on a bicycle airbag, used widely all over the Netherlands, that can be fitted to car bonnets.* TNO developed the idea of an airbag for cyclists by studying the Swedish airbag manufacturer Autoliv, which developed an airbag for pedestrians. The results were promising and TNO decided to develop a car airbag that protects both pedestrians and cyclists. The cyclist's airbag covers the whole front exterior of the windscreen in a crash—instead of just the bottom 25% of the lower window, as with the pedestrian airbag being developed by Autoliv—extending it to the front frame and the front part of the car roof. Such a car would also be equipped with a camera with a wide-angle lens positioned under the rearview mirror and an intelligent system that classifies any objects entering into the camera's view. If they are pedestrians or cyclists, the system puts itself on alert. When the sensors in the bumper of the car are signaling contact with the object, the airbag immediately activates. It is expected that this will decrease the number of road fatalities among pedestrians and cyclists by more than 20%. In order to test this, the detection system has been fitted to selected test cars in various European cities, such as Amsterdam, from 2012. The cyclist's protection airbag will probably be built into commercial vehicles in 2015. The cost will be around U.S. $225 per airbag. In 2013, the Swedish carmaker Volvo released the V40 model with the first pedestrian airbag.†

---

* This is done by order of the Ministry of Transport under the name SAVECAP (*SAver VEhicles for Cyclists And Pedestrians*). Zie www.savecap.org/.
† http://www.cbsnews.com/news/volvo-adds-pedestrian-airbag-to-v40-model/.

5.3.4 Pre-Crash Systems

In 2003, DENSO Corporation, a global automotive components manufacturer, developed the Pre-Crash System (PCS). PCS can detect an impending unavoidable accident just before the collision would occur. The system recognizes via a microwave radar object that a collision will occur and, based on that, determines the force of the impact and the probability of a collision. If the system detects that the probability of a collision is high, the brakes are automatically activated. This assures the driver of the maximum braking force to possibly avoid a collision. If a collision has become inevitable, PCS activates a number of other safety systems to reduce the impact of the collision in order to reduce the chance of injury to a minimum. When braking automatically or braking hard using the foot, seatbelts are automatically tightened.* The car manufacturer Lexus has already installed this system in some of its models.

Similar developments, although slightly less advanced, are the Ford Active City Stop and City Safety by Volvo. These systems check the road for other traffic just in front of the car using sensors placed behind the windscreen of the car. When a collision is imminent with another vehicle in front, which is braking, slow moving, or stationary, the car automatically activates the brakes. With an initial speed of 15 km/h (or 9 mph) the car comes to a complete stop with no damage being done, and at initial speeds of between 15 and 30 km/h (or 9 and 19 mph) the damage is kept to a minimum.†

5.4 Limited Self-Driving Automation (Level 3)

Three elements of limited self-driving automation (level 3) can be distinguished: traffic management (Section 5.4.1), cooperative systems (Section 5.4.2), and cooperative driving (Section 5.4.3). Traffic management has the objective of guiding the driver (or vehicle) and communicating with the driver (or vehicle) in order to communicate, which is necessary for cooperative systems. Cooperative systems contain vehicle-to-vehicle communication (V2V) and vehicle-to-infrastructure communication (V2I) that is enabled by traffic management. We call

---

* www.lexus.nl/range/is/key-features/safety/safety-pre-crash-safety.aspx.
† www.euroncap.com/.

these cars with cooperative systems *connected cars*: cars connected to numerous information and communication networks through navigation systems, roadside systems, and traffic services. An example of cooperative driving is "platooning," in which cars are mutually linked through electronic communication by cooperative systems. These cars drive close to each other, exchanging information about their speed, position, and acceleration, and then become autonomous.

5.4.1 Traffic Management

In order to reduce congestion on highways and in order to keep traffic moving, clever and flexible ways of using the optimal road capacity are being developed. This is being done in part by management of traffic, influencing supply and demand in traffic at a given time and place and aiming for the best possible functioning road system. Traffic management guides the choice of route, time of travel, and driving behavior. It works with real-time traffic information, alerting the driver about traffic congestion, roadworks, and accidents. Examples of traffic management are ramp metering, that is, limiting the number of cars flowing onto a road and dynamic speed limits. Ramp metering can prevent congestion and optimize the flow of traffic, for example, by installing on a ramp a "green light, one car" system. Dynamic Route Information Panels (DRIP) provide information by means of variable message signs enabling motorists to make a better route choice, based upon the current traffic situation, and regulate the inflow of cars to existing congestion, thereby reducing and then ending them. Several studies on the effectiveness of DRIPs have observed a significant increase in traffic flow (see, e.g., Edara, Sun, Keller, & Hou, 2012).

The introduction of navigation systems has changed the way we drive cars. The first navigation systems were built into the more expensive cars. Only after the arrival of portable navigation systems in 2004 did the market really start to take off, and nowadays most drivers use an in-car navigation system or a navigation app. Through these systems, motorists receive traffic information, including alerts about route, congestion, and incidents, as well as warnings about speed limits and dangerous curves in the road. Through such connected navigation systems, a vehicle is visible and the traffic manager

can communicate with the user. This causes a shift from only providing information toward advising and guiding motorists. Navigation
determines the fastest route—or the shortest and safest—and guides
the motorist. According to a survey study into the effects of driving with a navigation system on traffic safety performed by TNO,
navigation systems have a positive effect on traffic safety, such as a
positive effect on driver awareness and behavior; a reduction in the
driver's stress; and a reduction of traveling distances and traveling
time when driving through an unknown area (Feenstra, Hogema, &
Vonk, 2008).

TomTom is the world's leading player in navigation and is inventive
in expanding paid services. At the end of 2007, it launched *TomTom
HD Traffic*—now called TomTom Traffic—based on actual information.* Until then, traffic information was determined by the data
calculated by electric loops within the upper layers of the highway.
Instead, TomTom Traffic uses anonymous location data from mobile
phones to estimate local traffic flow. Using the movements of mobile
phones, one can make statistically reliable statements about traffic
flow on highways and in the city. TomTom Traffic devices communicate with a database every 2 minutes to receive new locations via
built-in SIM cards. Meanwhile, TomTom navigation systems have
an integrated GPRS modem and SIM card built in to them that send
data to and from the car. In addition, the system makes use of other
sources of information, including the traffic centers. From a case study
by Van't Hof, Van Est, and Kolman (2011), it follows that such a system helps to bring down the travel distance and time and saves fuel
and other costs. TomTom expects, for example, a reduction in travel
time for all road users of 5% if 10% of all road users have switched
on their TomTom Traffic equipment.† Navigation also has a positive
effect on safety: the device heightens the attention of the driver, while
it lowers stress on being lost (Van't Hof et al., 2011). To secure the
privacy, TomTom cuts off the beginning and end of the trip, so that it
cannot be traced to a specific address. Furthermore, cars are given a

---

* In 2011, Google launched Google Live Traffic. This is an extension of Google Maps
  with live traffic. The traffic element is an extra layer of information on top of the
  existing Google Maps. Google Live Traffic gives the planned route, taking into
  account traffic congestion and delays on major routes and highways.
† www.tomtom.com/manifesto.

random number, which changes every hour. TomTom only needs the amount, speed, and location of vehicles, and claims that no one can be identified through the system, and therefore options for road taxing or law enforcement are excluded in the architecture of the system (Van't Hof et al., 2011).

Predictions about travel information are produced by combining actual traffic measurements with a traffic model (Mahmassani, 2011). Development of these prediction models is still in its infancy, and is watched with great interest by traffic engineers. These traffic models will accurately predict traffic, in both short-term and long-term forecasting, and will yield advice when adjustments are required. A traffic manager presents information in such a way to motorists that it allows them to take the optimal route and drive along a specific road during the most convenient and safest time. The effectiveness of information on predictions of travel time is expected to increase even more if we also take the reaction of other road users into account (Van Lint, Valkenberg, Binsbergen, & Bigazzi, 2010). This requires that cars communicate with each other. Cooperative systems will enable this in the near future.

5.4.2 Cooperative Systems

The next phase that presents itself in car robotization is that of cooperative systems. To achieve the maximum benefits in safety, energy, and land use, as well as time, the vehicles of tomorrow would be networked. In the future, this will allow cars to communicate with other cars and with the infrastructure through V2V and V2I. Actually, the cars would be networked, both with each other and with the surrounding infrastructure. This would engender a whole new range of applications and, in this way, one would be able to take into account all kinds of circumstances located far outside the car's field of vision.

With V2I, communication exists between, for example, the vehicle and roadside systems with transmission and control apparatus located in roadside units: a motorist approaching an intersection will be informed that a cyclist is approaching on a priority road. An electric loop in the road or another detection system, video, radar, or GPS, records this approaching cyclist, and then the roadside system communicates this information to the car. At an intersection, these roadside systems can easily be integrated with existing traffic lights. In addition, via a coupling of different traffic junctions on ring roads, it will independently yield an advisory speed for cars, creating an optimal green wave in the future. Another example of V2I communication is *eCall*, an electronic safety system that automatically calls emergency services if there is a serious accident (see Box 5.2). Examples of V2V are alerting traffic after detecting slippery road surfaces and mutual communication in traffic congestion in order to optimize traffic flow. Often, V2V and V2I will operate in conjunction with each other.

In Ann Arbor, near Detroit, some 3000 residents of the area have allowed researchers from the University of Michigan Transportation Research Institute to install V2V and V2I communications equipment in their vehicles so that they can exchange data with other vehicles, as well as nodes at traffic lights, intersections, and roadway curves. The experiment spans roughly 73 lane miles in the northeast part of the city. This world's largest street-level connected-vehicle experiment, called *Safety Pilot*, is being conducted to find out how well connected-vehicle safety technologies and systems work in a real-life environment with real drivers and vehicles. It will test performance and usability and will collect data to better understand the safety benefit of a larger-scale deployment.[*] Connected-vehicle safety applications will enable drivers to have 360° awareness of hazards and situations they cannot even see. Through in-car warnings, drivers will be alerted to imminent crash situations, such as cars in the driver's blind spot, or a vehicle ahead braking suddenly. By communicating with the roadside infrastructure, drivers will be alerted when they are entering a school zone, when road menders are working, and when a traffic light they are approaching is about to change. At present, it is the human drivers who receive and act on the alerts, but in the near future the researchers will experiment with driverless cars that will be able to respond automatically. The researchers hope to demonstrate that fully driverless vehicles can operate within the whole infrastructure of the city by 2021 and to show that these can be safe, effective, and commercially successful.[†]

---

[*] http://safetypilot.umtri.umich.edu/.
[†] http://phys.org/news/2013-11-driverless-networked-cars-ann-arbor.html.

BOX 5.2   eCall

The V2I communication system *eCall* is a warning system that automatically alerts emergency services when an accident occurs based on precise GPS-based positioning and the provision of eCall prioritization within the mobile communication network.* The eCall system is activated when the airbags in a vehicle are activated or when impact sensors detect a collision. Over the cellular network, an SMS is sent to the local emergency phone center, and that message includes the GPS coordinates, time, direction of travel, and vehicle identification. Also, the driver gets a voice connection with the emergency desk at the push of a button, and a voice connection is also established when an automated system kicks in. Thus, a desk helper can determine whether the car occupant is still able to communicate. This system has been developed with European funding and should significantly lower the number of victims of road accidents by bringing emergency services faster to the scene of the accident: "Times saved translates into lives saved."[†] The European Commission estimates that eCall will reduce crash response time by about 50% in rural areas and up to 40% in urban areas.[‡] When medical care for the severely injured is available soon after the accident, the death rate and severity of trauma can be significantly reduced. It was shown in a Finnish study that a reduction of 5%–15% of fatalities and a reduction of 10%–15% of serious injuries can be expected when eCall is fully deployed (Virtanen, Schirokoff, Luoma, & Kulmala, 2006). Moreover, the system will result in less congestion, because crashed vehicles can be recovered faster. This system has been made compulsory by the European Commission and will be installed in passenger cars sold from 2018. The European Commission believes that a pan-European eCall is estimated to have the potential to save up to 2500 fatalities annually in the EU when fully deployed.[§]

---

* www.esafetysupport.org/download/ecall_toolbox/049-eCall.pdf.
[†] http://ec.europa.eu/digital-agenda/en/ecall-time-saved-lives-saved.
[‡] http://www.europarl.europa.eu/news/en/news-room/content/20140224IPR36860/html/Parliament-supports-life-saving-eCall-system-in-cars.
[§] http://www.europarl.europa.eu/news/en/news-room/content/20140224IPR36860/html/Parliament-supports-life-saving-eCall-system-in-cars.

5.4.3 Cooperative Driving

The Grand Cooperative Driving Challenge took place in May 2011 and was organized by the Dutch innovation program HTAS, *High Tech Automotive Systems*, and TNO. Ten international teams demonstrated their ideas about cooperative driving (see Figure 5.1). According to the researchers, cooperative driving has the advantages of increasing traffic safety, helping traffic flow, lowering harmful emissions, and making driving easier.[*] The idea of vehicles communicating with each other is not new, but until recently the necessary technical components were either lacking or were too expensive for commercial application. However, more and more cars are equipped today with satellite navigation, communication devices, and radar, or even a camera that monitors the distance to the car just in front. These constitute the main requirements for cooperative driving. Thus, a large number of companies, research institutes, and universities show interest in this concept and have established research programs. Cooperative systems are also high on the EU's agenda, given the number of co-funded research projects in this field, such as *Safespot* (2006–2010),[†] *CO-OPerative SystEms for Intelligent Road Safety* (COOPERS, 2006–2010),[‡] *Cooperative Vehicle-Infrastructure Systems* (CVIS, 2006–2010),[§] and *Safe Road Trains for the Environment* (SATRE, 2009–2013).[¶] These projects mainly focus on improving the communications between vehicles and roadside systems. Box 5.3 contains a brief description of the SATRE project.

In principle, the driver is not required to remain attentive when the car is in an autonomous mode, for example, when driving in a "platoon," but must be available to take control of the car within a certain amount of time after receiving an alert. Cooperative driving is technically feasible. All necessary systems are commercially available and are even being fitted today as standard equipment in many cars. Research such as that of SATRE shows that it will become feasible in the short term. Even standardization—essential for the practical application of

---

[*] www.htas.nl/index.php?pid=127.

[†] www.safespot-eu.org/.

[‡] www.coopers-ip.eu/.

[§] www.cvisproject.org/.

[¶] www.sartre-project.eu/.

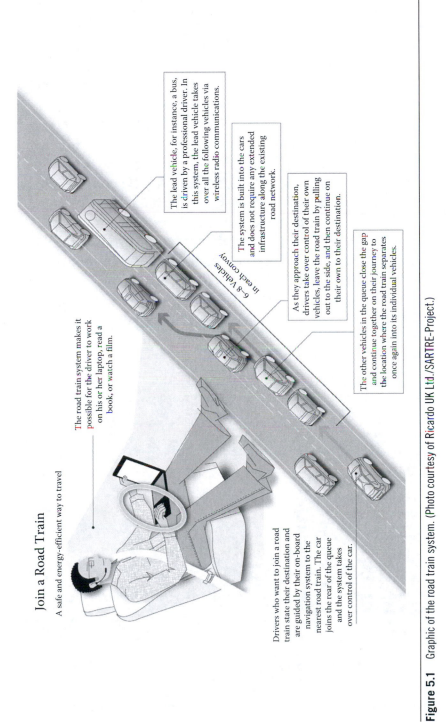

Join a Road Train

A safe and energy-efficient way to travel

The road train system makes it possible for the driver to work on his or her laptop, read a book, or watch a film.

Drivers who want to join a road train state their destination and are guided by their on-board navigation system to the nearest road train. The car joins the rear of the queue and the system takes over control of the car.

The lead vehicle, for instance, a bus, is driven by a professional driver. In this system, the lead vehicle takes over all the following vehicles via wireless radio communications.

The system is built into the cars and does not require any extended infrastructure along the existing road network.

6–8 vehicles in each convoy

As they approach their destination, drivers take over control of their own vehicles, leave the road train by pulling out to the side, and then continue on their own to their destination.

The other vehicles in the queue close the gap and continue together on their journey to the location where the road train separates once again into its individual vehicles.

Figure 5.1   Graphic of the road train system. (Photo courtesy of Ricardo UK Ltd./SARTRE-Project.)

BOX 5.3   DESCRIPTION OF THE SATRE PROJECT

Started in 2009, SATRE is a research program funded by the European Union in which car manufacturer Volvo also cooperates. The aim of the research is to develop a technology that allows the development of so-called car trains (see Figure 5.2). The idea is not new. In 1997, General Motors carried out tests on a series of Buicks that could travel at a short distance behind each other, taking their guidance from magnetic sensors in the road surface (V2I). In the SATRE project, however, the cars are mutually linked by electronic communication (V2V). According to the researchers, these "platoons," or "road trains," will become reality within 10 years. Since 2011, the first vehicles with this technology have been tried on test-drive circuits. When "platooning," cars have a mutual following distance of 0.2 seconds instead of the nearly 2 seconds regular cars have on average as their following distance. These cars mutually exchange information about their speed, position, and acceleration. When one car brakes, the vehicles following it also reduce their speed almost immediately. This is in fact a form of cooperative adaptive cruise control. Because of

Figure 5.2   Cooperative driving. (Photo courtesy of Ricardo UK Ltd./SARTRE-Project.)

this immediate response, cars can be packed more densely, and will drive without safety implications. Such a platoon consists of six to eight cars, with a "leader," which is a car that has an experienced driver who knows the route. A particular driver who wants to take an exit retakes control of the car again and leaves the convoy by steering toward the exit. The other cars in the platoon close the gap and continue, until the convoy splits again. Researchers see great benefits for long-distance commuters on a highway: while the "platoon" moves forward, they are free to read, watch TV, make phone calls, etc. Platooning has a positive effect on road capacity and yields a reduction in fuel consumption because as a result of the exchange of information road users maintain a constant speed. Platooning also has fewer problems in relation to drivers' overreaction. For example, if one motorist brakes slightly, the car to the rear brakes a little harder. This overreaction spreads like a shock wave through traffic and is the cause of so-called ghost traffic jams, in which traffic suddenly halts, seemingly without a clear cause, such as merging traffic or a narrowing road.

communication—is a minor problem, since there is already consensus: most current projects make use of a Wi-Fi transmitter containing a specific standard for the exchange of information. However, legal authorization, as a legal framework for cooperative driving, must be created before the launch, and acceptance by the general public could be major factors in delaying the launch.

It is expected that the most realistic introduction for the commercial adoption of the cooperative driving will start with long-distance haulage. Since road haulage is a huge industry, the impact of cooperative driving could be significant. In Europe, for example, there are 6.5 million heavy goods vehicles in operation,[*] and in the United States, there are 15.5 million.[†] Lorries (of more than 3.5 tons) in the EU account for more than one-quarter of road transportation CO_2 emissions, that is, 6% of total EU emissions of CO_2.[‡] Since 2014, the

---

[*] http://www.cepi.org/topic/transport/positionpaper/roadfreighttransport.
[†] http://www.truckinfo.net/trucking/stats.htm.
[‡] http://ec.europa.eu/clima/policies/transport/vehicles/index_en.htm.

European Commission is working on a comprehensive strategy to reduce CO_2 emissions from lorries. Experiments with driverless truck convoys have been under way for years. In 2013, for example, Japan's New Energy and Industrial Technology Development Organization (NEDO) tested a caravan of self-driving lorries.* They put four lorries on the road, with the first truck being driven by a human, followed by three autonomous lorries. The caravan successfully used technologies for steering, for maintaining speed, and for staying in close formation at a speed of 80 km/h (or 50 mph), and kept a 4 meter (or 13 foot) distance between each truck to create a slipstream of lower air resistance. The main goal of the researchers is to investigate what can be accomplished in terms of fuel efficiency. They say that running convoys of lorries in this manner could contribute to lower air resistance, helping to reduce fuel consumption by 15% or more. This corresponds with the results of the tests done with two lorries platoons by Peloton Technology in California[†] and with three lorries platoons by Lu and Shladover (2014). According to Mike Baker, the chief engineer at Ricardo UK Ltd. and involved in the SATRE Project, "[T]he long-haul vehicles have the most to gain, in terms of both safety and economic benefits. The fuel savings witnessed by lorries in a platoon has a significant impact on the operating profits of the operator, not to mention the environmental impact of reduced CO_2 and emissions."[‡]

With the aid of traffic management, it becomes possible to not only guide the movement at the level of individual vehicles but also manage traffic flow. This offers the prospect of improved safety levels, better traffic performance, and less impact on the environment. In addition, the development of in-car technologies will increase in the coming years. Several experts consider these developments as harbingers of further robotization of driving, which will eventually lead to the autonomous or self-propelled vehicle. In this development, the prefix "auto" in the word automotive would really take on its true meaning of "self-propelling."

---

* http://phys.org/news/2013–03-japan-group-fuel-saving-driverless-trucks.html. In the United States, for example, Peloton Technology in California has already carried out tests with two truck platoons.

† http://www.bbc.com/future/story/20140610-the-trucks-which-drive-themselves.

‡ http://www.techhive.com/article/2046262/the-first-driverless-cars-will-actually-be-a-bunch-of-trucks.html.

5.5 Autonomous Cars (Level 4)

The autonomous car was first promised in 1939 by Bel Geddes during the *Futurama* exhibition he designed for General Motors for New York World's Fair. At Futurama, Geddes speculated about what society would look like in the future. In his book *Magic Motorways* (1940), he writes: "These cars of 1960 and the highways on which they drive will have in them devices which will correct the faults of human beings as drivers. They will prevent the driver from committing errors." In 1958, General Motors engineers demonstrated the first "autonomous car." This car was autonomously driven over a stretch of highway by way of magnets attached to the car and wiring in the roadway, also called "automatic highways" (V2I technology). In a press release, General Motors proudly announced the result:

> An automatically guided automobile cruised along a one-mile check road at the General Motors Technical Center today, steered by an electric cable beneath the concrete surface. It was the first demonstration of its kind with a full-size passenger car, indicating the possibility of a built-in guidance system for tomorrow's highways.... The car rolled along the two-lane check road and negotiated the banked turn-around loops at either end without the driver's hands on the steering wheel.

> **Wetmore (2003, p. 7)**

The first real autonomous car was developed in 1977 by the Tsukuba Mechanical Engineering Laboratory in Japan. It tracked white street markers and achieved speeds of up to 30 km/h (or 20 mph).*

In 1974, 46 researchers predicted that automatic highways would become a reality between 2000 and 2020 (Underwood, Ervin, & Chen, 1989). Since then, several researchers have been working on the development of autonomous cars, but with relatively little success. Developments have been given a boost by the initiative of the U.S. military agency Defense Advanced Research Projects Agency (DARPA), which in 2004 took the initiative and organized a contest for autonomous vehicles. This DARPA Grand Challenge competition implied that autonomous cars had to travel a distance of over

---

* http://www.computerhistory.org/atchm/where-to-a-history-of-autonomous-vehicles/.

200 kilometers (or 120 miles) in the desert between California and Nevada. The result, however, turned out to be pathetic, because the team that had traveled the furthest only covered about 10 kilometers (or 6 miles). In 2005, the competition was organized again. Out of the 23 participating teams, 5 made it to the finish line. The winning team of Stanford covered the distance with an average speed of 30 km/h (or 20 mph). After this success, the bar was raised in 2007 with the so-called DARPA Urban Challenge, in which the car had to travel 100 kilometers (or 60 miles) in a simulated urban environment. Six of the eleven participated teams completed the task. Although this competition was a huge achievement, Urmson and Whittaker (2008) concluded that one could not yet speak of a fully autonomous car, because none of the participating cars could respond to traffic lights, and most cars had a hard time recognizing pedestrians. Partly in response to this competition, General Motors predicts that the autonomous car will be ready for the commercial market in 2020.

The development of autonomous cars continues steadily, and researchers and those who develop these cars constantly emphasize the benefits of them. The societal and economic benefits of autonomous cars include a decrease in the number of crashes, decreased loss of life, increased mobility for the elderly, disabled, and blind, and decreases in fuel usage. The large potential savings, which Morgan Stanley (2013) estimates at U.S. $1.3 trillion/year, should accelerate the adoption of self-driving vehicles. The researchers outline four key areas in which the U.S. savings will come:

1. Fuel cost savings of $158 billion ("an autonomous car can be 30% more efficient than an equivalent non-autonomous car").
2. Annual savings of $488 billion will come through a reduction of accident costs ("[i]f 90% of accidents are caused by driver error, taking the driver out of the equation could theoretically reduce the cost of accidents by 90%").
3. Increased productivity is likely to result in a gain of $507 billion ("people can work in their cars while commuting to work or at any other time").
4. A further $149 billion ("[a]utonomous cars should be able to largely eliminate congestion due to smoother driving style and actively managed intersections and traffic patterns").

Eight percent of the GDP of the United States is valued at $1.3 trillion. The company Morgan Stanley, a financial advisor to companies, governments, and investors from around the world, stated that "[e]xtrapolating these savings to a global level … we estimate global savings from autonomous vehicles to be in the region of 5.6 trillion per year" (Morgan Stanley, 2013, p. 7). These savings, however, can only be achieved at level 4 (full self-driving automation).

A benefit is also that with the introduction of autonomous cars, disabled persons or the elderly will be enabled to drive their own cars. Another possible benefit is related to the fact that autonomous cars are less prone to crashing. They need fewer safety features and can therefore be smaller and lighter than current vehicles, making them 10 times more energy efficient and making them better suited to using electric power (Burns, Jordan, & Scarborough, 2013). According to Litman (2014), however, "the ability to work and rest while travelling may induce some motorists to choose larger vehicles that can serve as mobile offices and bedrooms ('commuter sex' may be a marketing strategy for such vehicles) and drive more annual miles" (p. 6). This corresponds with the vision of Dieter Zetsche, the chairman of Daimler and head of Mercedes-Benz Cars: "The car is growing beyond its role as a mere means of transport and will ultimately become a mobile living space."* The new research vehicle F015 Luxury in Motion of Mercedes-Benz is a concrete example of this vision of autonomous driving of the future (see Figure 5.3). The autonomous car will ultimately become a private retreating space, where passengers have the freedom to use their valuable time on the road in manifold ways. As a consequence, the autonomous car will be very large: the F015 Luxury in Motion has an unusual length of more than 5 meters (or 16 feet) and a width of more than 2 meters (or 6.5 feet). This is in contrast to the vision of Burns et al. (2013) that autonomous cars will be smaller and lighter.

But before the autonomous car can be introduced into the commercial market, it is essential that autonomous cars can be tested in

---

* http://www.cnet.com/news/mercedes-benz-unveils-luxury-concept-self-driving-car/.

Figure 5.3 The F015 Luxury in Motion of Mercedes-Benz. (Photo courtesy of Daimler AG.)

real life to show their actual performance. In such practice, Google (Section 5.5.1) and the Free University of Berlin, with its project AutoNOMOS (Section 5.5.2), are already experimenting with autonomous cars.

5.5.1 Google

In 2010, it was announced that Google would undertake research on autonomous vehicles (see Figure 5.4). According to Google executives, the goal of the Google car was to "help prevent traffic accidents, free up people's time and reduce carbon emissions by fundamental changing car use."* Google hopes that the development of the car will ultimately contribute to better traffic flow and a reduction in the number of accidents and estimates that the annual number of 1.2 million casualties in the entire world can be reduced to half by the autonomous car.

---

* http://googleblog.blogspot.nl/2010/10/what-were-driving-at.html.

Figure 5.4 The autonomous Google car. (Photo courtesy of Google.)

Meanwhile, the company has driven seven autonomous cars on Californian public roads, six Toyota Prius and one Audi TT. Legal fines were avoided by the positioning of the drivers' hands just over the steering wheel, ready to intervene in case of problems. The vehicles have driven over 1.13 million accident-free kilometers, apart from one incident that occurred while a car was being driven manually,* since 2011, and have mastered staying in a lane and maintaining speed on the highway. Now that the Google cars have mastered highway travel, researchers are addressing the complexities of driving on a city street.† Before putting these cars on the road, Google engineers drive along the route one or more times to gather data about the environment. The autonomous cars compare the data they are acquiring to the previously recorded data to differentiate between objects, for example, pedestrians from stationary objects such as poles and mailboxes.‡ Although we can only speculate, Google is likely to maintain a continually updated database of road data that car manufacturers will have to subscribe to in order for their cars to drive autonomously.§

---

* http://www.businessinsider.com/googles-self-driving-cars-get-in-their-first-accident-2011-8.
† https://www.linkedin.com/today/post/article/20140528072025-142059068-don-t-laugh-the-new-google-prototype-car-has-implications-for-your-business.
‡ http://spectrum.ieee.org/automaton/robotics/artificial-intelligence/how-google-self-driving-car-works.
§ http://spectrum.ieee.org/cars-that-think/transportation/self-driving/google-autonomous-cars-are-smarter-than-ever.

Early in 2011, Google started to lobby in the U.S. states of Nevada and California about adjusting road traffic regulations. According to Google, autonomously driven vehicles should be legalized and the ban on text messaging from moving autonomous cars should be lifted. Meanwhile, four states in the United States have passed a law that allows one to take to the road in a self-steering car.* In Nevada, the Federal Department of Transportation Law has provided a designated area in which Google may perform these tests. A second law that allows people to make a phone call or send a text message while in the driver's seat has, however, not been adopted.

On the roof of each car robot is a Velodyne 64-beam laser, "the heart of the system," that generates a detailed 3D map of the environment. A computer combines these data with data from *Google Street View*, a GPS system, and a radar fitted within the car. In addition, next to the rearview mirror there is a camera that recognizes stop signs and traffic signals and helps detect pedestrians and cyclists. In addition, the car features a number of in-car technologies such as a stop-and-go system.

In June 2014, Google officially announced that they had built their very own prototype self-driving car. It is a radical design in which the normal configuration of a car is changed, since it lacks a steering wheel or pedals and is operated by pushing a button to start and stop (see Figure 5.5). It is designed to be used in urban settings at up to 40 km/h (or 25 mph).† Google had planned to run a small pilot with this prototype in California, but it has been blocked by the new testing rules of California's Department of Motor Vehicles, which require a driver to be able to take "immediate physical control" of a vehicle on public roads if needed. That means the car must have a steering wheel and brake and accelerator pedals.‡

While Google has been at the forefront of developing and testing self-driving technologies, it is not alone in its driverless vision for the future. Nissan, General Motors, Ford Motor Co., and automotive supplier Continental expect self-driving cars on the road by 2020. Tesla Motors

---

* http://cyberlaw.stanford.edu/wiki/index.php/Automated_Driving:_Legislative_
   and_Regulatory_Action.
† http://googleblog.blogspot.nl/2014/05/just-press-go-designing-self-driving.html.
‡ http://blogs.wsj.com/digits/2014/08/21/a-car-without-a-steering-wheel-or-pedals-
   not-so-fast-california-says/.

Figure 5.5 Google's new prototype autonomous car. (Photo courtesy of Google.)

even claims that it will develop a mostly autonomous car, which can handle 90% of driving duties by 2016.* In the following section, we will discuss another initiative of an autonomous car developed by the Free University of Berlin as a remotely controlled community taxi.

5.5.2 AutoNOMOS and the Remotely Controlled Community Taxi

Since 2011, an autonomous car has also been driven in Berlin; it is called *Made in Germany* (see Figure 5.6) and is the successor to the

Figure 5.6 The self-steering car Made in Germany. (Photo courtesy of Autonomos-Labs, Freie Universität Berlin.)

---

* http://www.usnews.com/news/articles/2014/10/09/will-teslas-d-be-a-self-driving-car.

Spirit of Berlin, which participated in the DARPA Urban Challenge in 2007. The car, a modified Volkswagen Passat, the result of the AutoNOMOS car project, was subsidized by the German government and implemented by the Artificial Intelligence Group of the Free University of Berlin.* The car has six stationary and one rotating laser scanner, seven radar sensors, four video cameras, and an infrared camera in order to get the fullest view of the environment. The developers have been awarded a license to carry out car tests on the roads in the states of Berlin and Brandenburg. The next goal of the developers is to drive the car across Europe.

A breakthrough in the AutoNOMOS project is that the car now drives smoothly, giving the impression that a person is driving the car, and the initial problem of sensitivity loss that robots typically have, which results in a jerky ride, has been remedied.

The most notable development is that you can order the car with your smartphone. The developers thus demonstrate a clear vision of the future. The idea is that cars should vanish from the road when they are not driving. In the developers' view, future cars should remain in central parking lots until an order call is made. As soon as the call is received, the car, a driverless taxi, sets off for the customer's location and then picks up the customer and takes him/her to a destination specified by the customer by smartphone. During the trip, customers can read, work, eat, talk on the phone, watch a film, or send e-mails. The car's system may be able to decide whether it picks up other customers it encounters with a matching destination on the planned route. There is no need to park—the vehicle zooms off to pick up another customer (see also Burns, 2013).

According to the researchers, in a city like Berlin, given the tie-in with existing public transportation, driverless taxi mobility can meet the personal mobility needs of the population with a fleet whose size is approximately 10% of the total number of passenger vehicles currently in operation in Berlin. This corresponds with the results of Burns et al. (2013). In a Singapore simulation, Spieser et al. (2014) found that a fleet of 250,000 robotaxis could replace all modes of personal transportation and fulfill the transportation needs (including private and public cars, taxis, scooters, buses and trains, etc.) of the entire population of Singapore.

---

* http://autonomos.inf.fu-berlin.de/.

There are currently about 800,000 passenger vehicles in total. These researchers' analysis showed that the maximum waiting time with this fleet size is about 30 minutes during rush hour, and is significantly lower during nonpeak periods. This also reduces emissions and congestion and reduces the need for parking lots. Hence, the researchers see this development as part of a trend for "greener" cars.

The German researchers' vision for the future fits with the vision of urban designer and futurist Michael Arth (2010), who predicts that the private car owner, whose car is stationary about 90% of the time and takes up space, will be swapped for public, self-propelled taxis that are constantly in use. In this system, there will no longer be any individual car ownership, and the current public transportation of trains and subways will have had its day. Instead, there will be a highly tailored public transportation system: for each type of transportation need, there will be an appropriate solution, such as an autonomous bus for a group, an autonomous limousine for a night out, and an autonomous transport for children. According to Arth, the development of the autonomous car therefore yields new prospects for the development of public transportation.

The scenario of Arth assumes that the motorist gives up his or her private car, because cars will be on call and will stand ready at the door, thereby taking away the necessity of owning a car. This argument is not entirely convincing, as current taxis (with a driver) also make owning cars less necessary. In order for it to succeed, the pricing for such an autonomous taxi will have to compete with public transportation, so that the consumer is willing to waive the right to use the private car. A useful thought experiment would be to see whether motorists would do away with the private car if all taxis were free. More motorists would probably take a taxi more often, but they would not do away with their car because cars are more convenient in many cases: they can get into the car without having to wait (as one has to for a taxi), they can leave equipment or luggage in the car and they can carry dirty loads when they need to. Furthermore, autonomous taxis have additional costs: they will require cleaning when passengers smoke, spill food and drinks or bring pets, and so on; they are often victims of vandalism, so frequently repairs are needed; and people will have to give up some privacy, since the cars will probably record their activities to minimize the cleaning and repairs (Litman, 2014).

A conceivable scenario is that people will purchase their own autonomous car and use it for all sorts of extra things that might even lead to heavier traffic in cities. The autonomous car could be used for multiple businesses and for tasks that would not be possible with an ordinary self-driven private car. For example, if a woman drives to work by car, the car is no longer available for other family members. But after an autonomous car has taken this woman to her workplace, it can return home empty, then take the children safely to school. Taking the children to school by bicycle and supervising them on this journey will not be necessary any more. Upon its return, the car will be available again for, for example, the errands of other family members, and at 6 o'clock the car will pick up the woman from work. This scenario will actually increase traffic because of these empty autonomous cars in particular. This is a scenario diametrically opposed to the one presented by Arth and the researchers from the AutoNOMOS project. The solution may be to ensure that autonomous cars become the new public transportation, rather than private possessions. Instead of being owned by individuals, they should be owned by cities or by rental car-and-ride-sharing companies, reducing ownership and parking costs (see also Schonberger & Gutmann, 2013).

5.6 Social and Ethical Issues Surrounding Car Robotization

Clearly, car robotization or the introduction of the intelligent car is surrounded by a large number of social and ethical issues: What are drivers' attitudes toward their driving tasks being taken over? What about the privacy of the motorist? How reliable are these advanced systems? Are autonomous cars really better drivers than humans? To what extent can the robot car lead to negative behavioral adaptations? What are the consequences of this development for liability in accidents? Does legislation allow cooperative or autonomous driving? We will raise these issues in the following subsections.

5.6.1 Acceptance

In several European research projects, research is being carried out into the acceptance of the robotic control of the car—and in particular the acceptance of driver support and cooperative systems, such as in the projects *European Field Operational Test on Active Safety Systems* (euroFOT)* and *Adaptive Integrated Driver-vehicle InterfacE* (AIDE).[†] It focuses on two questions: (1) How do motorists feel about technology taking over the driving task? and (2) Will motorists accept interference from these systems?

In principle, drivers are hesitant about accepting systems that take over driving tasks, because they often feel initial discomfort in a machine-dominated environment, and because they may see these systems as a restriction on their freedom, since many people consider the car to be the symbol of true freedom.[‡] However, acceptance grows as motorists have driven in a car with these systems and have come to trust the systems (Van Driel, Hoedemaeker, & Van Arem, 2007). Donner and Schollinski (2004) have shown that for a successful market introduction of driver assistance systems, the focus should be on showing the public that the systems are effective and safe. When ABS and ESC were first introduced, negative publicity and poor consumer education delayed mass-market adoption. Consumers did not fully understand how to make use of the technology. On the road, however, these systems delivered a clear, quantifiable reduction in fatalities. Once consumers understood how these systems worked, widespread adoption of ABS and more effective use of ESC followed (KPMG and the Center for Automative Research, 2012). More large-scale field studies are needed to determine how trust and therefore acceptance of different types of systems can be increased and used over time to achieve the desired effect of cooperative driver assistance systems. A first step could be to make being able to drive using driver assistance systems a compulsory part of the requirements for gaining a driving license and in the future to expand this with cooperative driving systems. In addition, an international ISO standard for driver assistance and cooperative systems could contribute to gaining public trust and hence public acceptance.

---

* www.eurofot-ip.eu/.

[†] www.aide-eu.org/.

[‡] According to the technology writer Andy Boxall at Digital Trends, "[c]ars are about adventure, excitement, and seeing the world; but for that to happen a person needs to be behind the wheel" (http://www.digitaltrends.com/mobile/google-driverless-cars-dt-debate/).

Autonomous vehicle technology will revolutionize the driving experience, and drivers will need time to learn how to use and manage the new features. They will need to feel comfortable with the new technology's functionality and interface, and they will probably have to overcome a psychological hurdle before they cede control and let the car drive autonomously (KPMG and the Center for Automative Research, 2012). According to the large international consulting firm KPMG (2013), consumers also seem open to the idea that they may opt for an autonomous car one day. In fact, "[t]he marketplace will not merely accept self-driving vehicles; it will be the engine pulling the industry forward. Consumers are eager for new mobility alternatives that would allow them to stay connected and recapture the time and psychic energy they squander in traffic jams and defensive driving" (KPMG and the Center for Automative Research, 2012, p. 6). In a study, however, conducted by Seapine Software among 2000 U.S. adults, 88% would be worried about riding in a driverless car, especially with regard to failing equipment, such as a glitch in braking software or a faulty warning sensor. In addition, 52% fear a hacker could breach the driverless car's system and gain control over the car.* Consumers will not relinquish control until they are certain their vehicles and the mobile environment are 100% safe and reliable. Furthermore, the ramifications of an early autonomous or connected-vehicle traffic crash could be calamitous: "Bad publicity is a significant risk for the deployment of innovative automotive technology, even if the technology itself is not the cause" (KPMG and the Center for Automative Research, 2012, p. 19). Acceptance may be growing, just as with driving assistance systems, as consumers have driven these cars. More research is needed on the acceptance of autonomous cars. Now that ever more autonomous cars take to the road, the first small-scale field studies on the acceptance of this car and confidence in an autonomous car will be carried out within the foreseeable future.

5.6.2 Privacy

As mentioned earlier in this chapter, the eCall system becomes mandatory in the EU for all new cars sold from 2018. Besides this eCall

---

* http://www.seapine.com/pr.php?id=217.

system, there are obviously several systems inside and outside the car producing and receiving information relating to location and identification, such as navigation systems and cooperative systems. These systems relay personal data, and because of that one has to take into account the privacy laws in both European directives and national legislation on privacy. The task force *Driving Group on eCall* (2006), set by the European Commission, is worried about the fact that data could be used for other purposes or even misused. Data related to time and locations could be used in road pricing or congestion pricing—or for traffic violations such as incidents of speeding. It is conceivable that eCall information could be very useful for insurers. The road authority could use this information for traffic management. In addition, these data increase the possibility of detecting criminal suspects. This is a real consideration, because in recent years the powers of law enforcement agencies have been considerably expanded in this area. In Amsterdam, for example, it was announced in 2011 that cameras installed to monitor polluting vehicles within the city would also be used for other purposes as well. The images collected are now used to see if car license plate owners have something to answer for—or still have an unpaid fine. In fact, these cameras now function as "a digital moat around the city."* It is highly desirable to get clarity in terms of laws and regulations relating to powers of investigative agencies in dealing with localization-related services surrounding motorists.

In order to meet privacy concerns, the eCall system will lead a "dormant life," that is, the eCall system will only switch on a connection to the data network when an accident has occurred, if the motorist activates the system, or if the car is reported stolen. This is in accordance with the opinion of the task force Driving Group on eCall. The question is whether they can keep their promise that eCall will remain only a dormant system.

Navigation systems that include a SIM card allow a data flow to and from the car—and thus are not dormant systems. A common method of addressing privacy concerns relating to these systems is anonymizing travel data by encryption, by which third-party abuse is prevented. Having anonymous information, however, can lead to controversy. In the Netherlands, the police had been accessing TomTom

---

* *Het Parool* (September 7, 2011), Digital ring scans all cars.

data to determine where to station traffic cops—not in areas with slow traffic, but in locales where people were speeding. And even though the company makes the data anonymous and aggregates it for the purposes of providing traffic information, the controversy led to a public outcry and a government investigation in which the company was cleared of violating Dutch data protection laws.*

Data can be decrypted by the suppliers of navigation systems at the demand of investigators when a car is found at a crime scene. But what should be done with the information when a driver flees after a traffic accident, for example? Should the police have the ability to trace this car? After all, it goes one step further than just being able to follow a motorist on a map. Or does privacy prevail and so the system guarantees anonymity, such as the TomTom navigation system (see Section 5.4)?

In cooperative systems, another problem arises because different parties have to exchange data with each other. The questions are who should own or control the data, how should time data and location data relating to the driver be stored, to what ends will the data be used and how can it be ensured that the privacy of the motorist is guaranteed. Now that cooperative systems have proven their worthwhile application in real life, it is high time that politicians and information lawyers started discussing these questions.

As the use of autonomous cars expands, maintaining individual privacy within the transportation system may become even more arduous (cf. Glancy, 2012). A lot of data would be gathered by autonomous cars in order that cars could be driven safely, but these data can be misused. Numerous stakeholders have commented on the high value of data that would be gathered by in-vehicle communications platforms about the vehicle itself and its driver. For example, insurance companies would be interested in individual driving habits and retailers in attracting motorists to their locations (Anderson et al., 2014). As we have already mentioned, law enforcement agencies have a considerable interest in using these data. In California, the Consumer Watchdog's Privacy Project director John Simpson states that the autonomous vehicles legislation

---

* http://data-informed.com/profitable-lessons-tomtoms-brush-data-privacy-controversy/.

should "provide that driverless cars gather only the data necessary to operate the vehicle and retain that data only as long as necessary for their operation. It should not be used for any additional purpose, such as marketing or advertising, without the consumer's explicit opt-in consent."[*]

5.6.3 Security and Safety

As we have already mentioned, advocates state that autonomous cars will eliminate 90% of crashes, but autonomous cars will introduce new risks with regard to security and safety, such as reliability, negative behavioral adaptation, and cyberattacks.

5.6.3.1 Reliability System failures, when a car is operating in an autonomous mode, could be fatal to vehicle passengers and other traffic participants. All critical components will need to meet high manufacturing installation, repair, testing, and maintenance standards, similar to aircraft components. This will probably be relatively expensive (Litman, 2014). Furthermore, the special equipment, including sensors and computers, needed for an autonomous car is already very expensive.

Because of the fear of liability, producers will only put these systems on the market when they are found to be perfectly safe. ISO standards are important to reduce liability, since then car manufacturers can argue that they are operating at the state-of-the-art level of the industry and have observed mechanisms for functional safety (Anderson et al., 2014). For most driving assistance systems, ISO standards are developed, such as the ISO standard 15622:2010 for adaptive cruise control systems.[†] Only after such a norm is created and much research will producers dare to take this plunge. A lot of the research in recent years has focused on increasing the safety of driver assistance systems, and therefore, these benefits are presented as unique selling points. We can see a huge increase in these systems in all car models. This trend is also expected for cooperative systems,

---

[*] http://www.consumerwatchdog.org/blog/dmv%E2%80%99s-autonomous-vehicle-regulations-must-protect-users%E2%80%99-privacy.

[†] http://www.iso.org/iso/catalogue_detail.htm?csnumber=50024.

which are relatively new. For that purpose, large-scale studies and ISO standards are needed.

5.6.3.2 Negative Behavioral Adaptation   Cooperative driver assistance systems can also lead to negative behavioral adaptations, such as a reduced level of attention and overestimating the system's advantages, with the result that some of the positive effects are nullified. It appears that the ACC system not only has a major positive effect on road safety but also leads to higher speeds and shorter headways (Dragutinovic, Brookhuis, Hagenzieker, & Marchau, 2005). Another possible unintended negative effect—with respect to mixed traffic—may be that road users without such a system will anticipate an assumed behavior of cars with such a system or that by guesswork or imitation they behave as if they do have such a system, for example, drivers of nonautonomous cars may be tempted to join autonomous car platoons. In addition, these systems can lead to de-skilling, so that driving ability may deteriorate. This can lead to dangerous situations at times when the (semiautonomous) car does not respond autonomously and control should be taken over by a driver who has become less road savvy.

Research conducted by Virginia Tech Transportation Institute and General Motors in cooperation with the U.S. Department of Transportation Federal Highway Administration in 2011 to understand the factors that impact the effectiveness of alternative concepts of operation for cars with limited ability autonomous driving features (specifically, adaptive cruise control capable of maintaining a set speed and headway, cooperative driving and lane centering capable of autonomously following a single lane on a highway) has addressed concerns that drivers (1) have become overreliant upon the systems; (2) operate the systems outside of design parameters; or (3) are not aware when the systems are not operating as intended. Evidence also suggests that drivers tend to spend less time looking at the roadway in front of them, have longer off-road glances when operating the semiautonomous car compared to the nonautonomous car, and do not adequately anticipate when the car needs to operate in a nonautonomous mode (Llaneras, Salinger, & Green, 2013). According to a study by Jamson, Merat, Carsten, and Lai (2013), drivers demonstrate increasing symptoms of fatigue with a car in an autonomous mode and become more heavily involved with the in-vehicle entertainment. Drivers, however, do demonstrate additional attention to the road as traffic conditions become more congested, implying that these responsibilities are taken more seriously as the supervisory demand of a car in an autonomous mode increases.

5.6.3.3 Cyber Security Cooperative systems (and autonomous cars) have to deal with the security of the information and communication network. Cooperative driving, for example, necessitates both communications hardware and a link to the engine management system so that the vehicle can control its own speed. A disadvantage of this is that the system is fragile and the car could become the victim of hacking attempts. U.S. researchers at CAESS (*Center for Automotive Embedded Systems Security*) have shown that it is possible to hijack and take over full control of the car (Checkoway et al., 2011). In theory, malicious people could take over a highway node, causing collisions and traffic disruptions. The European research project Preserve (Preparing Secure Vehicle-to-X Communication Systems), started in 2011, deals with the development and testing of a security system.* Securing data is complicated because cryptology increases the information flow, and the available bandwidth is restricted. In the United States, the National Institute of Standards and Technology is currently developing a framework to improve the critical infrastructure of cyber security, and recommendations that stem from this framework may be incorporated into automated- and connected-vehicle technologies (Fagnant & Kockelman, 2013).

5.6.4 Better Drivers

Autonomous cars developed by, for example, Google and the Free University of Berlin make the driver redundant. Many researchers see the autonomous car as a method of preventing traffic accidents, for conscious or unconscious human error is involved in almost all

---

* www.preserve-project.eu/.

traffic accidents. Several studies show that in more than 90% of cases, accidents occur due to human error and that only 5%–10% are the result of deficiencies in the vehicle or the driving environment (see, e.g., Broggi, Zelinsky, Parent, & Thorpe, 2008; Dewar & Olson, 2007; National Highway Traffic Safety Administration, 2008). Autonomous vehicles have continuous complete attention and focus, keep within the speed limit, do not get drunk, abstain from aggressive behavior, and so on. In addition, Peter Sweatman, director of the University of Michigan Transportation Research Institute, states that "[h]umans are not suited to monitoring tasks like driving [because] human attention is easily diverted."* In fact, the researchers at the institute state that the solution to preventing traffic accidents is not more speed limits, airbags or distracted driving policies, and so on, but getting humans out of the loop. But before the human factor can be switched off in traffic, the autonomous vehicle must be thoroughly tested in the actual dynamic traffic before safely functioning on the road. Levy and Murmane state that "[a]s the driver makes his left turn against traffic, he confronts a wall of images and sounds generated by oncoming cars, traffic lights, storefronts, billboards, trees, and a traffic policeman. Using his knowledge, he must estimate the size and position of each of these objects and the likelihood that they pose a hazard. ... Articulating this knowledge and embedding it in software for all but highly structured situations are at present enormously difficult tasks" (Levy & Murmane, 2004, p. 28). Or, if an autonomous car is programmed to faithfully follow the traffic regulations, then it might refuse to drive in automatic mode if, for example, a headlight is broken, or it might come to a complete stop when a small tree branch pokes out onto a highway because crossing a continuous line is prohibited whereas humans would simply drift a little into the opposite lane and drive around it (see also Lin, 2013).

According to Smith (2012), autonomous cars cannot yet be claimed to be significantly safer than those with human drivers. In his analysis, which uses a Poisson distribution and assumes the accuracy of the crash and mileage estimates, he concludes that for autonomous cars to be declared safer, they would need to drive 1.17 million kilometers

---

* http://www.engin.umich.edu/college/about/news/stories/2013/november/driverless-connected-cars.

(or 727,000 miles) on representative roadways without incident and without human assistance with, say, 99% confidence that they crash less frequently that a human driver, and 483 million kilometers (or 300 million miles) if considering only fatal crashes. Most autonomous cars, including the Google cars, have mainly traveled safely so far on dual carriageways and highways. Although preliminary evidence does not prove that autonomous cars are safer than those with human drivers, it seems likely that autonomous cars will eventually reduce the crash rate (Goodall, 2014b).

Having an autonomous car that deals in a completely safe way with the dynamic environment could take many years; predictions range from 5 to 30 years (cf. Knight, 2013). The question is, however, whether an autonomous car should be 100% safe. According to Marchant and Lindor, "it may be better to have autonomous vehicles sooner rather than later even if they are imperfect, given that even imperfect autonomous vehicles will be safer than vehicles on the road today" (Marchant & Lindor, 2012, p. 1340). According to Lin (2014b), a reduction in overall fatalities may be considered unethical when there is a sacrifice or "trading" of lives. Lin provides the following example. Suppose that an autonomous car is faced with a terrible decision of whether to crash into a sport utility vehicle (SUV) or a Mini Cooper. As a matter of physics, the autonomous car chooses a collision with the heavier vehicle, that is, the SUV, which can better absorb the impact of the crash. Such a crash-optimization algorithm would require the deliberate and systematic discrimination against SUVs as the preferred vehicles to collide with. The owners and passengers of these SUVs would bear the burden through no fault of their own.

Even perfect autonomous cars without human-driven traffic will crash. These cars would still face threats from wildlife, unexpected obstacles, pedestrians, and cyclists (Goodall, 2014b). Because of this unpredictability of real-life situations in traffic, there will always be a need for some type of ethical decision-making system. Autonomous cars can make split-second choices to optimize crashes by minimizing harm; however, it is unclear how to do that in difficult cases. For example, an autonomous car is driving on a public road when another car, approaching from the opposite direction, suddenly veers into its lane (e.g., because it is malfunctioning). The autonomous car must decide what to do and has two alternatives: (1) crash head-on into the other car and (2) veer right and have a collision with two cyclists. In such a case, an autonomous car needs a method to determine an ethical action. In the literature, there is no agreement about such a method. A solution could be that the autonomous car makes a random decision in a difficult case like this. Such a random decision mimics human driving insofar as split-second emergency reactions are mostly unpredictable. Lin (2014b) states that this randomness may be inadequate for several reasons. First, it is not an obvious benefit to mimic human driving, since a key reason for developing autonomous cars is that they should be able to make better decisions than humans do. Second, although drivers may be forgiven for making a poor split-second decision, programmers have all the time in the world to get this right. According to Lin (2014b), "[i]t is the difference between premeditated murder and involuntary manslaughter." Third, randomness evades the responsibility to make thoughtful decisions and to be able to defend them.

Another solution could be that customers choose their own personalized ethics setting (Lin, 2014a), so ethics would become a selling point for an autonomous car, which customers could change depending on their personal taste. In a survey, 44% of the respondents said they would prefer to have a personalized ethics setting, whereas 12% thought that manufacturers and 33% that lawmakers should predetermine the ethical standard.* For example, people could then choose that the autonomous car sacrifices their life to avoid crashing into a child or that the car saves their life and kills the child. Or it could choose that the car saves insured cars over uninsured cars or men over women.

But even if there is an agreement about the "best" method to determine an ethical decision, there will be ethical issues to consider (Lin, 2013). Take, for example, a variant of the classic trolley problem: on a narrow road, your autonomous car detects an imminent head-on crash with another car. Your car does not swerve to avoid the crash, because if it did it would knock down three pedestrians. Not swerving is probably the right thing to do: to sacrifice yourself to save the three pedestrians. Should such an ethical decision feature be something

---

* http://rca.mchrbn.net/ref/2014/09/if-death-by-autonomous-car-is-unavoidable-who-should-die-poll/.

explicitly disclosed to the passengers of an autonomous car, and should informed consent be required to ride in something that may purposely cause your own death? According to Hevelke and Nida-Rümelin (2014), the aforementioned problem of the autonomous car is quite different from the classic trolley problem. In contrast to the classic trolley problem, we should not focus on the actual damage done in the end when we try to determine which decision of an autonomous car is morally acceptable: "[i]t is a fallacy to take the real consequences of a decision into account when confronted with probabilistic phenomena. What counts when we decide if a possible action is in the interest of a person is the probability of that decision's consequence, not the actual consequence itself" (Hevelke & Nida-Rümelin, 2014, p. 5).

Goodall (2014a) proposes a three-stage strategy for developing the "best" method to determine an ethical decision: (1) having a rationalistic moral system that will take action to minimize the impact of a crash based on generally agreed principles, such as that injuries are preferable to fatalities; (2) using machine-learning techniques to study human decisions across a range of real-world and simulated crash scenarios to develop similar values; and (3) having an autonomous car that expresses its decision using natural language, so that its highly complex and potentially incomprehensible-to-humans logic may be understood and corrected.

It is clear that autonomous cars have to deal with hard life-and-death choices, and these decisions should be considered thoughtfully and openly to ensure a responsible product that millions will buy, ride in, and possibly be injured by. The fact that the autonomous car may injure us is all the more reason to focus on ethics in steering the future of transportation in the right direction. Creating a debate about the ethical autonomous car is necessary to help raise public, industry, and political awareness and to ensure that autonomous cars make moral sense. Before autonomous cars are introduced into the commercial market, a legal framework concerning life-and-death choices based on ethics should be developed.

5.6.5 Liability

Driver assistance systems, cooperative driving, and autonomous cars have and will reduce crashes. However, crashes will happen, and even autonomous cars will almost certainly crash (Goodall, 2014b). Therefore, we have to look at the consequences for liability in traffic accidents with regard to car robotization. In this section, we will briefly discuss the consequences of the introduction of driver assistance systems, cooperative systems, and autonomous cars for the liability of the following three actors: manufacturers, operators, and drivers.

5.6.5.1 Liability of the Manufacturers   The manufacturer is liable when damage is caused by a defect in its product. There is a defect if the product does not provide the safety which might have been expected of it under the given circumstances. Legally, there is no vacuum regarding the liability of the manufacturer for driver assistance systems. In assessing any lack of safety features in driver support systems, the law addresses the following factors: the expectations of the public, the presentation of the product, the reasonably expected use, and compliance with safety requirements. With regard to the last factor, there is a lack of legal or other safety regulations. Meeting safety standards will then be concretized by assessing whether the system meets the safety level one would expect.

However, the field becomes more problematic when we consider the liability of manufacturers in relation to cooperative systems. In cooperative systems and autonomous cars, the cause of the product's injurious behavior could be traced back to an external cause. For example, it is possible that a cooperative system has missed a warning for a local hazard because of a defect in a roadside system. In principle, it is up to the injured party to prove this defect. For consumers this is an almost impossible task, so the judge accommodates the consumer by allowing suggestive evidence. In driver assistance systems, this is workable, because a stop-and-go system that fails to break for traffic congestion shows a system defect. It becomes more difficult with cooperative systems and autonomous cars, because the incorrect functioning of a system may be due to data communication with other road users and infrastructure of the road. Evidence suggestions may then be insufficient. A solution to this evidence problem is to use an electronic data recorder, such as the eCall system or a black box, to capture information about the functioning of the system and driver behavior. This allows a manufacturer or road administrator to provide proof that the product was functioning properly. An additional problem relates to privacy: which parties have access to the data and under what conditions and for what purposes?

Autonomous cars will shift the responsibility for avoiding accidents from the driver to the manufacturer (see, e.g., Marchant & Lindor, 2012). According to LeValley (2014), autonomous car manufacturers should be held to the same standard as common carriers, since autonomous car manufacturers share many traits with common carriers, such as exclusive control over the vehicle's operation. Generally, common carriers are held to owe passengers the highest standard of care and must take extraordinary diligence to protect the lives and physical safety of his passengers. So, autonomous car manufacturers would then be liable for even the slightest negligence. His conclusion is further reinforced by the fact that policy justifications for holding common carriers to the highest standard of care are similarly applicable to autonomous car manufacturers. This liability burden on the manufacturer, however, could impede the development and introduction of autonomous cars.

5.6.5.2 Liability of the Road Authorities   The road authority is liable if the road does not meet the requirements that may be expected in the given circumstances. The public highway also includes equipment: all of those objects placed next to or above the traffic lanes, which are put there to design the road for traffic use, such as crash barriers, reflector poles, road signs, and traffic lights. The question here is whether digital roadside systems also fall under the heading of road equipment, because the operation of these systems depends on remote, centralized information processing systems. This may give rise to discussion about the scope of the term "road equipment" and thus the applicability of the liability of the road authority. The most obvious solution is to define roadside systems and their underlying computer systems as road equipment.

5.6.5.3 Liability of the Driver   In most countries, the owner of a motor vehicle is liable for damage caused by his vehicle, unless it is likely that the accident is due to *force majeure*. The interpretation of *force majeure* is very restrictive: it is limited to cases in which the accident is solely attributable to the behavior of the victim or that of a third party. If there is a suddenly occurring defect in the car, so even with a failing driver assistance system, *force majeure* cannot be invoked. The idea is that those who enjoy the benefits of car use should also bear the burden. The resulting question is whether *force majeure* would be able to succeed in the event of the failure of a roadside system or a lack of data communication with other vehicles because the cause can be considered as coming from outside and is due to the behavior of a third party.

The liability of a driver against other motorized road users is assessed on the basis of tort, which is attributable if this is the driver's fault. When a driving assistance system is not adequately functioning, the driver alone cannot be blamed if and when particular hazards are encountered, such as sudden, unforeseeable circumstances that the driver did not reasonably have to take into consideration. An example would be sudden—unprovoked—full automatic braking due to a collision avoidance system, which then caused a collision.

The fact that well-functioned driver assistance systems allow drivers to better respond to hazards may affect their liability. If a system has warned of a slippery surface, the driver cannot claim that this was an unforeseen circumstance. A case becomes a more difficult one when a driver has failed to activate the collision avoidance system and when it is established that the system could have prevented or limited damage. In that case, the court will probably decide whether a perfect driver—the standard generally used in law—would have activated the system.

As we already have seen, drivers may become over-reliant upon driver assistance and cooperative systems, with the result that drivers will concentrate less. If an accident occurs, the relevant question for the court is, again, whether a perfect driver could have prevented the accident.

In relation to cooperative driving and autonomous cars in particular, there are unique features that should be considered with respect to liability. Therefore, specific liability regulation will be needed for cooperative driving and autonomous cars. Hevelke and Nida-Rümelin (2014) propose a liability regulation based on strict liability. These authors argue that it is justifiable to hold users of autonomous cars collectively liable for any damage caused by autonomous cars. An owner of an autonomous car could only be held responsible for taking the risk of using the car, which would be a risk taken daily by millions of people and highly comparable. The owner would, therefore, share this responsibility with every other person who owns an autonomous car: "[f]rom this perspective, they did participate in a practice which carries risks and costs for others and it therefore is their responsibility to shoulder that burden" (Hevelke & Nida-Rümelin, 2014, p. 6).* A tax or mandatory insurance would be the easiest and most practical way to achieve collective liability.

5.6.6 Legislation for Limited and Full Self-Driving

The 1949 Geneva Convention on Road Traffic, to which the United States and most European countries are parties, probably does not prohibit automated driving. The treaty promotes road safety by establishing uniform rules, including Article 8, which requires that every vehicle or combination thereof must have a driver who is "at all times ... able to control" it (see also Smith, 2014). The 1968 Vienna Convention on Road Traffic, to which the United States is not a party but most European countries are, contains similar language in Article 8. In comparison to the Geneva Convention, it requires a more extensive obligation that "every driver shall possess the necessary physical and mental ability and be in a fit physical and mental condition to drive." A sixth paragraph on distracted driving was added to this article in 2006:

> A driver of a vehicle shall at all times minimize any activity other than driving. Domestic legislation should lay down rules on the use of phones by drivers of vehicles. In any case, legislation shall prohibit the use by a driver of a motor vehicle or moped of a hand-held phone while the vehicle is in motion.

The idea of an autonomous car with the human-out-of-the-loop is discouraged by these conventions. The key issue in these conventions is the concept of control, and, probably, amending these conventions is needed to ultimately realize the potential of autonomous cars.

---

* An exception is when a car does not satisfy the safety regulations, since then the manufacturer may be liable.

To more accurately capturing this potential, a legal framework for testing and implementing autonomous car technologies beyond level 2 is needed. These frameworks are lacking in most countries—which could impact the rate of adoption of autonomous cars (Mogos, Wang, & DiClemente, 2014)—with the exception of some states in the United States, the United Kingdom, and some ad hoc legal permits passed by regional authorities in some EU countries. The authority of Gothenburg (in Sweden), for example, has given Volvo permission to test 100 driverless Volvos on public roads in everyday driving conditions. This Volvo Car Group's project "Drive Me" started in 2014.* In the United Kingdom, testing of autonomous cars has been allowed on public roads since 2015.† Tests on public roads, instead of on private or specialized roads, are necessary to ensure safe testing and development of autonomous cars and their integration into the public sphere.

Several countries in Europe and some states in the United States have expressed interest in developing legal frameworks for testing and implementing autonomous vehicles, and some have even started taking active steps to examine their current laws (Kim, Heled, Asher, & Thompson, 2014).

5.7 Concluding Observations

In this chapter, we have explored the robot car against the background of developments in the field of driver assistance systems, traffic management, and cooperative systems. We have shown that the robotization of the car is in full swing. There is clearly a recognizable trend of driver assistance systems that inform only toward systems that also warn and even intervene. The next step in this trend is taking over driving tasks through cooperative systems—in conjunction with traffic management. Ultimately, it is expected that this will lead to autonomous vehicles being introduced. In this final section, we look back on the social issues that have arisen from driver assistance systems and cooperative systems in conjunction with traffic and autonomous cars.

---

* https://www.media.volvocars.com/global/en-gb/media/pressreleases/145619/volvo-car-groups-first-self-driving-autopilot-cars-test-on-public-roads-around-gothenburg.
† http://www.bbc.com/news/technology-28551069.

On this basis, we identify some issues that require attention in the public and political domains.

5.7.1 Short Term: Driver Assistance Systems (Levels 1 and 2)

5.7.1.1 Expectations   ADAS support the driver, but do not yet allow fully automated driving in traffic. The application of driver assistance systems is rapidly developing and is being fully stimulated by the car manufacturing industry, research institutions, and government. There are high expectations of these systems regarding safety effects. The available driver assistance systems are probably only harbingers of a major development that will lead to a progressive "automation" of the driving task. This trend can now be observed. Systems that in principle only advised or warned, as in alerting the driver about speeding or unintentionally veering off the roadway, are further developed into systems that actually intervene, causing the car to return to the right lane when the driver unintentionally leaves the roadway. Given technological developments, it is expected that in the near future more self-correcting driver assistance systems will be able to intervene in driving behavior instead of only warning or advising the driver.

5.7.1.2 Social, Ethical, and Regulatory Issues   Although the development of driver assistance systems is stimulated, much of the research shows that many consumers have insufficient knowledge about what driver assistance systems can do. This problem has now been addressed by the European Commission, which has set up the eSafety Forum, which includes the aim of increasing the public awareness of driver assistance systems.* In addition, these kinds of systems require new driving skills. It is important that attention is paid to this, and a possible solution could be that driving with driver assistance systems becomes a mandatory part of the requirements for being granted a driving license. The downside of this development is that many drivers come to rely on these systems, making them less alert.

---

* www.esafetysupport.org/.

The greatest benefit of these systems, sought by the European Commission in particular, is in traffic safety. The Commission aims to halve the total number of road deaths in the EU by 2020 as compared to 2010. This is a very ambitious goal, which in our opinion can only be achieved by rigorous measures such as mandating a number of driver assistance systems. The Commission has already made the ABS, ESC, and eCall compulsory. The next systems that could qualify for such an obligation are the ACC system and the Lane Departure Warning (LDW). Research shows that these systems have a great effect on road safety. According to a study by Rijkswaterstaat, part of the Dutch Ministry of Infrastructure and the Environment, 47% of accidents on highways are caused by keeping too little distance—respectively, 24% and 18% of accidents on country roads and urban roads (Rijkswaterstaat, 2007). With an ACC system, this can be prevented, especially in combination with a PCS. The European Commission is seeking to achieve the strategic goal of halving the number of road casualties by stronger enforcement of traffic regulations. Speed is also a basic risk factor in traffic and has a major impact on the number of traffic accidents. According to the Royal Society for the Prevention of Accidents (2011), the number of fatalities would decrease by between 10% and 26%, and the number of serious injury accidents by between 6% and 21%, if all drivers adhered to speed limits. This could easily be addressed by using a far-reaching variant of the Intelligent Speed Adaption systems: the intelligent speed authority, so that drivers cannot violate the speed limit. It is not expected that this variant will be implemented easily, because public acceptance of such an intrusive system is quite low (Morsink et al., 2007). The car is considered in our culture as a symbol of freedom (Husak, 2001), and an intrusive system restricts the freedom of the driver, since the driver would be forced to have his or her car fitted with technology that keeps them within the speed limit. The question is whether a moral duty exists to mitigate this freedom in the interest of road safety. The introduction of the intelligent speed authority will reduce the yearly toll in traffic victims which is a powerful moral reason for the European Commission to make this intelligent speed authority compulsory, even though this will be met by a lot of resistance from both drivers and political parties. Furthermore, the argument that it restricts the freedom of the driver is not a strong one, since there is no legal (and even moral) freedom to violate speed limits that endanger other road users.

Policy makers should be aware that if they would like to introduce an intervention driver assistance system that controls automatically, the penetration level should be sufficient from the start to convince others to adopt such a system. For example, in the Ghent ISA trial in 2002,* it was noted that most of the drivers were convinced of the effectiveness and were highly in favor of the intelligent speed authority but they stated that they would only use this system further when more or certain groups of drivers would also be forced to use the system. According to Vlassenroot (2011), promoting such a system by implementing it in certain types of vehicles, for instance, busses, taxis, and lorries, may be helpful for the acceptance of such a system.

5.7.2 Medium Term: Cooperative Systems (Level 3)

5.7.2.1 Expectations   In addition to measures that focus on driver assistance systems, science also points to the potential of cooperative systems—in conjunction with traffic management—through connected navigation. Nowadays, much research is being executed and many of the European research projects are funded by the EU. The first pilot projects have now been realized, and it is expected that these cooperative systems will lead to less congestion and better use of the road network. The European Commission will propose short-term technical specifications required to exchange data and information between vehicles (V2V) and Vehicle to Infrastructure (V2I) (European Commission, 2010). This proposed standardization should become a push for the further implementation of these systems. "Platooning" requires a cooperative driving electronics standardization, so that different car brands can click into the "platoon." Despite the expected positive contribution of cooperative systems—in contrast to driver assistance systems—little research has been done on the safety and possible side effects of cooperative systems. It will take some years before V2V and V2I communication is

---

* https://ideas.repec.org/a/eee/transa/v41y2007i3p267-279.html.

safe and reliable enough to be used in cooperative driving, starting with long-distance haulage on highways.

5.7.2.2 Social, Ethical, and Regulatory Issues   It is high time that parties (governments, industry, research institutes, and interest groups) considered the technical and legal aspects of cooperative driving, such as legal authorization, system security, standardization, and liability in case of malfunction. These aspects require time to be dealt with and, after all, would needlessly slow the introduction of cooperative systems in a few years' time if no moves are made now.

Some of the most pressing problems are going to arise in the short term in the field of privacy. Increasing the enforcement of road traffic regulations can be easily accomplished via V2I systems that can monitor a driver's behavior, allowing the owner to be fined automatically for violations of traffic rules. In addition, insurance firms may introduce premiums for drivers who drive safely and use monitoring. This is still considered an invasion of privacy, but the question is whether in the long term traffic safety concerns will take priority over privacy concerns. Therefore, the remaining question is whether politicians can keep their promise that the eCall system will remain "dormant" and that this system will not be used as an electronic data recorder for tracking criminals or for fining drivers who violate traffic regulations. This danger is real, as shown by the example in Section 5.4.2 of the CCTV cameras in Amsterdam that were only supposed to be keeping an eye on polluting lorries, and were installed for monitoring only those, but that were subsequently also used for other purposes.

5.7.3 Long Term: Autonomous Cars (Level 4)

There are many challenges ahead before the autonomous car can be introduced into the commercial market: improving the reliability of the cars and addressing daunting legal and liability issues, and so on.

5.7.3.1 Expectations   The development of cooperative systems will contribute to the further implementation of autonomous driving. Autonomous driving may have to be applied on highways with cooperative ACC, for which V2V communication will be necessary. The infrastructure will not need to change much, because drivers can already get information about local traffic regulations, traffic congestion, roadwork, and the like via the navigation system or any other information source. Perhaps roadside systems could be placed on the road to guide autonomous driving, especially on highway ramps. This semiautonomous driving allows autonomous driving of the car on certain roads with noncomplex traffic situations, such as highways, but not in places with more complex traffic situations, such as cities. Scientists consider that this will be realistic by about 2020 (Visbeek & Van Renswouw, 2008, see also the SATRE Project). The expected result of this semiautonomous driving is that road safety on highways will increase, that traffic congestion will be partly mitigated, especially in shock wave traffic congestion, and that cars will become more fuel efficient. During autonomous driving, the driver can read a book, use the Internet, have breakfast, and so on.

Fully autonomous driving will not be a realistic picture before 2020, even though this is predicted by several car manufacturers. For example, Nissan promised in 2013 to deliver the first "commercially viable" self-driving cars by 2020.* Given the development in the field of car robotics, it seems inevitable that the autonomous car will become commonplace, but a more likely estimate is that these systems will function around 2030. Significant technical obstacles must be overcome before the autonomous car can safely drive on public roads. Operating an autonomous car on public roads is more complex than flying an airplane, because there are more and closer interactions with often unpredictable objects, such as nonautonomous cars, pedestrians, cyclists, animals, trash, and potholes (Litman, 2014), and developers have to solve the weather-related problems. A natural move toward the introduction of fully autonomous vehicles will be the launch of short-range vehicles that provide local mobility at low speeds and in relatively controlled environments. Such an approach is planned by Milton Keynes, a British city in which 100 electric autonomous vehicles will be installed between 2015 and 2017 to run between the city's

---

* http://nissannews.com/en-US/nissan/usa/releases/nissan-announces-unprecedented-autonomous-drive-benchmarks.

central train station, shopping center and office parks. The autonomous pods will carry two passengers, plus shopping bags, luggage, or a baby stroller, and will travel up to 20 km/h (or 12 mph) in dedicated lanes inside the city. Passengers will pay £2 (or U.S. $3) per trip and summon their rides through a smartphone app.*

According to a study conducted by the automotive industry consultant IHS Automotive (2014), autonomous cars that include driver control are expected to hit highways around the globe before 2025 and self-driving "only" cars (with no human control) are anticipated around 2030. In this study, IHS Automotive forecasts that total worldwide sales of autonomous cars will grow from nearly 230,000 in 2025 (less than 1% of the cars expected to be sold globally that year) to 11.8 million in 2035 (9% of the global automotive sales expected that year). In all, there should be nearly 54 million autonomous cars in use globally by 2035. The study anticipates that nearly all of the vehicles in use are likely to be autonomous sometime after 2050. The premium for the autonomous car technology will add U.S. $7000–$10,000 to a car's sticker price in 2025, a figure that will drop to around $5000 in 2030 and about $3000 in 2035 when no driver controls are available. Most of these sales will be in well-established automotive markets such as the United States, Western Europe, and Japan.

5.7.3.2 Social, Ethical, and Regulatory Issues The study by the automotive industry consultant IHS Automotive also notes some potential barriers to autonomous car deployment and two major technology risks: software reliability and cyber security. The barriers include implementation of a legal framework for self-driving cars and the establishment of government rules and regulations.

The social impact of the introduction of the autonomous car could be very significant. Visions of the future of the autonomous car now lead to different scenarios, sometimes even diametrically opposed ones. The benefits of autonomous cars could be smaller and their costs greater than the optimistic scenarios. Therefore, policy makers and politicians must anticipate these possible scenarios. What exactly will the implications be for public transportation, privacy,

---

* http://www.wired.com/2013/11/milton-keynes-autonomous-pods/.

security, car ownership, road safety and road use, and so on? This should be investigated for each scenario, so that policy makers and politicians can design the road of the future and discourage undesirable developments promptly.

The autonomous car will force regulators to rethink the car and driver relationship and will possibly place more emphasis on the regulation of the car than of the driver. For instance, instead of certifying that a driver can pass a road test, the state might certify that a car can pass a test, upending the traditional drivers' licensing system. Questions also arise about liability for accidents, since the technology that makes an autonomous car is an after-market system. So if it hits something, does the fault lie with the manufacturer or the technology company that modified the car?

Furthermore, the introduction of autonomous cars would become a job-eliminating technology (Rotman, 2013). With the introduction of autonomous cars, individuals whose income depends on driving, such as taxi drivers or long-distance lorry drivers, may see reduced opportunities for employment.

Finally, cars crash, even autonomous ones. A major challenge is to effectively encode complex ethical reasoning into software so that it can make decisions, in particular in relation to complex driving situations. Countries beginning to pass legislation concerning autonomous driving should consider not only the technical pre-crash behavior of autonomous cars but also the method of ethical decision making in life-and-death choices, including issues such as discrimination and justice. This raises the question of whether we want to leave decisions about life and death to a machine. So, the question is whether we even want to have autonomous cars, since they will make choices that have moral consequences. The answer first depends on the technological possibility of building moral machines. This will be difficult, since an autonomous car, for example, should provide in advance potential responses to all kinds of emergency situations. If this attempt does succeed, the successful introduction of the moral autonomous car would then depend on whether we are able to deal with very difficult ethical questions, such as which ethical algorithm should be implemented in an autonomous car and who should make this decision?

Interview with Bryant Walker Smith (Assistant Professor
of Law, University of South Carolina, United States)

"Automated cars are much more likely to yield to pedestrians."
Working on the interface of law and technology, lawyer-cum-engi-
neer Bryant Walker Smith has written extensively about automated
driving.

"Increasing automation of vehicles is now a reality. What cars can
do now is nothing like what they could do 5 or 10 years ago, and I
expect that development to accelerate. But the holy grail of a car that
will drive you from your home to work while you sleep, or that can
pick up groceries for you—that car is probably further away than most
people think. Only in geographically restricted areas such as cam-
puses will that be possible any time soon."

How safe should autonomous cars become so as to be acceptable?

"That's a social question. Law, technology and cost-benefit analysis
all have bearing on it, but ultimately it's social. Most of us are far
more comfortable with the status quo, in which many people die in
human-driven vehicles, than with the idea of robot-driven vehicles
crashing and killing people. The pressure on such autonomous vehicles to perform well will be very high, and the backlash of a weird
crash, particularly one that a human could have prevented, might be
significant."

*If autonomous cars acquire at some point an unquestionable safety advan-
tage, is it conceivable that human driving will be banned?*

"What is considered safe and reasonable will naturally evolve along
with these technologies. You can't take a horse and cart on the highway today, and certainly we can reach a point where human-driven
vehicles are unacceptable in some or all situations, or are confined to
certain areas. In the meantime, human driving will be increasingly
assisted and monitored by technology."

*While autonomous cars may become safer under normal conditions, there is
the risk of their systems being hacked.*

"Yes, that's a risk of all information technology, including—it has
been shown—our cars today. We need to realize that automation
will never be perfect. Therefore it's important that we design forms
of graceful degradation, ways of limiting the danger and the potential

for harm, regardless of whether the cause of the trouble is a hack or a more direct failure. I'll give a stylized example. Say we design an automated vehicle so that it will never accelerate rapidly on its own, not even to avoid a crisis situation. On the one hand, this means that every now and again it fails to avoid a crash by going really fast, and somebody will get injured. But it also means that if there is a bug or a hack, it still won't accelerate rapidly. It would have a hard block on the system to prevent this greater danger."

What would the autonomous car mean for the safety of other, more vulnerable road users?

"Automated vehicles could completely change the dynamics between pedestrians and motor vehicles. Today, many people hesitate to cross a street because they're not sure, and rightly so, that the driver is going to see them and stop and yield to the right of way. An automated vehicle is much more likely to see you, perhaps signal that and then actually yield. We could have a much more pedestrian-friendly downtown."

The motor car as we know it has completely transformed our urban planning. Could the autonomous car do that all over again?

"It could. What we see historically is that people will move to the extent that their commuting time remains reasonable. They are willing to spend some 20 minutes of their time unproductively and under stressful conditions, driving into work, if that means they can afford big houses and good schools. If automation makes driving time more productive and if highway flows and highway capacity would increase, as seems likely in the long run, this might result—if left unchecked—in more remote suburbs, more urban sprawl.

The biggest positive urban design potential of autonomous vehicles is probably in parking. Many of our cities now have huge parking lots in central business districts. Automation would allow a more efficient allocation of vehicles and vehicle space, through robotic taxis, transportation on demand, and so on, which would reduce the need for parking spaces.

More broadly, highways are getting smarter, through sensors, monitoring, and so on. The integration of these things with the vehicle fleet could really change the way that roads function, even if at first they don't look any different."

Potential advantages of the autonomous cars include a reduction in crashes and in pollution. But a strong increase in traffic could undo those effects. What can regulators do to make sure society reaps the potential benefits of autonomous cars and avoids the negatives?

"I think the key for a good transportation policy would be a two-step: internalize the costs of driving and allow them to be charged on the basis of 'pay as you drive.' This could be done in several ways: with fuel taxes, travel taxes or more factorability for insurance, so that the insurance company can charge based on the distance driven, the time of day, the type of road or even the type of weather.

Internalizing costs does several things. First, it would make people more rational about their travel behavior. If driving 40 miles is more expensive, maybe you won't do it every day. In itself, automation could reduce some of the other costs of driving, so it would make driving more attractive. That effect will be reduced if costs are kept high, for good policy reasons. Second, internalizing costs will make automation potentially even more attractive compared to human driving. So, if automated driving is safer and you're paying for your insurance by the trip or by the kilometer [or mile], then automated driving is going to be the cheaper option. Likewise, if automated driving is more fuel-efficient, then you're going to be more attracted by it if you're actually paying for the true cost of your fuel."

You have done research into the legal aspects of automated driving, and concluded that in most American states autonomous vehicles are probably legal. But Europe and the USA are bound by different international conventions on road traffic. Does Europe face more of a legal obstacle here?

"I don't see it as an obstacle at all. Unlike the Geneva Convention, which applies in the United States, the Vienna Convention can fairly easily be modified. Moreover, these technologies are going to develop somewhat independently from these perceived legal barriers. At some point, the argument will be, 'States follow laws and therefore, if the Vienna Convention prohibited these vehicles we wouldn't have them, and since we do the Vienna Convention must be consistent with them.' It will introduce some conceptual challenges. Adapting national laws to these vehicles is a challenge. Understanding obligations of the human driver, which is something the Vienna Convention does touch on, is a source of some uncertainty, but not insurmountably so."

Another legal aspect is liability. Who can be held responsible when an autonomous car causes a crash?

"Companies are frequently liable for products they have sold and for injuries that these have caused, and yet they continue to sell them. The legal perspective is that, if automated vehicles operate well, not perfectly but well, then companies are going to be liable for a greater share of those crash injuries, but their absolute number will be lower.

It has been suggested that even when people are not operating the vehicle at the time of an accident, they might still be liable. This is not as odd as it may sound. In many states, when I lend you my car and you crash it and injure somebody, I am still liable for those injuries. As a result, I get insurance that covers not just the injuries that I directly cause, but the injuries that are caused with my car. So with the use of insurance, it's just another way of passing along costs.

I think it's important to realize that everything about the world in which we have automated vehicles is going to be different. It's not simply a case of extrapolating vehicle development and holding the rest of our world the same. We're going to have a host of other technologies that will be relevant. So we *might* send our driverless car to the store to get a toothbrush, but then, we might have it delivered by drone or we might print it in our 3D printer at home, or there is something else entirely that we cannot even conceive at this point. Likewise, our values are going to be different. Our expectations for safety will be higher. Expectations for privacy could look wholly different than they do now, either higher or lower. So that's a real challenge when we think about the role of automated vehicles in the future: it will be in a future that we don't really understand yet."

References

Anderson, J. M., Kalra, N., Karlyn, D., Sorensen, P., Samaras, C., & Oluwatola, O. (2014). *Autonomous vehicle technology. A guide for policy-makers.* Santa Monica, CA: RAND Corporation.

Arth, M. (2010). *Democracy and the common wealth: Breaking the stranglehold of the special interests.* DeLand, FL: Golden Apples Media.

Broggi, A., Zelinsky, A., Parent, M., & Thorpe, Ch. E. (2008). Intelligent vehicles. In B. Siciliano, & O. Khatib (Eds.), *Springer handbook of robotics* (pp. 1175–1198). Berlin, Germany: Springer.

Burns, L. D. (2013). Sustainable mobility: A vision of our transport future. *Nature, 497*(7448), 181–182.

Burns, L. D., Jordan, W. C., & Scarborough, B. A. (2013). *Transforming personal mobility.* New York: Earth Institute, Columbia University.

Checkoway, S., McCoy, D., Kantor, B., Anderson, D., Shacham, H., Savage, S., ..., Kohno, T. (2011). Comprehensive experimental analyses of automotive attack surfaces. In D. Wagner (Ed.), *Proceedings of the 20th USENIX on Security (SEC'11)* (pp. 77–91), San Francisco, CA, August 10–12. Berkeley, CA: USENIX Association.

Dewar, R. E., & Olson, P. L. (2007). *Human factors in traffic safety* (2nd ed.). Tucson, AZ: Lawyers & Judges Publishing Company.

Donner, E., & Schollinski, H. L. (2004). *Deliverable D1, ADAS: Market introduction scenarios and proper realisation. Response 2—Advanced driver assistance systems: From introduction scenarios towards a code of practice for development and testing.* Contract Number: ST 2001-37528. Köln, Germany.

Dragutinovic, N., Brookhuis, K. A., Hagenzieker, M. P., & Marchau, V. A. W. J. (2005). Behavioural effects of advanced cruise control use. A meta-analytic approach. *European Journal of Transport and Infrastructure Research, 5*(4), 267–280.

eCall Driving Group. (2006). *Recommendations of the DG eCall for the introduction of the pan-European eCall.* Brussels, Belgium: eSafety Support.

Edara, P., Sun, C., Keller, C., & Hou, Y. (2012). *Evaluating the benefits of dynamic message signs on Missouri's rural corridors* (prepared for Missouri Department of Transportation). Columbia, MO: University of Missouri—Columbia.

Erke, A. (2008). Effects of electronic stability control (ESC) on accidents: A review of empirical evidence. *Accident Analysis and Prevention, 40*(1), 167–173.

European Commission. (2010). *Towards a European road safety area: Policy orientations on road safety 2011–2020* (SEC(2010) 903). Brussels, Belgium: European Commission. http://ec.europa.eu/transport/road_safety/pdf/com_20072010_en.pdf (accessed September 4, 2014).

European Commission. (2013). *The European Union explained: Transport.* Brussels, Belgium: European Commission. http://europa.eu/pol/trans/index_en.htm (accessed September 4, 2014).

European Commission. (2014). *Road safety vademecum. Road safety trends, statistics and challenges in the EU 2010–2013.* Brussels, Belgium: European Commission. http://ec.europa.eu/transport/road_safety/pdf/vademecum_2014.pdf (accessed September 4, 2014).

European Environment Agency. (2012). *Towards low carbon transport in Europe.* Copenhagen, Denmark: European Environment Agency.

Fagnant, D. J., & Kockelman, K. M. (2013). *Preparing a nation for autonomous vehicles. Opportunities, barriers and policy recommendations.* Washington, DC: Eno Center for Transportation.

Feenstra, P. J., Hogema, J. H., & Vonk, T. (2008). Traffic safety effects on navigation systems. *IEEE intelligent vehicles symposium* (pp. 1203–1208), Eindhoven, the Netherlands, June 4–6. Washington, DC: IEEE.

Ferguson, S. A. (2007). The effectiveness of electronic stability control in reducing real-world crashes: A literature review. *Traffic Injury Prevention*, *8*(4), 329–338.

Geddes, N. B. (1940). *Magic motorways*. New York: Random House.

Glancy, D. J. (2012). Privacy in autonomous vehicles. *Santa Clara Law Review*, *52*(4), 1171–1239.

Goodall, N. J. (2014a). Ethical decision making during automated vehicle crashes. *Transportation Research Record: Journal of the Transportation Research Board*, *2424*, 58–65.

Goodall, N. J. (2014b). Machine ethics and automated vehicles. In G. Meyer & S. Beiker (Eds.), *Road vehicle automation. Lecture notes in mobility* (pp. 93–102). Berlin, Germany: Springer.

Gusikhin, O., Filev, D., & Rychtyckyj, N. (2008). Intelligent vehicle systems: Applications and new trends. *Informatics in Control Automation and Robotic. Lecture Notes in Electrical Engineering*, *15*, 3–14.

HEI Panel on the Health Effects of Traffic-Related Air Pollution (2010). *Traffic-related air pollution: A critical review of the literature on emissions, exposure, and health effects* (HEI special report). Boston, MA: Health Effects Institute.

Hevelke, A., & Nida-Rümelin, J. (2014). Responsibility for crashes of autonomous vehicles: An ethical analysis. *Science and Engineering Ethics*. http://link.springer.com/article/10.1007/s11948-014-9565-5 (accessed November 16, 2014).

Husak, D. (2001). Vehicles and crashes: Why is this moral issue overlooked? *Social Theory and Practice*, *30*(3), 351–370.

IHS Automotive. (2014). *Autonomous cars—Not if, but when*. http://orfe.princeton.edu/~alaink/SmartDrivingCars/PDFs/IHS%20_EmergingTechnologies_AutonomousCars.pdf (accessed October 26, 2014).

Jamson, A. H., Merat, N., Carsten, O. M. J., & Lai, F. C. H. (2013). Behavioural changes in drivers experiences highly-automated vehicle control in varying traffic conditions. *Transportation Research Part C: Emerging Technologies*, *30*, 116–125.

Jost, G., Allsop, R., & Steriu, M. (2013). *Back on track to reach the EU 2020 road safety target?* (7th road safety PIN report). Brussels, Belgium: European Transport Safety Council. http://archive.etsc.eu/documents/PIN_Annual_report_2013_web.pdf (accessed August 1, 2014).

Kim, K. M., Heled, Y., Asher, I., & Thompson, M. (2014). *Comparative analysis of laws on autonomous vehicles in the U.S. and Europe*. http://www.auvsishow.org/auvsi2014/Custom/Handout/Speaker8657_Session789_1.pdf (accessed August 25, 2014).

Knight, W. (2013). Driverless cars are further away than you think. *MIT Technology Review*, *116*(6), 44–47.

KPMG. (2013). *Self-driving cars: Are we ready?* http://www.kpmg.com/us/en/issuesandinsights/articlespublications/pages/self-driving-cars-are-we-ready.aspx (accessed August 25, 2014).

KPMG and the Center for Automative Research. (2012). *Self-driving cars: The next revolution.* http://www.kpmg.com/US/en/IssuesAndInsights/ self-driving-cars-next-revolution.pdf (accessed August 25, 2014).

LeValley, D. (2014). Autonomous vehicle liability—Application of common carrier liability. *Seattle University Law Review Online, 36*(5), 5–26.

Levy, F., & Murname, R. (2004). *The new division of labor.* New York: Russell Sage Foundation.

Lin, P. (2013, October 8). The ethics of autonomous cars. *The Atlantic.* http:// www.theatlantic.com/technology/archive/2013/10/the-ethics-of- autonomous-cars/280360/ (accessed September 3, 2014).

Lin, P. (2014a, August 18). Here's a terrible idea: Robots cars with adjustable ethics setting. *Wired.* http://www.wired.com/2014/08/heres-a-terrible- idea-robot-cars-with-adjustable-ethics-settings/ (accessed September 3, 2014).

Lin, P. (2014b, May 6). The robot car of tomorrow may just be programmed to hit you. *Wired.* http://www.wired.com/2014/05/the-robot-car-of- tomorrow-might-just-be-programmed-to-hit-you/ (accessed September 3, 2014).

Litman, T. (2014). *Autonomous vehicle implementation predictions: Implications for transport planning.* http://www.vtpi.org/avip.pdf (accessed September 4, 2014).

Llaneras, R. E., Salinger, J., & Green, C. A. (2013). Human factors issues associated with limited ability autonomous driving systems: Drivers' allocation of visual attention to the forward roadway. *Proceedings of the seventh international driving symposium on human factors in driver assessment, training and vehicle design* (pp. 92–98). http://drivingassessment. uiowa.edu/2013/proceedings (accessed March 17, 2014).

Lu, X.-Y., & Shladover, S. E. (2014). Automated truck platoon control and field test. In G. Meyer, & S. Beiker (Eds.), *Road vehicle automation. Lecture notes in mobility* (pp. 247–261). Berlin, Germany: Springer.

Mackay, M. (2005). Quirks of mass accident databases. *Traffic Injury Prevention, 6*(4), 308–310.

Marchant, G. E., & Lindor, R. A. (2012). The coming collision between autonomous vehicles and the liability system. *Santa Clara Law Review, 52*(4), 1321–1340.

Mahmassani, H. S. (2011). *Impact of information on traveler decisions.* Washington, DC: TRB.

Mogos, S., Wang, R., & DiClemente, J. (2014). *Autonomous car policy report.* http://www.epp.cmu.edu/Spring2014Reports/Autonomous%20 Car%20Final%20Report.pdf (accessed January 23, 2015).

Morgan Stanley. (2013). *Autonomous cars. Self-driving the new auto industry paradigm.* New York: Morgan Stanley.

Morsink, P., Goldenbeld, Ch., Dragutinovic, N., Marchau, V., Walta, L., & Brookhuis, K. (2007). *Speed support through the intelligent vehicle* (R-2006-25). Leidschendam, the Netherlands: SWOV.

National Highway Traffic Safety Administration. (2008). *Traffic safety facts 2008. A compilation of motor vehicle crash data from the fatality analysis reporting system and the general estimates system.* Washington, DC: U.S. Department of Transportation.

National Highway Traffic Safety Administration. (2013). *Preliminary statement of policy concerning automated vehicles.* http://www.nhtsa.gov/About+NHTSA/ Press+Releases/U.S.+Department+of+Transportation+Releases+Policy+on +Automated+Vehicle+Development (accessed October 25, 2014).

Pawsey, A., & Nath, C. (2013, September). Autonomous road vehicles. *POSTnote, 443,* 1–4. http://www.parliament.uk/briefing-papers/post-pn-443.pdf (accessed August 25, 2014).

Rijkswaterstaat. (2007). *De rij-assistant. Wegen naar de toekomst.* Delft, the Netherlands: Rijkswaterstaat.

Rotman, D. (2013, June 12). How technology is destroying jobs. *MIT Technology Review, 116*(4), 28–35.

Royal Society for the Prevention of Accidents. (2011). *Inappropriate speed.* http://www.rospa.com/road-safety/advice/drivers/speed/inappropriate/ (accessed October 5, 2014).

Schonberger, B., & Gutmann, S. (2013, June 4). A self-driving future. At the intersection of driverless cars and car sharing. *Sightline Daily.* http://daily.sightline.org/2013/06/04/a-self-driving-future/ (accessed November 16, 2014).

Schrank, D., Eisele, B., & Lomax, T. (2012). *TTI's 2012 urban mobility report.* College Station, TX: Texas A&M Transportation Institute.

Schwab, A. (2009). Type-approval requirements for the general safety of motor vehicles I. (P6_TA-PROV(2009)0092). *Texts adopted at the sitting of Tuesday 10 March 2009, P6_TA-PROV(2009)03-10* (pp. 41–76). Brussels, Belgium: European Parliament.

Smith, B. W. (2012). *Driving at perfection.* http://cyberlaw.stanford.edu/ blog/2012/03/driving-perfection (accessed October 14, 2014).

Smith, B. W. (2014). Automated vehicles are probably legal in the United States. *Texas A&M Law Review, 411*(1), 411–521.

Spieser, K., Treleaven, K., Zhang, R., Frazzoli, E., Morton, D., & Pavone, M. (2014). Toward a systematic approach to the design and evaluation of automated mobility-on-demand systems: A case study in Singapore. In G. Meyer & S. Beiker (Eds.), *Road vehicle automation. Lecture notes in mobility* (pp. 229–246). Berlin, Germany: Springer.

Underwood, S. E., Ervin, R. D., & Chen, K. (1989). *The future of intelligent vehicle-highway systems: A Delphi forecast of markets and sociotechnological determinants.* Ann Arbor, MI: University of Michigan, Transportation Research Institute.

Urmson, C., & Whittaker, W. R. (2008). Self-driving cars and the urban challenge. *IEEE Intelligent Systems, 23*(2), 66–68.

Van Driel, C. J. G., Hoedemaeker, M., & Van Arem, B. (2007). Impacts of a congestion assistant on driving behaviour and acceptance using a driving simulator. *Transportation Research Part F, 10*(2), 139–152.

Van Driel, C. J. G., & Van Arem, B. (2010). Using traffic flow simulation to investigate the impact of an active accelerator pedal and a stop-and-go system on traffic flow efficiency and safety in congested traffic caused by a lane drop. *Journal of Intelligent Transportation Systems*, *14*(4), 197–208.

Van Lint, J. W. C., Valkenberg, A. J., Binsbergen, A. J., & Bigazzi, A. (2010). Advanced traffic monitoring for sustainable traffic management: Experiences and results of five years of collaborative research in the Netherlands. *Intelligent Transport Systems*, *4*(4), 387–400.

Van't Hof, C., Van Est, R., & Kolman, S. (2011). Networked cars. In C. Van't Hof, R. Van Est, & F. Daemen (Eds.), *Check in/check out: The public space as an internet of things* (pp. 51–65). Rotterdam, the Netherlands: NAi Publishers.

Virtanen, N., Schirokoff, A., Luoma, J., & Kulmala, R. (2006). *Impacts of an automatic emergency call system on accident consequences.* Ministry of Transport and Communications Finland, Finnish R&D Programme on Real-Time Transport Information AINO. http://www.ecall.fi/eCall_Safety_Effects_Finland_Summary_final_06.pdf (accessed October 14, 2014).

Visbeek, M., & Van Renswouw, C. C. M. (2008). *C,mm,n. Your mobility, our future.* Enschede, the Netherlands: University of Twente.

Vlassenroot, S. H. M. (2011). *The acceptability of in-vehicle intelligent speed assistance (ISA) systems: From trial support to public support.* Delft, the Netherlands: TRAIL Research School.

Wetmore, J. M. (2003). Driving the dream. The history and motivations behind 60 years of automated highway systems in America. *Automotive History Review* (Summer), 4–19.

World Health Organization. (2013). *Global status report on road safety 2013: Supporting a decade of action.* Geneva, Switzerland: World Health Organization.

6

ARMED MILITARY DRONES

The Ethics behind Various Degrees of Autonomy

6.1 Focus on Teleoperated and Autonomous Armed Military Robots

Military robot technology has gained momentum. All over the world, military robots are currently being developed and thousands of military robots are already deployed during military operations. According to Peter Singer, this development forms the new "revolution in military affairs" (Singer, 2009b). The U.S. *Future Combat Systems*, a program for future weapons and communications systems costing in excess of U.S. $200 billion—commissioned by the Pentagon—has had a major impact. Military robots are a focal point in this program, and many industrialized countries have become inspired.

What stands out in many defense policy documents on military robots, such as the *Unmanned Systems Roadmap 2009–2034* (U.S. Department of Defense, 2009), is that scant attention is paid to the social and ethical aspects of military robotics. As is often the case with new technological innovations, the emphasis is on promoting technology itself, not on reflection on possible, unwanted side effects. The developers are not inclined to anticipate any negative aspects, probably from a fear that this would hamper further developments.

The use of military robots and, in particular, armed military robots will, however, entail a large number of ethical issues. Now that military robots are increasingly used in military operations and we seem to be heading for *the* next military revolution, many scientific publications have appeared that discuss the possible negative consequences and the ethical aspects of military robots (see, e.g., the special issue on armed military robots of the journal *Ethics and Information Technology* [*14*(2) (2013)], and the edited book *Killing by Remote Control: The Ethics of an Unmanned Military* (Strawser, 2013)). These publications show that *armed* robots present us with very different social and ethical questions compared to *unarmed* military robots. In order to study the ethical and social consequences of military robots (see Section 6.5), it is important to distinguish between them in this chapter.

6.1.1 Unarmed Military Robots

Unarmed military robots have an added value because they perform dirty, dangerous, and dull tasks to solve operational problems and for the effective and efficient performance of tasks. Examples include "around-the-clock" performance, detecting roadside bombs, conducting reconnaissance sorties, guarding military compounds or camps, improving situational awareness, and even carrying out potentially offensive tasks such as raiding a hostile building, where a robot forces open a door and explores who may be hiding inside. In particular, these tasks should contribute to increasing the safety of the military staff. Nowadays, these operational bottlenecks mainly occur in peace and reconstruction missions, when the armed forces deal with insurgents carrying out asymmetric operations, such as the use of roadside bombs and urban warfighting. In urban warfighting, it is difficult to distinguish the fighters from citizens, with the result that civilians are inextricably part of the battle space and at the same time the insurgents are becoming protected from the threats and dangers of war (Johnson, 2013). Robotics technology may offer a possible solution in response to these operations by insurgents, since robots, such as unmanned aerial vehicles (UAVs), the *Micro Air Vehicle* or the *Global Hawk* (see Box 6.1), can provide military personnel with crucial information at a distance, information that facilitates better decisions that are more consistent with the principles of proportionality and discrimination.

Unarmed military robots provide a positive contribution to the completion of soldiers' tasks. Few objections can be made against

BOX 6.1   UNARMED UNMANNED AERIAL VEHICLES

An example of an unarmed UAV is the unmanned reconnaissance helicopter, the Micro Air Vehicle. It is a remote-controlled propeller plane as small as a model airplane, with a weight of about 20 grams (or 0.7 ounce) to a few hundred grams and equipped with powerful regular or infrared cameras for autonomous observation tasks. The cameras' images are so sharp that people planting parcel bombs or roadside bombs can be detected and monitored, alerting local forces to act. Also, these aircraft can search for targets and communicate the position for conventional bombing. At the end of 2001, the United States deployed about 10 unmanned reconnaissance aircraft in Afghanistan, but by 2008 this number had already grown to more than 7000 (Singer, 2009b). In addition to these small aircraft, there is the reconnaissance *Global Hawk*, with a wingspan of nearly 40 meters (or 130 feet) and a maximum altitude of almost 20 kilometers (or 13 miles). The *Global Hawk* can survey large geographic areas with pinpoint accuracy, giving military decision makers real-time information regarding enemy locations, resources, and personnel.*

---

* http://www.northropgrumman.com/Capabilities/RQ4Block20GlobalHawk/Documents/HALE_Factsheet.pdf.

deploying these robots. A possible objection could be that the privacy of citizens would be affected (see also Chapter 5).

6.1.2 *Armed Military Robots*

In contrast, armed military robots raise many urgent and significant social and ethical issues (see, e.g., Krishnan, 2009; Sharkey, 2008b). We, therefore, focus on the ethical and social issues that are raised in the deployment and development of armed military robots, especially armed military drones, or unmanned combat aerial vehicles (UCAVs). The United States expects a great deal from UCAVs for the coming years considering the huge investment that has been made in this type of robots compared to other armed robots.* Armed military robots raise three broad questions:

1. Are armed military robots in breach of humanitarian war law? (Section 6.3)
2. Who can be held responsible when an autonomous armed robot is involved in an act of violence? (Section 6.4)
3. Does the proliferation of armed military robots cause irresponsible risks? (Section 6.5)

The answers to these questions often depend on the degree of the armed military robot's autonomy. In military robotics, there is a process in which the degree of autonomy of an unmanned vehicle is gradually being extended from limited autonomy by teleoperation to fully independent or autonomous functioning systems and, in the future, even to self-learning systems (see Section 6.2). This shift is phrased by Sharkey (2010) as "from in-the-loop to on-the-loop to out-of-the-loop." We distinguish three types of military robots with regard to their degree of autonomy: tele-guided, autonomous, and self-learning. This chapter focuses on teleoperated and autonomous armed military robots and especially on drones. The debate over the deployment and development of drones is a very timely one, since drones are linked to existing questions of the appropriateness, legitimacy, and potential illegality of the drone strikes. Furthermore, the United States is the major player in the development of military robots, and more than 90% of its budget for the development of military robots for the coming 5 years (2014–2018) has been made available for the development of drones. For the United States, drones have become one of the key elements of military operations abroad, since drones deliver precision strikes without the need for more intrusive military action. Finally, the question about the level of autonomy for military robots is most relevant for drones, since drones suffer the least from the frame problem because they fly, and the sky is not a very complex environment, in contrast to a mountainous environment for ground robots (see Section 1.3.4). As a consequence, drones could be a fully independent or autonomous functioning system in the near future.

Most military robots are still remotely controlled by human operators and thus are tele-guided systems. Based on input from sensors,

---

* http://robotenomics.com/2014/01/07/us-military-to-spend-23-9-billion-on-drones-and-unmanned-systems/.

these robots are programmed to perform or not to perform certain tasks, with an operator controlling activities to varying degrees. Nonautonomous robots require humans to authorize any decision to use lethal force, that is, they require a "man-in-the-loop."

When an armed military robot performs tasks and decisions completely independently that relate to whether to destroy military targets, thus without human intervention, we speak of autonomous systems. These autonomous systems have explicit task programming and act according to a certain fixed algorithm. This means that the acts of the autonomous military robot are predictable and can be traced afterward.

Learning military robots, based on neural networks, genetic algorithms, and agent architecture, are able to decide on a course of action and to act without human intervention. The rules by which they act are not fixed during the production process, but can be changed during the operation of the robot by the robot itself (Matthias, 2004). The problem with these robots is that there will be a class of actions in relation to which no one is capable of predicting the future behavior of these robots. So these robots would become a "black box" for difficult moral decisions, preventing any second-guessing of their decisions. The control is then transferred to the robot itself, but it is nonsensical to hold the robot responsible at that moment, since robots that will be built in the next two decades will not possess anything like intentionality or a real capability for agency. The deployment of armed learning military robots would constitute a responsibility gap (Matthias, 2004), since it would constitute the injustice of holding people responsible for the actions of robots over which they could not have any control.* Although learning armed military robots appear high on the U.S. military agenda (Sharkey, 2008a), the deployment of these robots is, at least under present and near-term research developments, not likely to happen within the next two decades (Arkin, 2009a).† We will not discuss this type of military robot in this chapter, because they will not be developed within the coming decades, and statements about these learning robots would be very speculative.

---

* For a discussion of possible mechanisms and principles for the assignment of moral responsibility to autonomous learning (intelligent) robots, we refer to Hellström (2013) and Sparrow (2007).
† Barring some significant breakthrough in artificial intelligence research, situational awareness cannot be incorporated in software for lethal military robots (Fitzsimonds & Mahnken, 2007; Gulam, 2006; Kenyon, 2006; Sharkey, 2008a; Sparrow, 2007).

6.2 Autonomy of Military Robots Is High on the Agenda

In recent years, many countries have started to invest in military robot technology. The United States still tops every other country with its budget of approximately U.S. $4.8 billion/year for military robots (mainly on drones, approximately $4.4 billion/year) for the coming 5 years (2014–2018)* and is the undisputed leader in the field of military robotics. The rest of the world is, however, not far behind. At least 76 countries have entered into the extensive development of UAV technology, including Russia, China, Pakistan, and India. China is reported to have at least 25 separate drone systems currently in development (Zenko, 2012). In about 2012, there were 680 drone programs in the world, an increase of over 400 since 2005 (Zenko, 2012). In this section, we take a look at the development history of military robots and see what future developments we may expect.

Initially, research into military robotics mainly took place in the United States, and in 1989, it received a major boost. In that year, as a result of Congressional pressure, research on unmanned ground vehicles of the various armed services was merged into the *Joint Robotics Program*. In addition to being ordered to establish a master plan for 1990, U.S. $22 million was set aside for robot research, development, testing, and evaluation. The U.S. budget for 2001 stated that in 2006 the U.S. military must have at least 4000 unmanned ground vehicles in operation, that by 2010 a third of all operational long-distance aircraft must be unmanned, and that by 2015 a third of all ground troop vehicles must also be unmanned (Committee on Armed Services, House of Representatives, 2000).

Based on these objectives, the U.S. military agency *Defense Advanced Research Projects Agency* (DARPA) launched a study to determine which future weapons and communications systems would be used by the U.S. military. This study is known as the *Future Combat Systems* program—costing U.S. $230 billion, it is the most expensive defense program ever—in which autonomous military robots form an important part (Agence France-Presse, 2008). So far, it looks as though all of the objectives formulated in the 2001 U.S. budget have been met, showing a sharp increase in the number of combat missions carried out using unmanned systems. Since 2009, the U.S. Air Force has been training more "pilots"

---

* http://robotenomics.com/2014/01/07/us-military-to-spend-23-9-billion-on-drones-and-unmanned-systems/.

(operators) for the unmanned armed aircraft Predator and Reaper (see Box 6.2) than combat and bomber pilots (Vanden Brook, 2009). In 2012, 2300 operators were on active service (Fitzsimmons & Sangha, 2013).

BOX 6.2 UNMANNED COMBAT AERIAL VEHICLES: THE PREDATOR AND THE REAPER

The two most famous UCAVs are the *Predator* and the *Reaper* (see Figure 6.1). The Predator can fire Hellfire* missiles and was widely used in Afghanistan. The Predator has a wingspan of 16.8 meters, is 7.6 meters (or 55.1 feet) long and 2.1 meters (or 6.9 feet) high, carries a payload of up to 204 kilograms (or 450 pounds) to an altitude of 7.6 kilometers (or 4.7 miles), and has a maximum speed of 217 km/h (or 135 mph).[†] It is able to visually penetrate a cloud with day and night cameras. Following the success of the Predator in combat, the defense contractor General Atomics anticipated the U.S. Air Force's desire for an upgraded UCAV by redesigning the Predator so that it had increased ability. Its successor

Figure 6.1 The Reaper. (Photo courtesy of Corbis/Hollandse Hoogte.)

---

[*] The Hellfire missile is a universal laser-guided missile system that can be placed on helicopters and on aircraft, naval ships, and land vehicles and which serves to disable enemy targets. A Hellfire missile is 163 centimeters (or 64 inches) long, weighs 45 kilograms (or 100 lb), and can reach a speed of 1530 km/h (or 950 mph). The Hellfire has a range of 8 kilometers (or 5 miles).

[†] http://www.af.mil/AboutUs/FactSheets/Display/tabid/224/Article/104469/mq-1b-predator.aspx.

was called the Reaper. The Reaper has a wingspan of 20 meters (or 66 feet), is 11 meters (or 36 feet) long and 3.8 meters (or 12.5 feet) high, carries a payload of up to 1700 kilograms (or 3750 pounds) to an altitude of 15 kilometers (or 9 miles), and has a maximum speed of 483 km/h (or 300 mph).* The Reaper was reported at first being used in Afghanistan in 2008. These UCAVs are operated by a pilot and a sensor operator located at a ground control station in Nevada. As of September 2013, there were 156 Predators and 104 Reapers in the U.S. Air Force inventory.† Some countries in Europe, such as Germany, France, Italy, the Netherlands, and the United Kingdom, have purchased or are considering purchasing some U.S. Reapers.

---

* http://www.af.mil/AboutUs/FactSheets/Display/tabid/224/Article/104470/mq-9-reaper.aspx.
† http://www.bga-aeroweb.com/Defense/MQ-1-Predator-MQ-9-Reaper.html.

Since 2006, the U.S. Predators and Reapers have clocked up more than 1 million flight hours and have flown 80,000 missions, of which 85% were in combat missions (Jennings, 2010).

The U.S. vision of further robotization assumes a linear technical development in which the degree of autonomy of an unmanned vehicle is gradually extended from limited autonomy by teleoperation to fully independent or autonomous functioning systems (U.S. Department of Defense, 2005). Although it is unclear what degree of autonomy these UAVs will have, "the eventual deployment of systems with ever increasing autonomy is inevitable" (Arkin, 2009a). In the U.S. vision, 2010 is seen as a kind of turning point when autonomy outstripped teleoperation. It is predicted that in 2020 robots will function autonomously (see Figure 6.2). This prediction will not come true, because it is clear that military robots will still not be able to operate autonomously in the coming years. The U.S. military now assumes that in 2047 the first autonomous armed aircraft will be deployed (United States Air Force, 2009).

Thus, the autonomy of military robots is high on the agenda of the U.S. Armed Forces. The U.S. independent advisory policy body, the National Research Council, states accordingly: "The Navy and Marine Corps should aggressively exploit the considerable warfighting benefits offered by autonomous vehicles" (National Research Council, 2005).

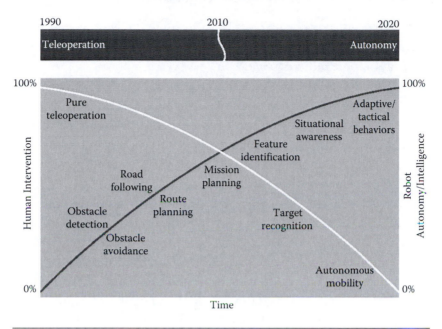

Figure 6.2   The robotics revolution. (Data from U.S. Department of Defense, *Joint robotics program master plan FY2005. Out front in harm's way,* OUSD (AT&L) Defense Systems, Washington, DC, 2005; Picture made by Textcetera commissioned by Boom Lemma uitgevers.)

In 2012, the U.S. Department of Defense issued a directive on autonomy in weapon systems (U.S. Department of Defense, 2012), and in 2013, the Department of Defense states in the *Unmanned Systems Integrated Roadmap FY2013–2038* that "[r]esearch and development in automation are advancing from a state of automatic systems requiring human control toward a state of autonomous systems able to make decisions and react without human interaction."*

The wish to deploy autonomous robots is mainly driven by the fact that tele-guided robots are more expensive, first because the production costs of tele-guided robots are higher and second because these robots incur personnel costs as they need human flight support. One of the main goals of the *Future Combat Systems* program, therefore, is to deploy military robots as "force multipliers" so that one military operator can run a multiple, large-scale robot attack (Sharkey, 2008b). To this end, robots are programmed to cooperate in swarms so that they can run coordinated missions. In 2003, the United States deployed the first test

---

* http://www.defense.gov/pubs/DOD-USRM-2013.pdf.

with 120 small reconnaissance planes in mutually coordinated flight (Krishnan, 2009). This swarm technology is developing rapidly and will probably become military practice in a few years.

In order to stimulate research into robotics, DARPA also initiated a competition in 2002 for autonomous robots, called The *DARPA Grand Challenge* (see also Chapter 4). This race was to yield useful information and technological innovations for research into new military robotics applications. In Europe, this has been copied on a much smaller scale within the *European Land Robotic Trial* (ELROB).

Besides this technology push, there is a demand pull for military applications. The call of U.S. society to reduce the number of military casualties has contributed to a huge boost for alternatives in robotics developments in the United States.* A few years ago, the number of U.S. soldiers killed in action rose to a high level because of insurgents operating in Iraq and Afghanistan using their popular homemade and deadly weapon: the improvised explosive device (IED), or the roadside bomb. Forty percent of the U.S. soldiers who were killed, died because of these IEDs (Iraq Coalition Casualty Count, 2008). During the invasion of Iraq in 2003, no use was made of robots, as conventional weapons were thought to yield enough "shock and awe." However, the thousands of U.S. soldiers and Iraqi civilians killed reduced popular support for the invasion and made the deployment of military robots desirable. According to Strawser (2010), in certain circumstances the use of armed military robots is not only desirable, but is even ethically mandatory under what he calls the "principle of unnecessary risk." From this principle, Strawser infers that it is morally reprehensible to command a soldier to run the risk of fatal injury if the task he or she is ordered to carry out could also have been carried out by a military robot.[†] By the end of 2008, there were 12,000 ground robots operating in Iraq, mostly being used to defuse roadside bombs, and 7000 reconnaissance planes or drones were deployed (Singer, 2009b).

---

* Former Senator John Glenn once coined the term "Dover Test": whether the public still supports a war is measured by responses to returning body bags. He called it the "Dover Test" because the coffins of killed U.S. soldiers came in from abroad to the air base in Dover, Delaware.

† The "principle of unnecessary risk" is disputed by several authors (see, e.g., Galliott, 2012; Steinhoff, 2013).

6.3 Military Robots and International Humanitarian Law

Before armed military robots can be used, they should be assessed against the principles of international law. In international law relating to warfare, it is common to distinguish between *jus ad bellum* and *jus in bello*. *Jus ad bellum* are the international rules pertaining to what extent the use of military force against another state is allowed. *Jus in bello* (or international humanitarian law) are the international rules pertaining to how armed conflict must be conducted, and thus, for example, how drones are deployed (see Box 6.3). With respect to *jus ad bellum*, it is debatable whether military robots will raise or lower the threshold for starting a war (Asaro, 2008; Kahn, 2002; Sharkey, 2008b). With regard to the deployment of both teleoperated and autonomous armed drones, a more elaborated debate has developed over the last few years. For this reason, we focus on the use of armed drones in the context of war and look at the extent to which they can be applied according to the principles of proportionality and discrimination. We will look first at tele-led armed drones and then at autonomous drones.

BOX 6.3 *Jus in Bello*

The law in war (*jus in bello*) governs the following aspects:

- *Proportionality*: The applied force must be proportional to the benefit gained. Civilian casualties are acceptable as collateral damage if this is proportionate to the military advantage.
- *Discrimination*: When selecting a target, a distinction must be made between combatants and noncombatants and between civilian objects, for example, hospitals and churches, and military objectives. Soldiers who are injured or have surrendered should cease to be targets.
- *Doctrine of double effect*: If an action has multiple effects, the bad effects, though foreseen, are outweighed by the good effects.

6.3.1 Tele-Led Drones

Tele-led armed military robots that are presently active may make a positive contribution to the principles of *jus in bello*. Drones, with their accurate perception achieved by target recognition sensors, are precise in hitting their target, so that unintended collateral damage is limited. However, we will show that these principles come under pressure with the deployment of tele-led drones and that targeted killing by tele-led drones may not compliance with international humanitarian law.

6.3.1.1 Proportionality Principle   The principle of "proportionality" may become compromised by the use of tele-led armed drones. Predators and Reapers can very accurately determine the target and relay the target's global positioning system (GPS) coordinates and images to the operator. Because of this precision, robotics can significantly reduce collateral damage. Using armed military robots can also allow greater risks to be taken than with manned systems. During the Kosovo War, for the sake of the pilots' safety, NATO aircraft flew not lower than 15,000 feet so that hostile fire could not touch them. During one specific air raid from this elevation, North Atlantic Treaty Organization (NATO) aircraft bombed a convoy of buses filled with refugees, although they thought they were hitting Serbian tanks.* This type of tragic mistake could be prevented by deploying unmanned aircraft that are able to fly lower because they are equipped with advanced sensors and cameras, allowing the operator to better identify the target. A possible consequence of this is that because of these advantages, armed drones will be deployed much sooner than, for example, a B-52 bomber, which would probably kill many people and would cause too much collateral damage (Sharkey, 2010). Thus, using drones will result in an expansion of the number of targets, and because of their robotic precision, they will also hit more targets near civilian locations. This sharp increase in air strikes by unmanned aircraft could ultimately lead to more casualties than before.

The U.S. government rarely acknowledges civilian casualties, but there is sufficient evidence that U.S. drone strikes have injured and killed civilians. According to estimates made by, for example, Bergen

---

* http://www.nytimes.com/1999/04/16/world/crisis-balkans-admission-nato-admits-mistaken-bombing-civilians.html?pagewanted=all&src=pm.

and Tiedemann (2011), the U.S. air strikes in Pakistan using UAVs increased from 10 strikes between 2004 and 2007, resulting in 175 victims, including 40 civilians (and 87 unknown), to 136 strikes in 2010, with 883 casualties, of whom 20 were civilians (and 160 unknown).* According to the estimates of four independent organizations,[†] we can conclude that every year after 2010 the number of strikes declines, as does the number of causalities. After 2009, we also see a decline in the proportion of civilian casualties.

However, the proportionality principle not only requires that the number of civilians killed by drones is much smaller than the number of (suspected) enemy combatants killed. It also needs the anticipated military advantage of an operation to be greater than the foreseeable civilian harm. Plaw states that "as long as drone strikes are required to suppress al-Qaeda command and control in Pakistan, a plausible argument can be made that the scale of the threat that al-Qaeda poses to civilians and the support it can potentially extend to regional franchise operations conduces to the proportionality of Central Intelligence Agency (CIA) drone strikes" (Plaw, 2013, pp. 146–147). His analysis shows that violations by U.S. drones have not been widespread or systematic, and he concludes that in terms of proportionality, "drone strikes should at least provisionally be permitted more time to achieve their objectives and to prove how precise and effective they can become" (Plaw, 2013, p. 153).

Besides civilian casualties, U.S. drone strike policies cause considerable and under-accounted-for harm to the daily lives of ordinary civilians, beyond death and physical injury, according to the report *Living under Drones* by the Stanford Law School and NYU School of Law (2012).[‡] The presence of drones terrorizes civilians, giving rise to anxiety and psychological trauma among civilian communities: "Those living under drones have to face the constant worry that a deadly strike may be fired at any moment" (Stanford Law School & NYU School of Law, 2012, p. vii). Furthermore, the report expresses

---

* umassdrone.org/stats.php?country=Pakistan.
† New America Foundation (natsec.newamerica.net/drones/pakistan/analysis), the *Long War Journal* (www.longwarjournal.org/pakistan-strikes.php), the University of Massachusetts (umassdrone.org/stats.php?country=Pakistan), and the Bureau of Investigate Journalism (www.thebureauinvestigates.com/).
‡ See also Amnesty International (2013).

serious concerns about the efficacy and counterproductive nature of drone strikes. Only 2% of the total casualties by drone strikes are "high-level" targets; the remaining 98% are lower-ranking targets, only some of whom are engaged in direct hostilities against the United States, and civilians (Brooks, 2012). Furthermore, the drone strikes lead to a loss of support for striving for peace: 74% of Pakistanis call the United States an enemy.* Instead of winning the hearts and minds of the civilian communities, drone strikes may create a breeding ground for terrorists. Baitullah Mehsud, the Pashtun commander of the Pakistani Taliban, claimed that for every citizen killed, 3–4 suicide terrorists emerged, being mostly members of victims' families (Ghosh & Thompson, 2009). Mehsud was killed in August 2009 by a U.S. drone. After this attack, his successor Hakimullah Mehsud sent a suicide bomber to a CIA base: seven CIA employees were killed. In October 2013, Hakimullah Mehsud gave an interview to the BBC in which he said he would be open to peace talks but that drone strikes must be brought to an end as a condition of any ceasefire. Hakimullah Mehsud was killed in November 2013 by a U.S. drone.†

The drone strikes may have a boomerang effect. Because of these drone strikes, the insurgents do not have enemy combatants within range of their conventional weaponry, which restricts their retaliatory options to an extent that drives them to utilize deplorable means they would not otherwise seek to use (Galliott, 2012). In addition, the available dual-use robotic technologies and the ability of skilled technicians working with radical movements will contribute to armed drones becoming part of their arsenal. U.S. researcher and former fighter pilot Mary Cummings outlines a doomsday scenario of a small, unmanned aircraft that deploys a biological weapon above a sports stadium in a terrorist action.‡

6.3.1.2 Discrimination Principle The principle of "discrimination" may also become compromised by the use of tele-led armed drones and the practice of geolocating. According to the top-secret NSA

---

* http://www.pewglobal.org/2012/06/27/pakistani-public-opinion-ever-more-critical-of-u-s/.
† http://www.theguardian.com/world/2013/nov/03/hakimullah-mehsud.
‡ http://harvardnsj.org/wp-content/uploads/2010/03/20100324_Forum_Cummings.pdf.

(National Security Agency) documents provided by whistle-blower Edward Snowden, it turns out that the NSA geolocates the SIM card or handset of a suspected terrorist's mobile phone by using a specially constructed device attached to the drone. This signals intelligence (SIGINT) enables the CIA or U.S. Air Force to conduct drone strikes to kill the individual in possession of the mobile phone. According to a former drone operator, about 90% of the drone strikes in Afghanistan relied on geolocation without knowing whether the individual in possession of a tracked mobile phone was in fact the intended target. This has the consequence that, for example, family members and friends who borrow a tracked mobile phone can be mistakenly targeted. As an operator put it:

> People get hung up that there's a targeted list of people. It's really like we're targeting a cell phone. We're not going after people – we're going after their phones, in the hopes that the person on the other end of that missile is the bad guy. ...But we don't know who's behind it, who's holding it. It's of course assumed that the phone belongs to a human being who is nefarious and considered an 'unlawful enemy combatant'. This is where it gets very shady.*

So, by geolocating, the distinction between combatants and noncombatants is unclear, since this distinction is based on who is in possession of a tracked mobile phone rather than the knowledge that this person is a combatant.

6.3.1.3 Targeted Killing The United States sees the drone as *the* weapon of choice in the fight against terrorism, and the association of such targeted killing using drones is clearly embedded in the public consciousness. Targeted killing is defined as the intentional killing of a specific civilian or unlawful combatant who cannot reasonably be apprehended and who is taking a direct part in hostilities; the targeting done at the direction of the state in the context of an international or non-international armed conflict (Solis, 2010, p. 538). There is growing criticism of these drone attacks on (alleged) terrorist suspects, because according to international law it is unlawful to

---

* https://firstlook.org/theintercept/2014/02/10/the-nsas-secret-role/.

authorize the killing of a citizen without a trial, and because there is doubt over whether transnational terrorism, which lacks a specific connection to an ongoing conflict, does in fact constitute an "armed conflict." The targeted killing by drones is justified on grounds of the so-called imminence of threat; however, according to Brooks (2012), many casualties pose no direct or imminent threats, but rather speculative ones, such as individuals who might 1 day attack the United States or its interests abroad.

Special attention given to extrajudicial executions was due to the death of the U.S. Muslim preacher Anwar al-Awlaki in Yemen in 2011, the first known time a U.S. drone strike deliberately targeted and killed a U.S. citizen, with the President's intent but without Anwar al-Awlaki having been charged formally with a crime or convicted at trial. Anwar al-Awlaki was suspected of conspiring to commit international terrorist acts.* The co-passenger in the car of al-Awlaki that was struck, the U.S. citizen Samir Khan, was not even on the "kill" list and was apparently in the wrong place at the wrong time when the attack began.† Two weeks later, the 16-year-old son of Anwar al-Awlaki was accidentally killed by a U.S. drone in Yemen.‡ Khan (2011) argues that these killings of suspected terrorists by drone attacks in Pakistan cannot be justified on moral grounds, because these attacks do not discriminate between terrorists and innocent women, children and the elderly. This tactical move of using the drones is counterproductive and is "unwittingly helping terrorists." He states that the U.S.'s international counterterrorism efforts can only be successful if a clear strategy is devised: adopting transparent, legitimate procedures with the help of Pakistan to bring the culprits to book and to achieve long-term results. More scholars (Alley, 2013; Camillero, 2013; Whetham, 2013) hold the opinion that the United States probably carries out illegal targeted killings in Pakistan, Yemen, Somalia, and elsewhere.

---

* In June 2014, a federal court released a legal memo that justified the drone killing of alleged U.S. terrorist Anwar al-Awlaki, after years of legal wrangling (http://www.nytimes.com/2014/06/24/us/justice-department-found-it-lawful-to-target-anwar-al-awlaki.html).

† http://www.nydailynews.com/news/national/anwar-al-awlaki-samir-khan-dead-al-qaeda-propagandists-killed-u-s-missile-strikes-yemen-article-1.958584.

‡ http://www.huffingtonpost.com/2013/04/23/obama-anwar-al-awlaki-son_n_3141688.html.

The government must be held to account when it carries out such killings in violation of the Constitution and international law.

The United States does not give access to information about the nature of the threat that the dead person represented, how imminent the threat was or how the "kill" lists are created, and so on. Former legal adviser John Bellinger who served under President George W. Bush called these drone attacks by the United States "Obama's Guantanamo Bay": killing citizens without any form of due process. Several authors and organizations have called for transparency from the U.S. government to avoid the lack of disclosure giving the United States a virtual and impermissible "licence to kill": "without disclosure of the legal rationale as well as the bases for the selection of specific targets (consistent with genuine security needs), States are operating in an 'accountability vacuum'" (Alston, 2010, p. 27). It is crucial that the United States consistently and strictly adheres to the existing international legal framework for the deployment of drones and provides the utmost clarity about the legal rationale that is involved in the use of armed drones before establishing a precedent that other nations may follow.

6.3.2 *Autonomous Drones*

On the basis of the discrimination principle, some scientists argue that autonomous armed military robots are not allowed to operate in a military force because they are not sufficiently able—without human intervention—to distinguish citizens from soldiers, and thus, the clear requisite definition of combatant is missing (Asaro, 2008; Gulam & Lee, 2006; Sharkey, 2008b; Sparrow, 2005). The problem, according to these scientists, is that no robot has the necessary competence to be able to make this distinction independently. The sensors, such as cameras, sonar, lasers, temperature-sensitive devices, etc., may indicate that something is a man, but a distinction between a civilian and a combatant cannot be made. This would require both situational awareness and empathy, that is, having the ability to understand the intentions of another human being and judge what someone is likely to do in a given situation. These qualities are all the more important given the fact that, increasingly, "asymmetric" and nontraditional warfare is carried out by a conventional army using modern technology against insurgents. Typically, the insurgents—consisting of irregular, often undisciplined, fighters led by warlords—are often not recognizable as combatants and perform operations that do not comply with the rules of international humanitarian law, such as planting roadside bombs.

Thus, making this distinction between civilians and military personnel is highly context dependent (Fielding, 2006). For example, a person who is recognized by a military robot as a combatant should cease to be a target if he intends to surrender or is seriously injured. It is, therefore, concluded by many scientists that unless there are very remarkable breakthroughs in the research on artificial intelligence, human intervention, or having a human-in-the-loop, will remain necessary to continue to meet the criteria of international humanitarian law (see also Fitzsimonds & Mahnken, 2007; Gulam & Lee, 2006; Kenyon, 2006; Sparrow, 2009). The same applies to the two principles "proportionality" and "doctrine of double effect." No algorithm exists today to determine whether the damage is proportional to the benefit gained. This, according to these scientists, also requires situational awareness and human judgment.

A solution to circumvent these so-called frame problems relating to autonomous armed military robots is to considerably reduce the possibility that they could attack civilians. One way to do this is to clearly define the boundaries within which a robot is allowed to kill. For example, by deploying them only in a particular war zone, the so-called kill boxes, which are shielded from citizens, citizens are clearly forewarned not to enter that kill area. An example of this application is the demilitarized zone between North and South Korea, where autonomous robots can shoot intruders (see Box 6.4). With this type of approach, the use of armed military robots becomes limited to places where civilians are not allowed. A different approach is taken by Canning (2006). He suggests simplifying the frame problem by programming robots so that they will only attack enemy fighters when they wear hostile weapon systems. However, in many circumstances this may lead to problems, such as with local tribesmen in Afghanistan, who usually carry a weapon while going about their daily activities and therefore are difficult to distinguish from armed enemy insurgents. An attack on an enemy's weapon system may then be disproportionate because of collateral damage—that is, civilian casualties.

BOX 6.4   THE DEMILITARIZED ZONE

To monitor the 250 kilometer (or 155 mile) long and 4 kilometer (or 2.5 mile) wide demilitarized zone between North and South Korea, South Korea uses the armed robot *SGR-A1*. The SGR-A1 is a stationary system equipped with machine guns developed by Samsung Techwin. These robots can autonomously detect intruders within a radius of 3.2 kilometers (or 2 miles) and shoot them, thus acting without human intervention. At the end of 2011, the first SGR-A1 was placed at a security post as a kind of test. In 2014, more SGR-A1s were installed to replace some of the thousands of patrolling soldiers at the South Korea border.*

---

* http://www.dailymail.co.uk/sciencetech/article-2756847/Who-goes-Samsung-reveals-robot-sentry-set-eye-North-Korea.html.

The fact that autonomous military robots cannot perfectly make the distinctions necessary to fulfill the principles of discrimination and proportionality, U.S. robot engineer Arkin (2010) argues, cannot be a good enough reason to ban these robots from appearing on the battlefield. Arkin predicts future possibilities for autonomous robots that become sufficiently reliable because of better target recognition algorithms to distinguish civilians from combatants. According to Arkin (2009a), the abilities of robot soldiers will surpass those of human ones in the future: "It is not my belief that an unmanned system will be able to be perfectly ethical in the battlefield, but I am convinced that they can perform more ethically than human soldiers are capable of." Robots will be the first to penetrate through the "fog of war" both literally, since the eventual development and use of a broad range of robotic sensors means that they will be better equipped for battlefield observations than human sensory abilities, and figuratively, because they have no feelings of fear or revenge and do not need to protect themselves when there is uncertainty about target identification. People make mistakes in military operations, especially if they are exposed to enormous stress.*

---

* At war soldiers are exposed to enormous stress, with all of its consequences. This is evident from a report that includes staggering figures on the morale of soldiers during the military operation Iraqi Freedom (Office of the Surgeon Multinational Force-Iraq & Office of the Surgeon General United States Army Medical Command, 2006).

Driven by this faith in the rationalization of warfare, Arkin (2009b) investigates the introduction of an adequate artificial ethical conscience into the armed military robot, which would program international humanitarian law and the rules of engagement into the robot. An advantage of these robots is that they "have the potential capability of independently and objectively monitoring ethical behaviour in the battlefield by all parties and reporting infractions that might be observed. This presence alone might possibly lead to a reduction in human ethical infractions" (Arkin, 2009a). The idea of the development of autonomous machines with an additional ethical dimension is quite a new emerging field of machine ethics. These robots, with the help of formal logic, are "able to calculate the best action in ethical dilemmas using ethical principles" (Anderson & Anderson, 2007). It is thus assumed that it is sufficient to represent an ethical theory in terms of a logical theory and to deduce the consequences of that theory. Beavers (2012) sees machine ethics as a threat to *ethical nihilism,* "the doctrine that states that morality needs no internal sanctions, that ethics can get by without moral 'weight,' that is, without some type of psychological force that restrains the satisfaction of our desire and that makes us care about our moral condition in the first place" (Beavers, 2012, p. 343).

This view that ethics can be made computable also misunderstands the unique—nonreducible—nature of ethics. Arkin (2009a) agrees that some ethical theories, such as virtue ethics, do not lend themselves well by definition to a model based on a strict ethical code. But he claims that the solution is simply to eliminate ethical approaches that refuse such reduction. However, the hallmark of ethics is its nonreducibility (Royakkers & Topolski, 2014). While many ethical situations may be reducible, it is the ability to act ethically in situations that call for a judgment that is distinctly human. A consequence of this approach is that ethical principles themselves will be modified to suit the needs of a technological imperative: "Technology perpetually threatens to coopt ethics. Efficient means tend to become ends in themselves by means of the 'technological imperative' in which it becomes perceived as morally permissible to use a tool merely because we have it" (Kaag & Kaufman, 2009, p. 604).

Another problem concerning the attempt to program the rules of the law of war and engagement into robots is that these rules are subject to interpretation and they do not provide a ready answer in any given situation. Moreover, the application of these rules is context dependent and requires the ability to understand complex social situations (Asaro, 2009). This ability is necessary in order to make ethical decisions, and as long as robots lack this ability, they will certainly not be able to take better ethical decisions than humans (Sullins, 2010).

Last but not least, Arkin does not address this essential moral question: should we relinquish the decision to kill a human to a nonhuman machine? (Johnson & Axinn, 2013). That a lethal decision should be made only by a human and not a machine was one of the main reasons that in 2009 four leading scientists (Noel Sharkey, Peter Asaro, Robert Sparrow, and Jürgen Altmann) established the *International Committee for Robot Arms Control*, ICRAC. The committee wants to restrict the use of armed robots for military purposes as much as possible (Altmann, Asaro, Sharkey, & Sparrow, 2013). One of the objectives is a ban on the development and deployment of autonomous armed robots, because the committee thinks that machines must never make decisions about life and death. In 2010 in Berlin, the committee organized the expert workshop *Limiting Armed Tele-Operated and Autonomous Systems* and invited scientists, politicians, and military delegates. Following the workshop, a statement was made—signed by a large majority of those present—which once again stressed the need to reduce armed military robots, and it was emphasized that a ban ought to be imposed on autonomous armed military robots because a human must always take life-and-death decisions. Part of the statement read, "it is unacceptable for machines to control, determine, or decide upon the application of force or violence in conflict or war. In all cases where such a decision must be made, at least one human being must be held personally responsible and legally accountable for the decision and its foreseeable consequences."* Robots can never relieve humans of their own responsibility for ethical decisions in time of war. In the words of the Dutch Major General (retd) Kees Homan (2011), "[T]he whole idea of automated killing is perverse. An action so serious in its consequences should not be left to mindless machines. Machines will never absolve mankind from its responsibility to make ethical decisions in peace and war." The United Nations, via the Acting

---

* For the statement see www.icrac.co.uk/Expert%20Workshop%20Statement.pdf.

Director-General of the United Nations Office, has also expressed its concerns about autonomous armed robots at an informal meeting of experts hosted by the UN Convention on Certain Conventional Weapons: "You have the opportunity to take pre-emptive action and ensure that the ultimate discussion to end life remains firmly under human control."* In November 2014, the United Nations organized a conference on this topic, and all of the 117 countries that took part in it had as their main issue a ban on or restricted use of armed autonomous robots.

6.4 Question of Responsibility

Lucas (2013, pp. 225–226) accuses Arkin of being needlessly provocative in citing the potential capacity for robots to behave "more ethically":

> they [autonomous robots] cannot, nor, in the final analysis, do we need to have them, 'behave ethically.' That is certainly asking for much more than we can currently deliver, and probably much more that we really require of them. We wish, instead, for military robots to be *safe and reliable* in their function, and to perform their assigned missions effectively, including following instructions that comply with the laws of armed conflict.

But whom should we hold responsible for how these robots act, or when a robot is involved in an act of violence that would normally be described as a war crime? Is it the designers or programmers, the manufacturer, the commanding officer, the Department of Defense, or the robot itself? Sparrow (2007) was one of the first authors to discuss this question, and he claims that no one can reasonably be held responsible for autonomous robots' behavior. Sparrow, however, uses a very strong definition of autonomy: intelligent robots that can make decisions independently with no more supervision than a human soldier would receive. Since it will (if possible) take several decades before such a robot will be deployed on the battlefield, we will discuss the question about responsibility for semiautonomous robots that are

---

* http://www.un.org/apps/news/story.asp?NewsID=47794#.VDkb8xZAphY.

tele-controlled or supervised for the following three actors: designers/ manufacturers, human operators, and commanding officers.

6.4.1 Responsibility of the Manufacturers

Casualties among citizens or a country's own military personnel can be the result of a malfunctioning military robot or of a programming error. There have already been instances of such malfunctioning: in October 2007 in South Africa, a malfunctioning robotic gun fired 500 high explosive rounds, killing 9 soldiers and seriously injuring 14 others (Shachtman, 2007). Although partly a result of their flying low and slowly, the relatively high loss rate of drones in military operations is perhaps also an indication of their unreliability. For example, some years ago the U.S. Air Force reported that it had lost 50% of its then 90 Predators (Jordan, 2007). The "race to the market," imposed by the U.S. Department of Defense, also increases the risk of faulty design or defective programming of military robots. These risks are often underestimated by policy makers and scientists in favor of other interests: "Public policy needs to recognize these dangers but to address them in a manner that does not unduly hold back research that could bring dramatic new capabilities to the market place and further national security" (Carafano & Gudgel, 2007, p. 6).

One could argue that designers have an (at least moral) obligation to make products that are safe and reliable and satisfy legal and ethical requirements. The stringent requirements for military weapons can be found in, for example, international humanitarian law (including the principle of discrimination, the principle of proportionality, and the principle of superfluous injury or unnecessary suffering). As it stands, there is, regarding the responsibility of designers of military robots, no lack of relevant legal and ethical concepts we can turn to, with liability seeming to be the most appropriate one (see also Lucas, 2011). A framework of liability would stimulate manufacturers to take their responsibility seriously (Van de Poel & Royakkers, 2011).

In military practice, however, designers are rarely held responsible for accidents caused by poor design (Thompson, 2007). It is, therefore, important that there is a regulatory framework that entails that "reasonable care" should be taken by designers of armed military robots.

Applying "reasonable care" could, for example, imply that the potential failures of military robots should be properly assessed before being introduced into the armed forces on a large scale. These assessments, mostly done through simulations and experiments, or small-scale field tests, however, do not always provide complete and reliable knowledge about the functioning of technological products and the potential hazards and risks involved. For example, laboratory and field tests are not always representative of the dynamic and complex circumstances in which military robots have to function. Evidently, it is difficult to capture the situations that occur in the "fog of war" in a test setting.

Nevertheless, it seems that the regulation of risk assessment can provide us with a regulatory framework for the design of military robots by formulating a set of boundary conditions for the design, production, and use of military robot technologies. If such a regulatory framework meets certain criteria, it could be considered an adequate way of dealing with the ethical issues that the design of a technology raises with regard to safety and responsibility. Such a regulatory framework for risk management should be laid down in specific norms, such as the ISO 10218 requirements for industrial robots. This norm "specifies requirements and guidelines for the inherent safe design, protective measures, and information for use of industrial robots. It describes basic hazards associated with robots, and provides requirements to eliminate or adequately reduce the risks associated with these hazards" (ISO, 2011). Although this ISO norm was developed for industrial robots, it can be used outside that context: "Examples of non-industrial robot applications include, but are not limited to: undersea, military and space robots" (ISO, 2011). Until now, however, such a regulatory framework specifically for military robots has been lacking (see also Lin, Bekey, & Abney, 2008). International legal safety guidelines can prevent neglecting the security aspect as a result of increasing pressure on the development and implementation of armed military robots. Furthermore, the potential use of military robots to save the lives of "friendly" human beings possibly explains why most of today's governments refrain from passing regulations with respect to the design of military robots, especially regulations that could lead to a ban on certain designs of military robots (Royakkers & Olsthoorn, 2014).

6.4.2 Responsibility of the Human Operators

Human drone operators, "the cubicle warriors," often work from their control center in Nevada, thousands of kilometers away from the actual battlefield (see Figure 6.3). Drones connect human operators with the war zone; they are the eyes—and sometimes hands—of the tele-soldiers. Based on the information shown on their computer screen, the tele-soldiers decide whether or not to engage a target. The things they see on screen can emotionally and psychologically affect these human operators and can thus influence their decision making. Although fighting from behind a computer is not as emotionally potent as being on the battlefield, killing from a distance remains stressful; various studies have reported physical and emotional fatigue and increased tensions in the private lives of military personnel operating the Predators in Iraq and Afghanistan (Donnelly, 2005; Kaplan, 2006; Lee, 2012). For example, a drone pilot may witness war crimes which he is unable to prevent, or he may even see how his own actions kill civilians.

The ensuing "residual stress" of human operators has led to proposals to diminish these tensions. In particular, the visual interface can play an important role in reducing stress; interfaces that only show abstract and indirect images of the battlefield will probably cause less stress than the more advanced real images (Singer, 2009a). From a technical perspective, this proposal is a feasible one, since it would not be hard to digitally recode the war scene in such a way that it induces less

Figure 6.3 The cubicle warrior. (Photo courtesy of Polaris/Hollandse Hoogte.)

psychological discomfort for the war operator. This cure may have some unwanted side effects though. Showing abstract images would in fact dehumanize the enemy, and as a result could desensitize military personnel operating unmanned systems. The danger of this development is that operators take decisions about life and death as if they are playing a video game (Sparrow, 2009), as evidenced by the words of an operator: "It's like a video game. It can get a little bloodthirsty. But it's fucking cool" (Singer, 2009b, pp. 308–309). The question this raises is whether operators can be held morally responsible for their decisions. Moral responsibility implies that they must have complete control over their behavior. That is, they fully know the consequences of their decisions and they take their decisions voluntarily (Fischer & Ravizza, 1998). This requires ethical reflection. Interfaces that only indirectly show abstract images of the enemy and military targets cause the operator to become less than fully aware of the consequences of his or her decisions. So in order to prevent stress, the ethical reflection of the operator is reduced or even eliminated (Royakkers & van Est, 2010). In short, these types of interfaces tend to dehumanize the operator.

Coeckelbergh (2013) shows that surveillance technologies can also enable a kind of "empathic bridging" between the operator and potential targets and thus can weaken the danger of dehumanization. Operators usually intensively watch potential targets, undertaking a wide range of normal human activities, including eating, smoking, and interacting with friends and family, before eventually attempting to kill them. By zooming in on the potential targets and watching what they are doing and what happens to them when they are bombed, the operator gains "a certain intimacy" (Bumiller, 2012): "I see mothers with children, I see fathers with children, I see fathers with mothers, I see kids playing soccer" (an operator quoted in Bumiller, 2012). As we said earlier, this, in turn, can lead to very stressful situations, since it is not so easy to kill a target who has become more of a person to the operator or whose image the operator can recall (see also Fitzsimmons & Sangha, 2013). In a survey of 900 drone crew members conducted by the U.S. Air Force in 2010 and 2011, 46% of drone pilots on active duty reported high levels of stress and 29% reported emotional exhaustion or burnout.*

---

* https://forums.gunbroker.com/topic.asp?TOPIC_ID=554910.

As Coeckelbergh (2013) concludes, more (empirical) research is needed to better understand drone fighting practices and the psychological experience of drone operators.

Another aspect is that decisions will be more and more mediated in the future by the armed military robot. Military robots are becoming more and more autonomous through artificial intelligence (AI) technology and often include "ethical governors" in which international humanitarian law has partially been programmed. Through ethical governors, the tele-led armed military robots may in the future correct the operator or may provide advice when the operator is deciding whether to use a weapon. Arkin's study (2009b) focuses on this. Ethical governors tell operators which type of ammunition should be used for the intended purpose and predict the resulting amount of damage. If the operator makes a decision that would result in too much collateral damage, the governor gives a warning and will, for example, block the bomber. The operator can override this, but he knows that there is a high probability that he is violating international humanitarian law and that he risks being tried for a war crime. The drawback of this technological mediation (see Verbeek, 2005) is that operators will no longer make moral choices but will simply exhibit influenced behavior—because subconsciously they will increasingly rely, and even over-rely, on the military robot (Cummings, 2006). As a consequence, a "moral buffer" may come into being between human operators and their actions, allowing human operators to tell themselves that it was the military robot that took the decision. This could blur the line between nonautonomous and autonomous systems, as the decision of a human operator is not the result of human deliberation, but is mainly determined or even enforced by a military robot. This would mean a shift from "controlling" to "supervising," and effectively the military robot would take autonomous decisions. This is provoked by the increase in the amount of information from different sources, which has to be integrated and then interpreted quickly in order to come to a decision. Military robots can do this more effectively and efficiently than people, for whom this is almost impossible. The operator will still be "on-the-loop" and will have the power to veto the system's firing actions.

According to Human Rights Watch (2012), "on-the-loop" will soon be rendered meaningless when the operator is given only a fraction of a second to make the veto decision, as is the case with several systems already in operation. This would imply that we could no longer hold a human operator reasonably responsible for his decisions, since it would not really be the operator taking the decisions but a military robot. This could have consequences for the question of responsibility in another way too. Detert, Treviño, and Sweitzer (2008) have argued that people who believe that they have little personal control in certain situations—such as those who carry out monitoring—are more likely to go along with rules, decisions, and situations even if they are unethical or have harmful effects.

In addition, Vallor (2013) argues that the shift from "in-the-loop" to "on-the-loop" leads to a dangerous *moral de-skilling* of the military profession. She states that an "expert supervisor of another's decision, in order to be worthy of the authority to override it, must have acquired expertise in making decisions of the very same or a similar kind" (Vallor, 2013, p. 483). The question then is "[H]ow would such a supervisor ever become qualified to make that judgement, in a professional setting where the decision under review is no longer regularly exercised by humans in the first place?" (Vallor, 2013, p. 483).

That the shift "from in-the-loop to on-the-loop to out-of-the-loop" has become an actual threat follows from geolocation, where drone strikes are dependent on electronic signal intelligence rather than on human intelligence and the decision about life and death does not actually need a man-in-the-loop or on-the-loop (see Section 6.3.1).

6.4.3 Responsibility of the Commanding Officer

Even if there is not a human operator directly controlling the robot, there is still a human agent that has decided whether or not to deploy this robot: "even if a system is fully autonomous, it does not mean that no humans are involved. Someone has to plan the operation, define the parameters, prescribe the rules of engagement, and deploy the system" (Quintana, 2008, p. 15). So, in the case of geolocation too, someone has to order that a drone is to be sent out to target a mobile phone by geolocation. The basis of this argument is the doctrine of command responsibility, and although this ancient doctrine is interpreted differently by different authors

(Garraway, 2009), it will usually cover the deployment of armed military robots. As Schulzke (2013, p. 215) puts it:

> Commanders should be held responsible for sending AWS [autonomous weapon systems] into combat with unjust or inadequately formulated ROE [rules of engagement], for failing to ensure that the weapons can be used safely, or for using AWS to fight in unjust conflicts, as all of these conditions that enable or constrain an AWS are controlled by the commanders.

For example, the possibility that an autonomous drone may engage the wrong targets could be an acknowledged limitation of the system. If the designers have made this clear to those who have purchased or deployed the system, then, Sparrow (2007) argues, they can no longer be held responsible should this occur; in that case, the responsibility should be assumed by the commander who (wilfully and knowingly) decided to send the drone into the battlefield despite its known limitations.

To conclude this section with respect to the question "who can be held responsible when an autonomous armed robot is involved in an act of violence?," we quote Krishnan (2009, p. 105): "the legal problems with regard to accountability might be smaller than some critics of military robots believe…. If the robot does not operate within the boundaries of its specified parameters, it is the manufacturer's fault. If the robot is used in circumstances that make its use illegal, then it is the commander's fault." That the future will include more autonomous systems seems almost a given, and although renouncing certain types of autonomous robots might be a good idea for many reasons, a lack of clarity as to who is responsible for their use is thus probably not among them (see also Kershnar, 2013). Whether one would want to have that responsibility is a different question altogether.

6.5 Proliferation and Security

The first signs of an international arms race in relation to military robotics technology are already visible. All over the world, significant amounts of money are being invested in the development of armed military robots. This is happening in countries such as the United States, Britain, Canada, China, South Korea, Russia, Israel, and Singapore. Proliferation to other countries, for example, by the transfer of robotics technology, materials, and knowledge, is almost inevitable. Many state and non-state actors that are hostile to the United States have also begun to enter the area of UAV technology. Iran has, for example, developed its own armed drone, called the *Ambassador of Death*, which has a range of up to 1000 kilometers (or 600 miles).* That drones are within the reach of many state and non-state actors is because, unlike other weapon systems, the research and development of armed military robots is fairly transparent and accessible. Furthermore, robotics technology is relatively easy to copy and the necessary equipment to make armed military robots can easily be bought and is not too expensive (Horton, 2009; Singer, 2009b) (see Box 6.5).

In addition, much of the robotics technology is in fact open-source technology and is a so-called dual-use technology; it is, thus, a technology that in future will potentially be geared toward applications in both the military and the civilian market. One threat is that in future certain commercial robotic devices, which can be bought on the open market, could be transformed relatively easily into robot weapons.

Chances are that unstable countries and terrorist organizations will deploy armed military robots. Singer (2009a) fears that armed military robots will become the ultimate weapon of struggle for ethnic rebels, fundamentalists, and terrorists. Noel Sharkey (2008c) also predicts that soon a robot will replace a suicide bomber. According to Sharkey (2008c), "the spirit [is] already out of the bottle." International regulations on the use of armed military robots will not solve this problem, as terrorists and insurgents disregard international humanitarian law.

An important tool to curb the proliferation of armed military robots is obviously controlling the production and purchase of these robots by implementing global arms control treaties. A major problem with this is that countries such as the United States and China are not parties to these treaties. In addition, legislation is needed in the field of the export of armed military robots in the UN

---

* http://www.dailymail.co.uk/news/article-1305221/Ahmadinejad-unveils-Irans-long-range-ambassador-death-bomber-drone.html.

BOX 6.5   U.S. DRONE IN IRANIAN HANDS

Toward the end of 2011, a U.S. *RQ-170 Sentinel* spy probe, an unmanned stealth aircraft, invisible to radar, which was kept secret until 2009, fell virtually intact into the hands of Iran (see Figure 6.4). Iran refused to give the device back and announced it was going to copy the stealth technology and software. Peter Singer said, "Flights from Moscow and Beijing to Tehran have I'm sure been full this week," knowing that with the help of Russia or China, which are very keen to replicate any U.S. technologies, it was possible to build an RQ-170 Sentinel. Moreover, the Pentagon wondered why the device had gone off-course. It was, therefore, speculated that the control of the RQ-170 Sentinel Iran had been disrupted with a cyberattack.

Figure 6.4   U.S. drone in Iranian hands. (Photo courtesy of Eyevine/Hollandse Hoogte.)

Source: From Blair, D. and Spillius, A., Iran shows off captured US drone, *The Telegraph*, December 8, 2011, http://www.telegraph.co.uk/news/worldnews/middleeast/iran/8944248/Iran-shows-off-captured-US-drone.html (accessed November 11, 2014).

framework to combat the illicit trafficking of armed military robots and to set up licenses for traders in armed military robotics technology. According to Boyle (2013), the ability of the United States to control the sale of drone technology will be diminished, because "it is only a matter of time before another supplier steps in to offer the drone technology to countries prohibited by export controls from buying U.S. drones" (Boyle, 2013, p. 23).

Another danger is that military robots will be hacked or may become infected by a virus. In October 2011, U.S. Predators and Reapers were infected by a mysterious virus (Shachtman, 2011). It logged every keystroke of the pilots in the control room on the base as they remotely flew drones on missions over Afghanistan and other battle zones. In 2012, U.S. researchers from the University of Texas repeatedly took control of a flying drone by hacking into its GPS system—acting on a thousand-dollar dare from the U.S. Department of Homeland Security.* The technique they used is called *spoofing*: a technique in which the drone mistakes the signal from hackers for the one sent from GPS satellites. In both cases, the situation was not immediately serious, but the danger of such viruses could become immense. Through hacking, others could take over unmanned combat aircraft or viruses could disrupt the drones in such a way that they become uncontrollable and could be hijacked and made to take unwanted actions, such as crashing into a building or turning on civilians.

6.6 Concluding Remarks

The robot as a technological development has a great influence on contemporary military operations, and this is seen as a new military revolution. Meanwhile, military robots have become part and parcel of the military context. Currently, tens of thousands of mostly unarmed military robots are operational. During the last decade, advances were made, especially in the development of armed military robots. From 2009, the U.S. Air Force has trained more operators—or cubicle warriors—than fighter pilots. The expectation is that unmanned aircraft will increasingly replace manned aircraft, and in the medium term will even make manned aircraft obsolete.

---

* http://www.bbc.com/news/technology-18643134.

A trend we are observing in military robotics development is a shift "from *in*-the-loop to *on*-the-loop to *out*-of-the-loop." We have seen that cubicle warriors are increasingly being assigned monitoring tasks rather than having a supervisory role. The next step would be for the cubicle warrior to become unnecessary and for the drone to function autonomously. The autonomous drone is high on the U.S. military agenda, and the U.S. Air Force assumes that by around 2050 it will be possible to deploy fully autonomous drones. Given current developments and investment in military robotics technology, this U.S. Air Force prediction seems not to be utopian but a real image of the future. This is a future in which the automation of death becomes reality.

In this final section, we look back on the social and ethical issues that come to the fore in relation to armed military robots and identify some regulatory issues that require attention in the public and political domains.

6.6.1 Social and Ethical Issues

6.6.1.1 Proliferation and Abuse The greatest, but most underrated, problem is the proliferation of armed military robots. Armed military robots are relatively cheap and easy to copy, and the first signs of an arms race are already visible. It does not require a lot of skill to make autonomous robot weapons. The idea that an autonomous robot weapon could become a standard terrorist weapon (e.g., a GPS-guided suicide bomber) that would replace the human suicide bomber is very disturbing.

In addition, there is a real and present danger that unmanned systems can be hacked, in that video images can be captured by hostile forces, as happened in 2009 with the Iraqis and a U.S. Predator (Gorman, Dreazen, & Cole, 2009), or can even be taken over by hostile troops.

6.6.1.2 Counterproductive Nature Although armed military robots, due to their precision robotics, are very accurate in hitting their targets, their use may eventually lead to more victims, because they will be deployed much faster and more frequently, even in civilian areas. In addition, the use of armed unmanned systems is often considered to be a cowardly act by locals, and every civilian victim of such an automated device will be used by insurgents for propaganda. All this leads to a loss of psychological support among the local population, and support is an essential tool for providing a positive contribution to stabilizing a conflict, for example (see, e.g., Khan, 2011). As Alley (2013, pp. 34–35) concludes,

> It will come through recognition that the undoubted weakening of al-Qaeda and Taliban leaderships has come at the price of inflamed anti-Western hostility, weakened authority of target state government, and renewed recruitment to militant networks. These conditions have given licence to internal and cross-border criminality and violence, as well as worsening sectarian conflict across the Middle East, South West Asia and North and Central Africa. These outcomes were no more than marginally envisaged when drone weaponisation began in 2004.

6.6.1.3 Humanization versus Dehumanization   Tele-led drones are remotely controlled by operators or cubicle warriors. The interface is the primary means by which these cubicle warriors gain access to information, on the basis of which they must make decisions about life and death. As this daily work can be psychologically very stressful, producers are developing interfaces that only show abstract and indirect images of the enemy and military targets. The result of this dehumanization is the creation of a moral and emotional distance between a fatal action and the implication of that action, which desensitizes the operator to the consequences of his or her decisions. It follows that the operator will not exercise ethical reflections on these decisions. Do we want operators to make decisions about life and death as though they are playing a video game? The key challenge for designers is to develop an interface that not only will reduce stress but will also overcome the issue of the moral distance, allowing operators to be fully conscious of actions so that they can actually be held responsible for their decisions, both morally and legally.

6.6.1.4 Autonomy   An important question regarding armed unmanned systems is whether they should be able to decide independently and thus autonomously on the use of deadly force. Of course, this automation of death raises ethical questions. Can autonomous military robots meet the principles of proportionality and discrimination?

Compliance with these principles often requires empathy and situational awareness. A tele-led military robot can be helpful for an operator because of its highly sophisticated sensors, but it does not seem feasible that in the next decade military robots will possess this ability to empathize and show common sense. Some scientists wonder if this will be at all possible because of the dynamic and complex battle environments in which these autonomous robots will have to operate (see, e.g., Sharkey, 2008b).

Another problem with autonomous robots is that they take the human moral agent entirely out of the firing loop. According to some authors, autonomous armed robots are objectionable because they can never relieve humans of their responsibility for ethical decisions in time of war.*

6.6.2 Regulation

6.6.2.1 Work on an International Ban on Autonomous Armed Robots   It is important that humans always remain in-the-loop and that decisions about life and death should not be made by robots. Therefore, it would be desirable to achieve an international ban on autonomous armed military robots. This must, however, be done quickly, because the current tendency is that the *man-in-the-loop* is being increasingly eroded and that technology is developing quickly in the direction of autonomous robots, which seems to be an inevitable outcome of the tremendous amount of research and political will that is being directed toward these robots. As Sullins (2010) remarks with regard to this rapid development of armed robots: "It is probably impossible to contemplate the alternative anymore, but we should have avoided arming robots in the first place" (p. 274).

Whether such a ban is feasible depends mainly on the attitude of the United States, which has set its sights on precisely this type of robot. It is, therefore, essential that other NATO countries and the United Nations get the United States to the negotiating table. This

---

* This moral objection against autonomous armed robots is made by Bolton, Nash, and Moyes, for example: "Decisions to kill and injure should not be made by machines and, even if at times it will be imperfect, the distinction between military and civilian is a determination for human beings to make" (http://www.article36. org/statements/ban-autonomous-armed-robots/).

will be difficult, because the United States regularly frustrates these developments. This is illustrated by the fact that the United States is not a party to arms control treaties and is not party to the Statute of the International Criminal Court. This court has jurisdiction in criminal matters in the fields of genocide, crimes against humanity, and war crimes. Boyle (2013, p. 28) concludes that

> [i]t is not realistic to suggest that the US stop using its drones altogether, or to assume that other countries will accept a moratorium on buying and using drones. The genie is out of the bottle: drones will be a fact of life for years to come. What remains to be done is to ensure that their use and sale are transparent, regulated and consistent with internationally recognized human rights standards.

However, the UN Convention on Certain Conventional Weapons (CCW) hosted an informal meeting about armed autonomous robots. Eighty-seven countries participated in this meeting, including countries with significantly advanced robot technology, such as the United States, China, Russia, and Israel. In November 2014, all 117 countries participated at the formal conference of the UN Convention on CCW. The objective of this convention was to ban or restrict the use of specific types of weapons that are considered to cause unnecessary or unjustifiable suffering to combatants or to affect civilians indiscriminately.* The future will tell whether the outcomes of the conference will really contribute to the achievement of the objective.

6.6.2.2 Curbing the Proliferation of Armed Military Robots   Curbing the proliferation of armed military robots by implementing global arms control treaties is a necessary condition that must be met—before armed military robots can be further developed in a responsible manner and can be deployed in military operations. The effects of armed military robots that are in the possession of fundamentalists or terrorists could well be devastating and would, by comparison, pale the impact of roadside bombs.

---

* http://www.un.org/apps/news/story.asp?NewsID=47794#.VDkb8xZAphY.

In the words of Homan (2011), "as the prices of robots falls and technology becomes easier, a robotic arms race can be expected, one that will be difficult to stop. It is of utmost importance that international legislation and a code of ethics for autonomous robots at war are developed before it is too late."

6.6.2.3 Broad International Debate on the Consequences of Military Robotics   For the development of military robotics technology, a broad international debate is thus required about the responsibilities of governments, industry, the scientific community, lawyers, nongovernmental organizations, and other stakeholders. Such a debate has not been realized because of the rapid development of military robotics so far. Fortunately, as we have seen, a start has been made with a debate during the informal meeting of experts hosted by the UN Convention of CCW in May 2014. The necessity of this is shown by the contemporary technological developments of military robotics, which cannot always be qualified as ethical.

The deployment of armed military robots affects the entire world, and it is therefore important that all stakeholders with a variety of interests and views enter into a mutual debate (see also Marchant et al., 2011). The starting point for this debate must be the development of a set of internationally recognized standards or norms governing the sale and responsible use of armed military robots. Otherwise, these robots will proliferate without control and will be misused by state and non-state actors.

Interview with Jürgen Altmann (Physicist and Peace Researcher at TU Dortmund University, Germany)

"I'm hopeful about a ban on autonomous arms."
Throughout his three-decade long career as a physicist at several German universities, Jürgen Altmann has been actively committed to the cause of arms control. In 2009, he was a co-founder, along with scientists from Australia, Britain, and the United States, of the International Committee for Robot Arms Control, which was among the initiators of the Campaign to Stop Killer Robots.

"ICRAC has fundamental objections against autonomous weapon systems. We think it's both ethically and legally unacceptable for arms to choose their own targets and to decide whom to kill. We are convinced that their deployment will lead to an increase in civil war victims, something which nations are under a legal obligation to avoid.

Personally, I think we should take a wider view of the effects of autonomous weapon systems, as to my mind they will increase the likelihood of war. Just imagine two fleets of these weapon systems facing each other across a national border or somewhere on the high seas. They may easily misinterpret some small event—a flash of light, an unexpected noise—as an attack and automatically open fire. Such hair-trigger systems could easily set off hostilities that neither party has an interest in or a wish for. Moreover, autonomous arms would lower the threshold for the use of international violence, given that the attacking side would expect few victims among its soldier. In recent years, remote-controlled armed drones—which are semi-autonomous systems—have given us a foretaste of that."

It has been argued that, quite on the contrary, autonomous arms may make morally superior decisions, reducing rather than increasing numbers of casualties among civilians. Unlike human soldiers, they always stay rational and do not run amok under stress.

"I can't rule that out once and for all, but in the foreseeable future, let's say for the next 20 to 40 years, that's extremely unlikely. Roboticist Ronald Arkin at Georgia Tech is working on software that will decide, on the basis of simple if-then conditions, whether an attack is justified or not. But it will be unable to adequately interpret complex battlefield situations, perhaps with civilians nearby, and then make ethically correct decisions. It simply can't be done without an understanding of the social context, without having emotional experience, knowledge of local cultural context, and so on. It's a very tall order to recreate these artificially. I fear that, without an international ban, military motives might well lead to the introduction of autonomous systems way before such conditions are met."

What have ICRAC and the Campaign to Stop Killer Robots achieved so far?

"The issue has been taken up very well not only by public opinion, but also by the United Nations. In 2013, many countries raised the issue in the General Assembly and the UN Special Rapporteur on

Extrajudicial, Summary or Arbitrary Executions called for national moratoria. In November of that year, the parties to the Convention on Certain Conventional Weapons (CCW) in Geneva began discussing the need for restrictions or a ban, resulting in an expert meeting 6 months later. In 2015, more detailed discussions are likely to take place, tackling practical issues such as the definition of what is to be banned. It looks as if the term 'meaningful human control' might discriminate usefully between what is and what is not to be banned under the CCW.

The good thing about a ban under this Convention is that all relevant parties are involved. The downside is that an agreement can only be reached by consensus, so each single country can block anything it doesn't like. There is a risk that negotiations will result in too weak a compromise. In that case, our campaign may change tack and go for a stronger separate treaty, even if that means losing some countries. That was what happened with other similar campaigns, which resulted in treaties banning anti-personnel mines and cluster munition. But for the time being, I'm hopeful. Even within the United States, many people are uneasy about autonomous arms."

"Even within the USA," you say. Is there a wide gap between the United States and Europe when it comes to the acceptance of these arms, and perhaps even of war in general?

"Well, I'm not a political scientist, but it is evident that the United States considers war as a potential reality at any given time. Also, its military-technology strategy states explicitly that it wants to maintain a technological lead that will enable it to defeat any enemy on any battleground. In Europe, with the possible exception of Britain, we feel very different about war, especially in Germany. The potential willingness to impose a limit on new arms technologies is somewhat stronger here.

In my view, the United States somewhat short-sightedly underestimates the capacity of several other countries to catch up in the autonomous-arms race quite fast, as a result of which U.S. citizens will not end up being safer but rather more at risk, abroad and even at home. I would advocate that the United States give up its claim to absolute military superiority and look rather more to a concept of cooperative security. Not that I think for a second that anyone in Washington will listen to some obscure scientist from Germany. It's up to U.S. scientists and U.S. public opinion to change the minds of military policy makers."

Talking about risk to ordinary people, do you think autonomous arms could become part of the terrorist arsenal?

"The smaller varieties, yes, definitely. Model airplanes could be mounted with bombs and outfitted with some degree of autonomous target recognition. But I don't think that any small group of people, wanted by the law, could build sophisticated weapons systems. On the other hand, given the arms race between technologically advanced states, sophisticated systems may well be exported to regions where terrorist groups are active. On the black or gray market they may then manage to acquire arms that they could never hope to build for themselves. This is actually an indirect benefit of preventive arms control agreements: while terrorists cannot participate in them, the fact that many nations do greatly reduces the risk of new types of arms falling into dangerous hands. Let me add that even without arms control, terrorists will not be able to use systems beyond those launched from the shoulder or a truck. I mean, things that require a 1000 meter runway or have a 30 meter wingspan are hardly practicable for them."

Returning to the negotiations on banning autonomous arms, how do you see the prospects for enforcement?

"Enforcement is always a difficult issue in a system of states with equal rights. Only the UN Security Council is entitled to it, and is often hampered by a veto. But verification of compliance is a necessary precondition, also for the nations entering an arms limitation agreement. Verification would be much easier—and, to my mind, peace and security would be greatly served—if all unmanned combat systems were banned. In that case, we could simply deploy inspection teams to visit military installations, to check whether all combat vehicles have a crew department and are manned during exercises. This would be similar to the practice under some other conventions. Unfortunately, such a blanket ban is politically unfeasible.

The trouble with enforcing a ban on autonomous arms only is that nations may well be able to turn a remote-controlled system into an autonomous one in a matter of minutes, simply by changing the software. This could even be done after the launch.

Enforcement would have to follow the example of some other treaties, such as the bans on dumdum bullets and anti-personnel mines. No advance inspections take place, so countries could have this type of ammunition or weapon in their arsenals. Only when the typical wounds have been observed will the Red Cross or a UN Commission start an investigation. The present case is more difficult: autonomous weapons would produce the same wounds as remotely controlled ones. Suspicion that a particular attack was carried out without 'meaningful human control' could arise from certain circumstances, such as the presence of a large swarm, which would be practically uncontrollable. In such a case, the operating state would have to prove the presence of a human operator.

To make this possible, it would have to be compulsory for armed forces to record the data stream into and out of remote-controlled weapons systems: those of their video and other sensors, their communications with base, the decision processes, the operator's hand movements, and so on. Such data, adequately authenticated, could prove that the attack was indeed under 'meaningful human control', under the direct command of Major so and so and operated by Captain such and such, as opposed to autonomous. My U.S. colleague Mark Gubrud and I have worked on this, but obviously, it has to be developed some more."

On a more personal note, do you consider yourself an activist scientist?

"I think that would be only half true. I do not have the time to organize demonstrations and the like, even though I would like to see more of them. Most of my time is taken up by professional activities such as keeping track of military R&D and reflecting on its potential long-term effects on peace and stability—obviously, this is what made me co-found ICRAC. Other than that, I'm also working on seismic and acoustic technologies that may be used to verify certain arms control agreements. I feel committed to using my knowledge as a physicist in the best interest of peace, and since this is a somewhat rare expertise, I think I may leave it to other people to organize the demonstrations."

References

Agence France-Presse (2008). *Automated killer robots: Threat to humanity.* http://www.commondreams.org/news/2008/02/27/automated-killer-robots-threat-humanity-expert (accessed April 22, 2014).

Alley, R. (2013). *The drone debate. Sudden bullet or slow boomerang* (discussion paper no. 14/13). Wellington, New Zealand: Centre for Strategic Studies. http://www.victoria.ac.nz/hppi/centres/strategic-studies/documents/DP1413OnlineVersion.pdf (accessed November 11, 2014).

Alston, P. (2010). *Report of the special rapporteur on extrajudicial, summary or arbitrary executions. Addendum, study on targeted killings.* Geneva, Switzerland: United Nations Human Rights Council. http://www2.ohchr.org/english/bodies/hrcouncil/docs/14session/A.HRC.14.24.Add6.pdf (accessed April 22, 2014).

Altmann, J., Asaro, P., Sharkey, N., & Sparrow, R. (2013). Armed military robots: Editorial. *Ethics and Information Technology, 15*(2), 73–76.

Amnesty International (2013). *Will I be next? U.S. drone strikes in Pakistan.* London, UK: Amnesty International. http://www.amnestyusa.org/our-work/issues/security-and-human-rights/drones/will-i-be-next (accessed January 23, 2014).

Anderson, M., & Anderson, S. L. (2007). Machine ethics: Creating an ethical intelligent agent. *AI Magazine, 28*(4): 15–26.

Arkin, R. C. (2009a). Ethical robots in warfare. *IEEE Technology and Society Magazine, 28*(1), 30–33.

Arkin, R. C. (2009b). *Governing lethal behavior in autonomous robots.* Boca Raton, FL: CRC Press.

Arkin, R. C. (2010). The case of ethical autonomy in unmanned systems. *Journal of Military Ethics, 9*(4), 332–341.

Asaro, P. M. (2008). How just could a robot war be? In A. Briggle, K. Waelbers, & Ph. Brey (Eds.), *Current issues in computing and philosophy* (pp. 50–64). Amsterdam, the Netherlands: IOS Press.

Asaro, P. M. (2009). Modeling the moral user. *IEEE Technology and Society, 28*(1), 20–24.

Beavers, A. F. (2012). Moral machines and the threat of ethical nihilism. In P. Lin, K. Abney, & G. A. Bekey (Eds.), *Robot ethics. The ethical and social implications of robotics* (pp. 333–344). Cambridge, UK: The MIT Press.

Bergen, P., & Tiedemann, K. (2011). *The year of the drone. An analysis of U.S. drone strikes in Pakistan, 2004–2011* (counterterrorism strategy initiative policy paper). http://newamerica.net/sites/newamerica.net/files/policydocs/bergentiedemann2.pdf (accessed September 3, 2014).

Blair, D. & Spillius, A. (2011, December 8). Iran shows off captured US drone. *The Telegraph.* http://www.telegraph.co.uk/news/worldnews/middleeast/iran/8944248/Iran-shows-off-captured-US-drone.html (accessed November 11, 2014).

Boyle, M. J. (2013). The costs and consequences of drone warfare. *International Affairs, 89*(1), 1–29.

Brooks, R. (2012, September 12). Take two drones and call me in the morning. *Foreign Policy*. http://www.foreignpolicy.com/articles/2012/09/12/take_two_drones_and_call_me_in_the_morning (accessed September 4, 2014).

Bumiller, E. (2012, July 29). A day job waiting for a kill shot a world away. *The New York Times*. http://www.nytimes.com/2012/07/30/us/drone-pilots-waiting-for-a-kill-shot-7000-miles-away.html (accessed October 26, 2014).

Camillero, J. A. (2013, July 20). Drone warfare: Defending the indefensible. *e-International Relations*. http://www.e-ir.info/2013/07/20/drone-warfare-defending-the-indefensible/ (accessed August 1, 2014).

Canning, J. S. (2006). A concept of operations for armed autonomous systems. Presented at the *third annual disruptive technology conference* (September 6–7), Washington, DC. www.dtic.mil/ndia/2006disruptive_tech/2006disruptive_tech.html (accessed November 16, 2014).

Carafano, J. J., & Gudgel, A. (2007). The Pentagon's robots: Arming the future. *Backgrounder, 2093*, 1–6.

Coeckelbergh, M. (2013). Drones, information technology, and distance: Mapping the moral epistemology of remote fighting. *Ethics and Information Technology, 15*(2), 87–98.

Committee on Armed Services, House of Representatives (2000). *Floyd D. Spence national defense authorization act for fiscal year 2001*. Washington, DC: U.S. Government Printing Office.

Cummings, M. L. (2006). Automation and accountability in decision support system interface design. *Journal of Technology Studies, 32*(1), 23–31.

Detert, J. R., Treviño, L. K., & Sweitzer, V. L. (2008). Moral disengagement in ethical decision making: A study of antecedents and outcomes. *Journal of Applied Psychology, 93*(2), 374–391.

Donnelly, S. B. (2005). Long-distance warriors. *Time, 166*(24), 42.

Fielding, M. (2006). Robotics in future land warfare. *Australian Army Journal, 3*(2), 99–108.

Fischer, J. M., & Ravizza, M. (1998). *Responsibility and control: A theory of moral responsibility*. Cambridge, UK: Cambridge University Press.

Fitzsimmons, S., & Sangha, K. (2013). Killing in high definition: Combat stress among operators of remotely piloted aircraft. *Proceedings of the international studies association annual convention*, San Francisco, CA, April 3–6. http://www.cpsa-acsp.ca/papers-2013/Fitzsimmons.pdf (accessed October 4, 2014).

Fitzsimonds, J. R., & Mahnken, T. G. (2007). Military officer attitudes towards UAV adoption: Exploring institutional impediments to innovation. *Joint Force Quarterly, 46*, 96–103.

Galliott, J. C. (2012). Uninhabited aerial vehicles and the asymmetry objection: A response to Strawser. *Journal of Military Ethics, 11*(1), 58–66.

Garraway, C. (2009). The doctrine of command responsibility. In J. Doria, H.-P. Gasser, & M. C. Bassiouni (Eds.), *The legal regime of the International Criminal Court* (pp. 703–725). Leiden, the Netherlands: Nijhoff.

Ghosh, B., & Thompson, M. (2009). The CIA's silent war in Pakistan. *Time*, *173*(21), 20.

Gorman, S., Dreazen, Y., & Cole, A. (2009, December 17). Insurgents hack U.S. drones. *The Wall Street Journal*. http://online.wsj.com/article/ SB126102247889095011.html (accessed October 5, 2014).

Gulam, H., & Lee, S. W. (2006). Uninhabited combat aerial vehicles and the law of armed conflicts. *Australian Army Journal*, *3*(2), 123–136.

Hellström, T. (2013). On the moral responsibility of military robots. *Ethics and Information Technology*, *15*(2), 99–107.

Homan, K. (2011, March 29). Will the human disappear in future warfare? Keynote speech at the Colloquium *The soldier in operations: Exploring the human factors*. Brussels, Belgium: The Belgian Royal Higher Institute for Defence (RHID). www.clingendael.nl/sites/default/files/20110300_ homan_speech.pdf (accessed March 17, 2014).

Horton, S. (2009, January 27). Prepare for the robot wars: Six questions for P.W. Singer, Author of *Wired for war*. *Harper's Magazine*. http://harpers. org/blog/2009/01/prepare-for-the-robot-wars-six-questions-for-pw-singer-author-of-_wired-for-war_/ (accessed January 23, 2015).

Human Rights Watch (2012). *Losing humanity: The case against killer robots*. http://www.hrw.org/reports/2012/11/19/losing-humanity-0 (accessed April 22, 2014).

International Organization for Standardization (2011). *Robots for industrial environments—Safety requirements*. http://www.iso.org/iso/iso_ catalogue/catalogue_tc/catalogue_detail.htm?csnumber=36322 (accessed April 22, 2014).

Iraq Coalition Casualty Count (2008). *Deaths caused by IEDs and U.S. deaths by month*. http://icasualties.org/oif/IED.aspx, http://icasualties.org/oif/ USDeathByMonth.aspx (accessed May 9, 2014).

Jennings, G. (2010). Predator-series UAVs surpass one million flight hours. *IHS Jane's: Defense & Security Intelligence & Analysis*. http://www.janes.com/products/janes/defence-security-report. aspx?ID=1065927694 (accessed August 8, 2014).

Johnson, A. J., & Axinn, S. (2013). The morality of autonomous robots. *Journal of Military Ethics*, *12*(2), 129–141.

Johnson, R. J. (2013). The wizard of Oz goes to war: Unmanned systems in counterinsurgency. In B. J. Strawser (Ed.), *Killing by remote control: The ethics of an unmanned military* (pp. 154–178). Oxford, UK: Oxford University Press.

Jordan, B. (2007, February 23). Half of predators fielded have been lost. *Air Force Times*. http://www.airforcetimes.com/news/2007/02/AFpredator losses070223 (accessed April 22, 2014).

Kaag, J., & Kaufman. W. (2009). Military frameworks: Technological know-how and the legitimization of warfare. *Cambridge Review of International Affairs*, *22*(4): 585–606.

Kahn, P. (2002). The paradoxes of riskless war. *Philosophy & Public Policy Quarterly*, *22*(3), 2–8.

Kaplan, R. D. (2006, September). Hunting the Taliban in Las Vegas. *Atlantic Monthly*. http://www.theatlantic.com/magazine/archive/2006/09/hunting-the-taliban-in-las-vegas/305116/ (accessed April 22, 2014).

Kenyon, H. S. (2006). Israel deploys robot guardians. *Signal*, *60*(7), 41–44.

Kershnar, S. (2013). Autonomous weapons pose no moral problem. In B. J. Strawser (Ed.), *Killing by remote control: The ethics of an unmanned military* (pp. 229–245). Oxford, UK: Oxford University Press.

Khan, A. N. (2011). The U.S. policy of targeted killings by drones in Pakistan. *IPRI Journal*, *11*(1), 21–40.

Krishnan, A. (2009). *Killer robots. Legality and ethicality of autonomous weapons*. Farnham, UK: Ashgate Publishing Limited.

Lee, P. (2012). Remoteness, risk and aircrew ethos. *Air Power Review*, *15*(1), 1–19.

Lin, P., Bekey, G., & Abney, K. (2008). *Autonomous military robotics: Risk, ethics, and design*. San Luis Obispo, CA: Ethics & Emerging Technologies Group.

Lucas, G. R. (2013). Engineering, ethics, and industry: The moral challenges of lethal autonomy. In B. J. Strawser (Ed.), *Killing by remote control: The ethics of an unmanned military* (pp. 212–228). Oxford, UK: Oxford University Press.

Marchant, G. E., Allenby, B., Arkin, R. C., Barrett, E. T., Borenstein, J., Gaudet, L. M., … Silberman, J. (2011). International governance of autonomous military robots. *Science and Technology Law Review*, *12*, 272–315.

Matthias, A. (2004). The responsibility gap: Ascribing responsibility for the actions of learning automata. *Ethics and Information Technology*, *6*(3), 175–183.

National Research Council (2005). *Autonomous vehicles in support of naval operations*. Washington, DC: The National Academies Press.

Office of the Surgeon Multinational Force-Iraq & Office of the Surgeon General United States Army Medical Command (2006). *Mental health advisory team (MHAT) IV. Operation Iraqi freedom 05-07* (final report). Washington, DC: Office of the Surgeon Multinational Force-Iraq & Office of the Surgeon General United States Army Medical Command. http://www.combatreform.org/MHAT_IV_Report_17NOV06.pdf (accessed April 22, 2014).

Plaw, A. (2013). Counting the dead: The proportionality of predation in Pakistan. In B. J. Strawser (Ed.), *Killing by remote control: The ethics of an unmanned military* (pp. 126–153). Oxford, UK: Oxford University Press.

Quintana, E. (2008). *The ethics and legal implications of military unmanned vehicles (occasional paper)*. London, UK: RUSI. https://www.rusi.org/downloads/assets/RUSI_ethics.pdf (accessed January 23, 2014).

Royakkers, L. M. M., & Olsthoorn, P. (2014). Military robots and the question of responsibility. *International Journal of Technoethics*, *5*(1), 1–14.

Royakkers, L. M. M., & Topolski, A. R. (2014). Military robotics & relationality: Criteria for ethical decision making. In J. van den Hoven, N. Doorn, T. Swierstra, B.-J. Koops, & H. Romijn (Eds.), *Responsible innovation 1. Innovative solutions for global issues* (pp. 351–367). Dordrecht, the Netherlands: Springer.

Royakkers, L. M. M., & van Est, Q. C. (2010). The cubicle warrior: The marionette of digitalized warfare. *Ethics and Information Technology*, *12*(3), 289–296.

Schulzke, M. (2013). Autonomous weapons and distributed responsibility. *Philosophy and Technology*, *26*(2), 203–291.

Shachtman, N. (2007, October 18). Robot cannon kills 9, wounds 14. *Wired com.* http://www.wired.com/2007/10/robot-cannon-ki/ (accessed March 17, 2014).

Shachtman, N. (2011, October 7). Computer virus hits U.S. drone fleet. *Wired.com.* http://www.wired.com/2011/10/virus-hits-drone-fleet/ (accessed March 17, 2014).

Sharkey, N. (2008a). Cassandra or false prophet of doom: AI robots and war. *IEEE Intelligent Systems*, *23*(4), 14–17.

Sharkey, N. (2008b). Grounds for discrimination: Autonomous robot weapons. *RUSI Defence Systems*, *11*(2), 86–89.

Sharkey, N. (2008c, February 27). *Killer military robots pose latest threat to humanity.* Keynote-presentation for the Royal United Services Institute, Whitehall, London, UK.

Sharkey, N. (2010). Saying "no!" to lethal autonomous targeting. *Journal of Military Ethics*, *9*(4), 369–383.

Singer, P. W. (2009a). Military robots and the laws of war. *The New Atlantis*, *23*, 25–45.

Singer, P. W. (2009b). *Wired for war: The robotics revolution and conflict in the twenty-first century.* New York: The Penguin Press.

Solis, G. D. (2010). *The law of armed conflict: International humanitarian law in war.* Cambridge, UK: Cambridge University Press.

Sparrow, R. (2005). Hands up who wants to die?: Primoratz on responsibility and civilian immunity in wartime. *Ethical Theory and Moral Practice*, *8*(3), 299–319.

Sparrow, R. (2007). Killer robots. *Journal of Applied Philosophy*, *24*(1), 62–77.

Sparrow, R. (2009). Building a better warbot: Ethical issues in the design of unmanned systems for military application. *Science and Engineering Ethics*, *15*(2), 169–187.

Stanford Law School & NYU School of Law (2012). *Living under drones. Death, injury, and trauma to civilians. From U.S. drone practices in Pakistan.* http://livingunderdrones.org/ (accessed April 22, 2014).

Steinhoff, U. (2013). Killing them safely: Extreme asymmetry and its discontents. In B. J. Strawser (Ed.), *Killing by remote control: The ethics of an unmanned military* (pp. 180–208). Oxford, UK: Oxford University Press.

Strawser, B. J. (2010). Moral predators: The duty to employ uninhabited aerial vehicles. *Journal of Military Ethics*, *9*(4), 342–368.

Strawser, B. J. (Ed.). (2013). *Killing by remote control: The ethics of an unmanned military.* Oxford, UK: Oxford University Press.

Sullins, J. P. (2010). RoboWarfare: Can robots be more ethical than humans on the battlefield? *Ethics and Information Technology*, *12*(3), 263–275.

Thompson, M. (2007, September 26). V-22 Osprey: A flying shame. *Time.* http://content.time.com/time/magazine/article/0,9171,1666282,00. html (accessed November 16, 2014).

United States Air Force (2009). *Unmanned aircraft systems flight plan 2009–2047.* Washington, DC: Headquarters, United States Air Force.

U.S. Department of Defense (2005). *Joint robotics program master plan FY2005. Out front in harm's way.* Washington, DC: OUSD (AT&L) Defense Systems.

U.S. Department of Defense (2009). *Unmanned systems roadmap 2009–2034.* Washington, DC: Government Printing Office.

U.S. Department of Defense (2012). *Autonomy in weapon systems.* Directive number 3000.09. http://www.dtic.mil/whs/directives/corres/pdf/300009p.pdf (accessed November 11, 2014).

Vallor, S. (2013). The future of military virtue: Autonomous systems and moral deskilling in the military profession. In K. Podens, J. Stinissen, & M. Maybaum (Eds.), *Proceedings of the fifth international conference on cyber conflict (CyCon 2013)* (pp. 471–486). Tallinn, Estonia: NATO CCDCOE.

Vanden Brook, T. (2009, June 16). More training on UAV's than bombers, fighters. *USA Today.* http://www.airforcetimes.com/news/2009/06/gns_airforce_uav_061609w/ (accessed March 17, 2014).

Van de Poel, I. R., & Royakkers, L. M. M. (2011). *Ethics, technology, and engineering. An introduction.* Oxford, UK: Wiley-Blackwell.

Verbeek, P. P. (2005). *What things do: Philosophical reflections on technology, agency, and design.* University Park, PA: Pennsylvania State University Press.

Whetham, D. (2013). Drones and targeting killing: Angels or assassins? In B. J. Strawser (Ed.), *Killing by remote control: The ethics of an unmanned military* (pp. 70–84). Oxford, UK: Oxford University Press.

Zenko, M. (2012, February 27). 10 things you didn't know about drones. *Foreign Policy.* http://foreignpolicy.com/2012/02/27/10-things-you-didnt-know-about-drones/ (accessed March 17, 2015).

7
AUTOMATION FROM LOVE TO WAR

Until recently, robots were mainly used in factories for automating production processes. In the 1970s, the appearance of factory robots led to much debate on their influence on employment. The mass unemployment that was feared did not come to pass. Still, robots have radically changed the work in countless factories. Driven by a belief in efficiency, factories and labor have been redesigned over the last century. The first half of the twentieth century saw a far-reaching simplification and specialization of the work. This paved the way for the mechanization and automation of the production process. As a result, robots have come to play a central role in this ongoing attempt to rationalize production. In essence, robotization presents a way to rationalize a certain social practice by reducing its dependence on people, ultimately by replacing them with machines.

Both rationalization and robotics no longer only concern factory applications. Currently, no aspect of people's lives is immune to rationalization any more. This book has explored the actual and intended use of robots in various social domains outside the factory. Our study looked at home robots, care robots, the use of drones in the city, car robots, and military robots. Thus, we began close to home and subsequently moved further away from home, ending up on the military battlefield. Accordingly, we have talked about the automation of numerous human activities, such as caring for the sick, driving a car, making love, and killing people. Robotics in the twenty-first century thus literally concerns automation from love to war.

This book is, therefore, about the use of robots in a complex and unstructured world: our social, everyday environment. It is important to realize, however, that before engineers began to rationalize it, the average factory was a chaotic place too. And nowadays, we even have fully automated factories that require no human presence at all.

We use these historical insights about industrial robots to reflect on the use of service and social robots outside the factory. One basic understanding is that the use of robots in messy social practices is only possible when these practices are organized around their technological limitations. So before robots can be applied in a certain social practice, that practice first has to be adapted to the limited capacities of robots. In other words, a basic level of rationalization of a certain social practice is required before robotization, as the next and further step toward rationalization, can take place.

This final chapter reflects on the societal meaning of the rationalization of society through robotization. Our analysis concerns the pros and cons of robotization as rationalization: both its reasonable and unreasonable aspects. The use of robots in our society not only offers numerous possibilities for making human life more pleasant and safe, but it also raises countless difficult societal and ethical issues. Based on the five preceding chapters, we first discuss the technical possibilities and future expectations surrounding the use of robots in society. Next, we review the social gains expected from robotics. Then we describe some social and ethical issues that are raised by the fact that robots are information technologies, have a lifelike physical or psychological appearance, and have a certain degree of autonomy. In addition, we look at to what extent robotic systems may lead to dehumanizing systems that may become antihuman or even destructive of human beings. To conclude, we touch upon a few political–administrative topics that enter the public agenda through the societal issues.

7.1 Future Expectations and Technical Possibilities

For better or for worse, the future of robotics is strongly connected to two long-term socio-technological imaginaries. First, there is the engineering dream of building multipurpose machines that can move and act autonomously in complex human environments. Related to this high level of autonomy is the horror scenario of the smart but evil and destructive machine that has gone out of control. On the bright side, there is the dream of the friendly robot that exhibits good social behavior and acts according to high moral standards. The notion that this will be feasible within a few decades is referred to as the strong artificial intelligence (AI) view.

At the start of this book, we introduced three types of concepts to reflect on the state of the art in relation to social and service robots and the expectations surrounding their future development. To indicate the level of autonomy of a robotic system, we used the terms "man-in-the-loop," "man-on-the-loop," and "man-out-of-the-loop." Second, we distinguished three categories of social robots: socially situated, socially embedded and socially intelligent. Third, four classes of machine ethics were defined: ethical impact agents, implicit and explicit ethical agents, and full ethical agents. We will use these concepts to address the following question: To what extent will the strong AI vision of autonomous, socially intelligent and morally competent robots become reality in a few decades?

7.1.1 Influential Strong AI Pipe Dream

In all the five fields that have been described in this book, socio-technical imaginaries exist in which autonomously operating robots play a role. The Netherlands Organization for Applied Scientific Research (TNO) (2008) expects that as soon as 25 years from now, autonomous cars will be driving around. And the U.S. Air Force expects that by around 2050, the first autonomous armed military airplanes will be in the air. At about that time, some also expect that there will be human-like robots with social and emotional behavior. The organizers of the RoboCup believe that a team of robots will beat the world champions in football by the middle of this century. And David Levy (2007) expects that by around 2050, the first marriage between a human and a robot will take place. Finally, Noel Sharkey (2008) expects human-like police robots to exist by 2084. However, these speculations are akin to reading tea leaves.

Yet, this is not about informal guesswork about the future; it is about expectations that strongly influence current techno-scientific agendas and research budgets. Speculations about the future therefore have actual political significance. They accurately convey what are seen today as important trends and also as scientific and technological challenges: more autonomous behavior and the development of artificial social and moral intelligence. This applies to every type of robot that we have researched. In Japan, hand in hand with the media, scientists and engineers are cultivating the image of multifunctional humanoid robots, which will replace human caretakers in the future (Lau, Van't Hof, & Van Est, 2009, p. 42). Such ambitious goals turn out to be successful in generating public funding for fundamental research in the field of robotics. Especially in the field of military robots, the United States sees similar dynamics. The U.S. Army wants autonomous robot armies by the middle of the twenty-first century and invests billions of dollars to reach this goal through specific subgoals.

The active goal of developing robotics for the domestic environment, care, the management of traffic, police, and the army goes hand in hand with a new research goal and thus a new societal goal. There is an aim for technology to enable an increasing number of autonomous moral and social actions of robots. Thus, a radical developmental path unfolds, namely, the modeling, digitization, and automation of human behavior, decisions, and actions in new social domains. This is at least partially legitimated by speculative socio-technical imaginaries, such as the multifunctional and autonomous machine that inhabits social intelligence and moral judgment.

7.1.2 Successful and Pragmatic Weak AI Approach

For the short and medium term, developments in the field of new robotics are mainly characterized by terms such as "man-in-the-loop" and "man-on-the-loop." Robot experts speak of *co-robotics.** First of all, this signifies the digitization of various previously low-technology fields. Robotics makes it possible to utilize machines in fields that traditionally involve mostly human labor, such as the sex industry and care for the elderly. Second, in high-technology practices, such as the automotive industry and the military, we see a shift from "man-in-the-loop" to "man-on-the-loop" and even "man-out-of-the-loop."

Multifunctional household and social care robots belong mostly to the future. During the last decade, the first robots that have entered the household on a relatively small scale are machines specialized in performing one task: *monomaniacal robots* such as the vacuum cleaning robot and the lawnmower robot. In the long term, though, it might

---

* http://www.nsf.gov/pubs/2014/nsf14500/nsf14500.htm.

be possible that robots will act as assistant caregivers for patients who wish to live independently at home for longer. Such care-supporting robots, however, will not make a large-scale entrance into care in the short term. The utilization of care robots is strongly linked to the development of home automation. Through home automation, part of the work of bedside caregivers can take place from a distance. If this happens, care of humans will be mediated through a technological system, and thus, caretakers will be placed in a technological loop. Although the introduction of home automation is a laborious process, the coming 10 years will probably see a breakthrough in telecare around the world.

Another currently relatively low-technology practice in which robotics can begin playing a role concerns sexual private life, particularly the sex industry. It is not known how sex robots will develop in the coming years. At the moment, this rather expensive technology is slowly being introduced in the sex industry. Some see a big future for sex robots. This is not unthinkable, because in the past, too, sex was a driver behind the introduction of new technologies, such as video recorders and the Internet.

Although in health care and the sex industry the "man-in-the-loop" is now being introduced, the military and cars have known a "man-in-the-loop" for a long time. Human drivers drive their cars, after all. In these fields, we see a shift from "man-in-the-loop" to "man-on-the-loop." In the automotive industry, there is a trend of moving from systems that support driving through informing to systems that also warn and even intervene. The next step in this trend is for driving tasks to be taken over through cooperative systems (in combination with traffic management). Considering the fast developments in the field of robotics, it is now likely that autonomous cars will gradually become common in the future, around 2030.

Just as with car robots, military robots have a trend that goes from "in-the-loop" to "on-the-loop" to "out-of-the-loop." The United States has invested in surveillance robots for years as well as in robots for distant roadside bomb cleanups and tele-controlled armed military robots. When cleaning up roadside bombs, military personnel control a cleanup robot from a safe distance. With fighter planes, we see a fast transition from airplanes manned by pilots to robot systems (*drones*) that are controlled remotely by operators. These operators no longer need the skills of a traditional air pilot to fly the aerial system, since the flying itself is done autonomously by the drone. The teleoperator can focus on other tasks, such as surveillance or targeting an enemy. In a similar way, operating a small recreational drone requires fewer skills than flying a traditional radio-controlled model airplane. Just as digital cameras have made it possible for nonprofessional people to make technically very good pictures with the help of robotics, people do not have to be fanatic hobbyists to teleoperate a model airplane. Thus, in the case of domestic drones operated through robotics, more people can act as the man-in-the-loop. Furthermore, technological systems are increasingly advising human operators on which action must be taken. This we call "man-on-the-loop." In other words, there is a shift from robots controlled remotely by operators to robot systems that advise operators and to more and more autonomous systems.

Our study shows that the level of autonomy reached by a robotic system strongly depends on the way the socio-technical environment it works in is structured. Current autonomous robots, such as the robotic vacuum cleaner, can only work successfully if we adapt our living space to its limited capacities. In industrial robotics, the factory was redesigned in such a way that robots could be used to take over simplified human tasks. Similarly, the successful use of robots outside the factory does not depend on the engineering ability to build a robot with human-like physical and mental capabilities, as promoted within the strong AI vision, as much as on smartly (re)framing a certain social practice and developing a socio-technical environment within which a "simple" robot can work. For example, trying to let a machine simulate the way human beings iron is currently a route to disaster. Instead, engineers redefined ironing so that it was suitable for a machine to do. In addition, our rapidly evolving technological systems, from communication to sensor networks, are increasingly enveloping our daily environment into an information and communication technology (ICT) friendly infosphere (cf. Floridi, 2014, p. 144). It is the gradual, steady development of these large technological systems (Hughes, 1983) that power and limit the use of robots in the social domains that were studied: global positioning systems (GPS) enable drones in the city and on the battlefield, and the use of care robots will hinge on the advancement of home automation systems.

7.1.3 Exploring Artificial Social Intelligence

With regard to the ambition to build socially intelligent and morally competent robots, a similar argument can be made. In the short or medium term, it is not expected that robots will display elaborate social and moral behavior. Developing socially intelligent robots, that is, "robots that show aspects of human style social intelligence, based on deep models of human cognition and social competence" (Fong, Nourbakhsh, & Dautenhahn, 2003, p. 145), is a very long-term goal. There is also strong doubt whether it is at all possible to build ethical decision making into machines in a human-like, nonreductionist way.

Despite this, many R&D efforts are put into making ICT more "social." This goal is part and parcel of the vision of ambient intelligence (AmI), which has been driving the ICT research agenda of, for example, the European Commission and many big ICT firms since the start of this century. In this vision, humans are surrounded by "smart environments," with computers that are aware of which person is present and what characteristics, emotions, needs, and intentions they have. A smart environment can adjust to, react to, and anticipate this. The AmI vision has a significant impact on the development of ICT in the fields of care, mobility, and the domestic environment. For example, the AmI vision offers a new view on the way in which we should deal with our health in the future: we should have personal health care that is automated as much as possible (Schuurman, Moelaert El-Hadidy, Krom, & Walhout, 2009). Emphasis within the AmI vision originally lay on information, communication, and amusement. Under the influence of robotics, the past few years have seen increased attention on the automation of physical tasks. The ultimate goal is a well-educated robot that can help humans with everyday tasks. Thus, robotics has become a supportive element of the AmI vision. As a consequence, the goals of developing socially situated robots (robots that can perceive and react to a certain social environment) and socially embedded robots (robots that are able to interact with agents and humans, and are at least partially aware of human interactional structures, such as taking turns) have become integral parts of the AmI vision.

Human–machine interaction is an important part of modern robotics research. Contemporary robots are still very limited and predictable in their social interaction. Again, the step-by-step improvement of the interaction between man and machine does not depend on strong AI breakthroughs. The current route forward tries to circumvent the currently limited capacities of ICT by making use of the social intelligence of human beings. For example, when the most developed chatbots, such as Cleverbot* and Eugene Goostman,† generate their answers, they make use of earlier answers to similar questions that have been put to people and that can be found on the Internet. This is known as human-based computation, since the computer makes use of the way in which groups of people have solved problems before, or data-driven (weak) AI (cf. Nielsen, 2011). The social performance of these chatbots thus depends on a large socio-technical system, namely, the Internet, or the social web. More importantly, as Floridi (2014, p. 149) explains, "[T]he innumerable hours that we spend and keep spending rating and ranking everything that comes our way are essential to help our smart yet stupid ICTs to handle the world in apparently meaningful ways." Besides playing a role in shaping the way the robot acts, human intelligence plays another important semantic role in the way we interact with robots. Namely, people are capable of filling in much of the social interaction with a robot themselves: we anthropomorphize the robot. Moreover, people have no clearly described expectations of such machines and consider them as toys to be played with and to have fun with.

7.1.4 Exploring Artificial Moral Intelligence

At the moment, a machine that acts like a full ethical agent, which can make explicit ethical judgments and is generally competent to reasonably justify them, belongs to the realm of science fiction. The current scientific debate concerns the feasibility and desirability of implicit and explicit artificial ethical agents. Whereas implicit ethical agents are designed so that they implicitly promote ethical behavior, explicit ethical agents have human morality encoded in their software.

---

* http://en.wikipedia.org/wiki/Cleverbot.
† http://en.wikipedia.org/wiki/Eugene_Goostman.

Designing information technology that accounts for human values is widely accepted as a legitimate activity. However, to include ethical aspects into the design of robotic systems, one must first acknowledge that developing and applying robots is normative and that all robots should, therefore, be considered ethical impact agents that should be evaluated in terms of their ethical consequences. This book tries to strengthen that awareness and gives an overview of the state of the art in relation to the debate on the potential ethical implications of robotics in a broad set of social practices. It is important to identify the significant moral values involved in the social practice that is being redesigned and then to follow this with an operationalization of these values in the robot design (cf. Van Wynsberghe, 2013).

The question of to what extent building ethical decision making into machines is feasible and/or morally acceptable is a hot topic for debate among a small group of ethicists. The difference between the two approaches appears to resemble the distinction between weak and strong AI. In the strong AI version, morality can be encoded in software even in such a way that the robot behaves "more ethically" than, for example, a human driver or soldier would if confronted with the same situation. In the pragmatic AI version, the issue of whether a machine can "behave ethically" is not considered an important one. Instead, it is crucial that robots can function safely and can perform their assigned missions effectively, including following instructions that comply with the law.

The expectation that robots will be better drivers than humans is widespread. Also, in the field of decision making concerning life-and-death situations on the battlefield, some believe that future robots will be more capable of this than humans. Autonomous action also implies that such machines are expected to act "morally." For example, autonomous car robots must heed traffic regulations and military robots must act according to the Geneva Conventions. Driven by these future socio-technical imaginaries, various attempts to build ethical decision making into machines have been made. The most well known is Ronald Arkin's army-funded work on the problem of how to make drones capable of dealing with the complicated ethics of wartime behavior. Arkin, Ulam, and Duncan (2009, p. 1) proposed the concept of an "ethical governor," which is supposed to be "capable of restricting lethal action of an autonomous system in a

manner consistent with the Laws of War and Rules of Engagement."
It is not yet clear, however, to what extent it is feasible to build such
explicit ethical agents. In particular, the effort to encode morality into
machine software seems to be in conflict with the essential nonreducibility of human ethical reasoning. In other words, there is strong
doubt whether computers will ever be able to deal in an appropriate
way with the ethical frame problem.

The core of the pragmatic approach toward building robotic systems that take account of human values and laws is finding ways to
circumvent this ethical frame problem. This approach does not so
much depend on the engineering ability to build artificial morality,
but on smartly (re)framing a certain socio-technical practice in such
a way that a "simple" robot can act as an implicit ethical agent. For
example, in the chapter on military robots, how to enable machines
to comply with the principle of discrimination in the law of war was
discussed. One option to considerably reduce the possibility that
armed autonomous robots could attack civilians was to deploy them
only in places where civilians are not allowed. Canning (2006) suggests building robots that only attack enemy fighters who wear hostile
weapon systems. However, reducing the frame problem in this way
can actually lead to ethical problems, for example, in circumstances
in which citizens usually carry a weapon and are therefore difficult to
distinguish from armed enemy insurgents. In fact, targeted killing
through tele-led armed drones by means of locating intended targets
by tracking their mobile phones is a topical example of reducing the
problem of distinguishing between an enemy and a civilian. Reducing
the frame problem in this way is also unethical, since it is not certain
whether the individual in possession of a tracked mobile phone is in
fact the intended target. To conclude, it is a real challenge to simplify
ethical frame problem in an ethically appropriate fashion. The danger
of a technological imperative—doing things because they are feasible
and not because they are desirable—always lurks around the corner.

7.2 Expected Social Gains

Robotization presents a way of rationalizing social practices
and reducing their dependence on people (cf. Ritzer, 1983).
Rationalization can have many benefits: greater efficiency, less

mistakes, cheaper products, higher safety levels, a higher quality of services, etc. The use of robots may be promoted because it is assumed that the robot can do things better than humans can. As indicated earlier, there is a widespread belief that robots in the near future will become better drivers than humans, and Arkin (2009) believe that in due course robots will be able to make better moral decisions on the hectic battleground than human soldiers. And Levy (2007) expects that sex robots will become better lovers than humans in the near term.

Another reason for investing public money in the development of robots is the expectation that in future there will not be enough people to do a certain socially relevant job, such as caring for the elderly. Besides saving labor and improving the quality, access, and efficiency of health care for the elderly, the promise is that robotics will also empower elderly people by allowing them to live longer at home and enabling them to exert more influence over their own care process. This empowerment outlook is also present in the discussion about the home robot. When a household robot is doing the cleaning, family members get more time to engage in activities that are more fun. Levy (2007) sees many advantages of the sex robot, for example, they can be sex teachers, saving the relationships of people with bad sex lives, and can be used to clarify one's sexual orientation.

Finally, the development and use of robotic systems are often legitimated by the fact that they will take over dirty, dull, and dangerous work done by human beings. Certain tasks can indeed be so disgusting, dangerous, or mind-numbingly dull that it would be inhumane to let such tasks be performed by humans if it is possible to have them done by machines. A good example of this is the utilization of robot jockeys at camel races in Qatar (see Box 7.1).

By replacing such inhuman tasks with technology, robotics can appeal to our moral duty to tackle certain societal problems. Some claim that the "principle of unnecessary risk" leads to an ethical obligation to apply robotics in certain situations. Our study shows various situations that raise the question of whether or not it is possible and desirable to robotize certain human activities.

Strawser (2010) believes that it is morally unacceptable to give a soldier an order that may lead them to suffer lethal injuries if a military

BOX 7.1   ROBOT JOCKEYS REPLACE CHILDREN AT CAMEL RACES

At the beginning of this century, it was common practice to employ children from the age of 4 as jockeys at camel races in various Gulf states (Lewis, 2005). These children were often kidnapped from nearby, poorer countries, such as Sudan and Pakistan, and were treated very badly. Most did not receive an education, and they were malnourished to keep their weight low. In the United Arab Emirates, this concerned 3000 children, and in Qatar, a few hundred. In 2003, Pakistani human rights activist Asnar Burney started a campaign against the use of child jockeys. His endeavor was supported by the United Nations and the United States, which accused Qatar of slavery and threatened it with economic sanctions. As a result, the emir of Qatar declared that all child jockeys be replaced by robots. The Swiss company K-Team developed the robot jockey *Kamal*, which has taken over the work of child jockeys ever since. These lightweight robots are remote controlled by operators that drive in cars alongside the racetracks.

robot could perform this same task. This principle of unnecessary risk leads to an ethical obligation to apply robotics in certain situations, for example, in searching for and dismantling roadside bombs.

In many parts of the world, prostitution is a dirty, dangerous, and degrading vocation. The extent of human suffering involved is huge: the International Labour Organization (2012) estimates that there are 4.5 million victims of sex trafficking worldwide. Dealing with this global issue is very complex, and both legalizing and criminalizing the selling and/or buying of sex at the national level often have unintended consequences. The Swedish law of 1999 that forbade the buying of sex has led to less prostitution in Sweden than in neighboring countries. Swedish sex workers, however, still face stigmatization and the threat of eviction, because landlords are vulnerable to pimping charges if they collect money earned from selling sex (Goldberg, 2014). Moreover, traffickers have chosen other more profitable destinations (Raymond, 2008). In the Netherlands, the welfare of prostitutes has suffered from the legalization of prostitution in 2000, and trafficking has increased. There, between 50% and 90% of the 25,000 women in prostitution work under coercion (KLPD, 2008). The current issue is how politics can fight such extensive and, for the women in question, traumatic cases of abuse. The development of sex robots, however, might bring in a new perspective to this discussion. It brings to light the question of whether or not sex robots could form a reasonable alternative to human prostitution. And should the presence of such a technological alternative, given the many degrading circumstances in prostitution, lead to an ethical obligation to utilize sex robots?

The "principle of unnecessary risk" is also applicable to driving cars. Contrary to other societal terrains, such as aviation and the work environment, in traffic we accept hundreds of fatalities per year. In 1997, the Swedish parliament renounced this pragmatic approach to traffic safety. Politics supported the so-called *Vision Zero*, which, as a matter of principle, refuses to accept that traffic accidents cause casualties (Whitelegg & Haq, 2006). This policy view is not based on a cost–benefit analysis, but on the ethical principle that it is not acceptable that humans die or become seriously wounded when they move through traffic (Tingvall & Haworth, 1999). Although most of the fatal accidents are caused by human errors, in this view road users are not to blame; the inadequate traffic system itself is regarded as the root source of these deadly accidents.

Robotics offers numerous possibilities for improving the current traffic system. This applies to, among other things, the existing Intelligent Speed Assistant (ISA). This system gives information on the speed limit and gives warnings or intervenes if there is speeding. According to experts, making compulsory the intervening aspect of the ISA system might contribute to halving the total number of traffic deaths in Europe. In 2009, more than 35,000 people died in traffic accidents (European Commission, 2010, p. 2). Over 140,000 people became permanently disabled through an accident, and 1.4 million people were seriously injured. The total cost of the accidents, medical care, and the loss of income amounts to €130 billion (or U.S. $145 billion) per year (European Commission, 2010). Here, too, new robotics sheds light on the question of whether or not we are morally obliged to make automatic speed adaptation a legal requirement.

The key message of this book is that the potential social gains of rationalization through robotization should be weighed against potential social costs. For example, one aspect that plays a role in the latter case of making automatic speed adaptation a legal requirement is paternalism, or the top-down government restriction of freedom. This awareness pressures us to take seriously questions such as: In what circumstances and to what extent can personal autonomy be affected? And if we apply technology on a large scale in this way, is there still space for people's personal decisions and their role of personal responsibility? In the remainder of this chapter, we will discuss various ethical and social issues that are raised by the fact that robots are information technologies, have a lifelike appearance, and have a certain level of autonomy. Moreover, we describe to what extent over-rationalization may lead to dehumanizing robotic systems.

7.3 Robots as Information Technology

The fact that robotic systems are information systems raises various familiar IT-related issues, such as privacy, informed consent, data ownership, cyber security, and hacking. In this section, we will summarize some of the privacy issues that relate to the fact that robots are monitoring devices. Moreover, we will consider to what extent the safe use of robots in society can be guaranteed and misuse can be prevented.

7.3.1 Monitoring and Privacy

Robotics can be applied in all sorts of ways to monitor certain situations, such as the patient's state of health, a car driver's focus of attention, and the safety situation in the street or on the battlefield. As a direct consequence of this, robot technologies can have an impact on our privacy in all sorts of ways. The right to privacy and the related right to data protection are not absolute, but they are often interpreted in a specific societal constellation. The utilization of robotics calls for an explicit and careful consideration of various potential conflicting interests in various fields, for example, between health and privacy, traffic safety and privacy, public safety and privacy or, in relation to drone journalism, between freedom of the press and privacy. It should also be taken into account that the utilization of this kind of information technology may go hand in hand with more intensive file keeping on the actions of caregivers, (professional) drivers, police officers, and/ or soldiers.

Citizens can lose their right to privacy temporarily when significant public security interests are at stake. At such a moment, the public security interest is more important than an individual's privacy. Surveillance drones, which can be equipped with high-definition cameras, infrared cameras, or smell detection sensors, offer the police a broad range of new possibilities for criminal investigation and law enforcement. A telling example is presented by the police in North Dakota (the United States), who used a Qube drone equipped with thermal-imaging equipment to find in only a few minutes someone suspected of drunk driving who ran away into a 2 m (or 6 feet) high corn field.* But using surveillance drones may also form an infringement of the privacy of citizens. It is important that law-abiding citizens' privacy in both private and public space is safeguarded. With respect to the latter, human beings should maintain a reasonable expectation of privacy, or anonymity, on public roads, parks, and other public settings.

As one of the oldest applications in the digitization of public space, closed-circuit television (CCTV) provides us with some guidelines on how to safeguard anonymity on the streets (Schreienberg, Van't Hof, & Koffijberg, 2011). History shows first of all that fierce public and political debates during the nineties were needed to encourage governments to set up specific legal frameworks. These frameworks were supposed to make clear who is allowed to watch and who can be watched under what circumstances. The point of departure is that people should have a reasonable expectation of privacy in public space, but that anonymity can be lifted if security is at stake. The use of surveillance drones, however, should be under democratic control; drone policies and the actual use of drones should be made public; drone image retention should be restricted; and a systematic examination of the costs and benefits of drone surveillance is needed (Cavoukian, 2012). Although, for example, the Dutch legal framework for applying CCTV is based on the principles mentioned earlier, in practice several flaws can

---

* http://motherboard.vice.com/read/police-used-a-drone-to-chase-down-and-arrest-four-dui-suspects-in-a-cornfield.

still be detected. In particular, this is caused by the blurring of the boundaries between public and private cameras, since the police are using private (security) camera footage and citizens actively disseminate self-made films of incidents (Schreienberg et al., 2011, p. 90). Drones equipped with cameras will definitely strengthen this trend. It might be able to regulate the deployment of surveillance drones by the police according to the aforementioned principles and enforce it. As a matter of fact, this is precisely the course that has been taken in the Netherlands, where the legislation for the use of CCTV has been amended to include surveillance cameras placed on drones. Under the new legislation, the city's mayor decides what kind of camera surveillance is used: fixed, vehicle mounted, or airborne. Privacy legislation, however, is expected to be far less effective in relation to privately owned drones, because the probability of detecting malicious use might be very low.

Home automation and robotics in telecare raise have already raised privacy issues in the short term. A balance must be found between protecting the personal living environment, on the one hand and, on the other hand, satisfying the need for patients to live independently. An important aspect of this is the degree to which people are in control of the data that are being collected about them and the way in which domotics enables caregivers to look into their private domestic sphere from a distance.

With the robotization of traffic, what is relevant is the comparative assessment of traffic safety and privacy. The European Commission has designed the eCall system, which is a system that automatically contacts the emergency services when there is an accident, and this is obligatory for private cars sold from 2015. The European Parliament demanded that the eCall system will have a "dormant existence" and will not take any action until an accident has taken place. This system, therefore, cannot be applied for tracking criminals. However, a shift in goals (function creep) continues to be a risk. The question is whether such possibilities in the future will also be applied to ensure (traffic) safety. For example, the enforcement of traffic rules could take place through "vehicle to infrastructure" systems, which monitor driver behavior or even enforce it. Previously, we even asked whether or not we are morally obliged to introduce such compelling systems from the perspective of traffic safety.

7.3.2 Safety, Cyber Security, and Misuse

The call for safety is an important driving force behind the development of robots. The technological safety and reliability of robotics itself, however, are also a constant point of concern. Especially because—contrary to factory robots—service and social robotics comprise extensive interaction with humans, the safety of robots is an important condition for public acceptance. It is important that (severe) accidents are avoided when robots are used. In 2014, ISO 13482, safety requirements for personal care robots, was released, which applies to the mobile servant robot, the physical assistant robot, and the person carrier robot. However, at this moment, there is a lack of internationally accepted safety norms for autonomous cars. This lack of norms stands in the way of social acceptance and often also of innovation. In the field of cars, an international ISO standard for task-supporting systems could significantly contribute to gaining public trust.

The use of public and civil drones presents the most urgent situation with regard to safety. In particular, the rapid increase of their use in the United States is causing major concerns in the Federal Aviation Administration (FAA) about aerial safety. The FAA especially fears midair collisions between a drone and an airplane. Although drones are not allowed to fly too high (lower than 120 meters [or 394 feet]) and not too close to commercial airports (not within 3.8 kilometers [or 2.4 miles]), many operators do not obey these regulations. Moreover, the current traffic avoidance systems that guarantee the safety of civil aviation often cannot detect small drones.

It is not only unsafe robot systems that breach public trust. The misuse of robotics, too, breaches trust, because it draws strongly on human fear. Think of the use of robotics by criminals and/or terrorists. Robots can be powerful weapons. Criminals and terrorists can utilize robots themselves and can also hack robots and thereby pose danger. The robotization of cars makes cars vulnerable to hacking attempts. Through hacking, sensitive data can fall into the hands of the wrong people. Through hacking cars, it is even possible to gain complete control over them. The same applies, even more so, to police and military robots. The vulnerability of such robot systems is connected to their networked character. This is because it is difficult to protect the information and communication networks on which the robot in question is dependent. The speed of communication and cyber security (usually dependent on the complex encryption of the message) are at odds with each other.

7.4 The Lifelike Appearance of Social Robots

> Friedman and Kahn (1992) raise another serious ethical concern posed by the human tendency to anthropomorphise technology: the harm that can result from imputing faculties to machines that they do not have.
>
> **Wallach and Allen (2010, p. 45)**

In addition to language, human interaction is very much about non-verbal communication. A major goal, therefore, is to engineer embodied types of communication, such as facial expressions, body warmth, and posture. To express these forms of nonverbal communication, a virtual or physical body is needed. To be successful, the robot body does not need to resemble a human body, since people have a strong tendency to attribute human motivation, characteristics, or behavior to animals and inanimate objects. Human beings thus easily anthropomorphize robots. At the same time, robots that do aim to have a strong animal- or human-like appearance raise high expectations in people with regard to their body and behavior. As a result, robots that seem almost, but not quite, human- or animal-like will repel us. This psychological effect has been known for a long time known as the "uncanny valley" (Mori, 1970). The engineering of social robots, therefore, is about controlling, as Patrick Duffy (2006) calls it, "the power of the fake."

As with any technology, this power can be used for good or bad to empower or disempower people, and sometimes there is just a thin line between the two. For example, the international children's aid organization *Terre des Hommes* put online child sex tourism on the public and on the political agenda in many Western countries by using an online avatar called Sweetie, a virtual 10-year-old girl.* Here, the power of the fake was used to ensnare more than 1000 online

---

* http://terredeshommesnl.org/en/sweetie.

pedophiles in 65 countries. In countries such as the United States, the United Kingdom, and the Netherlands, the possession of virtual child pornography is punishable. A Dutch court ruled that virtual child pornography could become part of a perverted subculture that promotes the sexual abuse of children. Similarly, the foreseeable development of sex robots that resemble children raises the question of whether this use of social machines with a childlike appearance for "recreational" goals should be criminalized or not. Before organizing a total legal ban, robot ethicist Ronald Arkin (Hill, 2014) and robot expert Ben Way,* however, plead for study and debate on the question of whether a state-controlled use of childlike sex robots, for example, as part of treatment plans for pedophile sex offenders, could reduce child abuse.

Since people easily perceive computers as social actors, computers can be used successfully to influence what we do and think (Fogg, 2003). This is even more true for a robot, which is a virtually or physically embodied computer. Social robots thus are persuasive technologies that may raise issues of autonomy and mental and physical integrity. To what extent these issues come into play depends on the actual application and many contextual factors, such as whether users are well informed about the way in which the robot tries to influence them and whether users willingly accept that the robot influences them or that it manipulates them on a subconscious level.

In most instances, people enter voluntarily into a relationship with companion robot for reasons relating to entertainment and having fun. The toy robot can become part of the fantasy world of a child. In a similar vein, the sex robot—a companion robot with a focus on the physical aspect of the relation—can only function if it stimulates and can play a part in the self-chosen imagination of the user. Roxxxy, the first female sex robot, for example, offers the user five different types of illusions, from the reserved Frigid Farah to SM-Susan. In these circumstances, users enter into a voluntary relationship with an artifact to build their dream world. Here, the power of the fake

---

* http://www.dailymail.co.uk/sciencetech/article-2695010/Could-child-sex-robots-used-treat-paedophiles-Researchers-say-sexbots-inevitable-used-like-methadone-drug-addicts.html.

enriches the imagination. Problems can occur when the dream world takes precedence over the real world, for example, if people become addicted to robots, just like many people have become addicted to playing video games. Another danger is that the artifact is going to dominate the fantasy world of people, and machines will "even dream your dreams" (Gorka, 2001).

Social interaction robots may be used for training and therapeutic purposes. For example, people who feel insecure about their sexual orientation could relatively easily experiment with male and female sex robots. The Japanese *Simroid* is a robotic dental patient that can be used to train dentists (Lau et al., 2009, p. 21). This android has been modeled on a 28-year-old female patient who is very scared of visits to the dentist. Through sensors in its mouth, the Simroid can react to incorrect maneuvers made by the trainee dentists and say, "Ouch … that hurts!" The female robot also has sensors in its chest to warn students if they accidentally touch inappropriate areas. In Belgium, a humanoid Nao robot* called *Zora*† is programmed to lead the weekly gym and bingo sessions in a home for the elderly. Interactive technology can also be applied for therapeutic reasons. A flight simulator, for example, can help to cure a fear of flying. The University of Hertfordshire has developed a child-sized robot called *KASPAR* that can be used as a therapeutic or educational tool to encourage social interaction skills in children with autism.‡ The fact that KASPAR has human features but is clearly not a human allows children to safely investigate the social robot, in a way (such as squeezing its nose) that would be difficult or inappropriate with a real person. Simulated social interaction thus can offer a safe route toward acquiring self-knowledge, confidence, and social skills.

The claimed effect of the therapeutic robot seal Paro on elderly patients is threefold: psychological (relieving feelings of loneliness and isolation, helping patients to be more relaxed, and encouraging patients to take care of the robot and thereby strengthening their self-esteem), physiological (e.g., improving vital signs), and social (e.g., encouraging patients to interact with the robot and being a

---

* http://en.wikipedia.org/wiki/Nao_(robot).
† http://www.zorarobot.be/?lg=en.
‡ http://www.herts.ac.uk/kaspar/supporting-children-with-autism.

topic of conversation between patients and their human caregivers) (cf. Borenstein & Pearson, 2010; Lau et al., 2009, p. 29). In the case of elderly people with dementia, it would often be hard to ask for their informed consent about the use of social robots like Paro. But if these effects could indeed be realized, the fact that they are to a certain extent "tricked" by an artificial social companion might well be regarded by their close family members as a permissible kind of white lie, at least when the use of Paro will not be at the expense of the care of their relatives that is carried out by humans.

Persuasive robot technology, however, also has the potential to deceive and manipulate people, in particular vulnerable groups, such as children and mentally less able people. For example, interaction technology is not limited to replicating or mimicking human behavior. Scientists in the field of human–technology interaction (HTI) are experimenting with "human responses to media technologies that transform or augment our capabilities beyond what was previously possible, creating and studying entirely new experiences and social phenomena" (IJsselsteijn, 2013, p. 20). Experiments with an avatar, for example, demonstrated that hearing somebody's heartbeat increased perceived intimacy even more than eye contact. This seems to indicate the strong, maybe even addictive, power of artificial heartbeats, for example, in sex robots. According to HTI professor IJsselsteijn (2013), such research illustrates "the potent transformational effects that media may have on our social interactions, deeply affecting people's interpersonal connectedness and behaviour" (p. 21).

7.5 Degree of Autonomy of Robots

The degree of autonomy of robots concerns the degree of control that is delegated by the user of the robotic system to that system. This section reflects on what this delegation of control from humans to robotic machines means for the distribution of responsibilities and liabilities. In particular, we focus on the automotive industry and the military, where a shift from "man-in-the-loop" to "man-on-the-loop" to "man-out-of-the-loop" (autonomous robots) is clearly visible. As a result, the introduction of robotics in these two socio-technical practices brings along a complex set of questions in the field of responsibility and liability.

7.5.1 Systems View on Responsibility and Liability

Addressing these issues requires a systems perspective. Let us take the example of cars. With cars, an integral legal–social and technical system has come into place for dealing with issues of liability. This concerns compensation rights, based on which it can be determined who is liable in what way, varying between liability based on fault and strict liability to product liability. Furthermore, there are legal definitions that ensure that those who are guilty can carry the burden of debt. Think of liability insurance, which protects insured parties against the risk of liability. Furthermore, the owner of a car is legally required to keep it in good repair (think of the obligatory periodic motor vehicle inspection test) and to have it insured. Furthermore, car manufacturers must meet all sorts of ISO standards at the product level. Robotization can pressurize this system of distributing responsibilities and liabilities in various ways. In the following, we indicate several issues that will come up for discussion if more and more control is delegated from humans to robot cars and/or armed military robots. We will look at the changing roles and responsibilities of manufacturers of cars and military equipment, car drivers, road authorities, drone operators, and commanding officers, depending on whether man is in, on, or out of the loop.

7.5.2 Man-in-the-Loop

Car drivers and plane pilots have always been in the loop, that is, part of a technological system. Robotization, however, is changing these socio-technological systems and challenging the existing way of organizing responsibilities and liabilities. For example, the introduction of task-supporting systems calls for new types of ISO standards. Section 7.3 has already mentioned the current lack of ISO standards for such robotic systems. Thus, although car manufacturers are liable when damage is caused by a defect in their products, there is uncertainty about what exactly they need to do to guarantee the safety of their products. The use of military robots faces similar problems. Although manufacturers have the moral obligation to make safe and reliable military products so that they satisfy various legal and ethical conditions, for example, as set out in international humanitarian law, they are rarely held responsible for accidents caused by poor design (Thompson, 2007). Specific international legal safety guidelines, such as ISO norms, can prevent neglect of the security aspects of armed military robots. Such a regulatory framework, however, is currently lacking.

The complexity of robotic systems increases the demand to record its functioning in order to keep track of whether failures are man-made or caused by technical deficiencies, and enables that recording to be done digitally. Cooperative driving systems provide an example of a complex robotic system. In addition to cars, vehicle-to-vehicle (V2V) and vehicle-to-infrastructure (V2I) communications are subsystems within that smart system. Such a complex system makes the establishment of producer liability complex. Road authorities are liable if the road does not meet the requirements that may be expected of it in the given circumstances. When roads become smarter there might be discussion about the scope of the term "road equipment" and thus the applicability of the liability of the road authority. It would be most obvious to also define intelligent wayside systems and their underlying computer systems as road equipment. But how can it be determined that the cause of damages can be traced to the failing of a product rather than to an external cause?

This same problem was faced with the robotization of airplanes. To determine who is liable for a particular accident, black boxes (flight data recorders) were installed in airplanes in which all flight data of the last part of the flight are stored electronically. This could also be done with other robotic systems. In the United States, the National Highway Traffic Safety Administration (NHTSA) demanded that producers incorporate a so-called event data recorder in all new cars as of September 2012.* Such a black box stores, among other things, information about the speed of the car before the accident, the use of the brakes, and the use of seatbelts. Just like with the European eCall system, the introduction of the event data recorder in the United States raises many questions in the field of privacy. While the NHTSA asserts that this system has been introduced to increase traffic safety, others state that this has been done to resolve liability issues in particular, especially those relating to the car producers themselves.† Besides

---

* http://www.nhtsa.gov/EDR, consulted on October 29, 2014.
† See, for example, http://www.youtube.com/watch?v=KzYLJHgUf0k, consulted on October 29, 2014.

recording the functioning of the car, there will also be a growing need to record the workings of other relevant parts of the smart robotic mobility system, such as the V2V and V2I systems.

In relation to the liability of the driver, measuring against the "perfect" motorist is usually the norm. The fact that driver assistance systems allow drivers to better respond to hazards may affect their liability. For example, if a system has warned about a slippery road surface ahead, the driver can no longer claim that the circumstances were unforeseeable. The trend, for example, in the European Union, is that more and more driver assistance systems are becoming mandatory. This leads to a thought-provoking tension. Guided by the norm of the perfect driver, driver assistance systems will increase the responsibilities of the driver. On the other hand, their reliance on such systems makes many drivers less alert. In addition, driver assistance systems can lead to de-skilling, so driving ability may deteriorate. Parallel to the introduction of these systems, it would be wise, therefore, to make driving with driver assistance systems a mandatory part of the driving license.

7.5.3 Man-on-the-Loop

Questions about the reliance of man on technology, or even over-reliance, become even more relevant when we move from "man-in-the-loop" to "man-on-the-loop." In this latter situation, decisions made by the human being will be mediated more and more by the robotic system. A crucial question is, therefore, in what way technological mediation influences the actions of the operator. In making decisions, drone operator is highly dependent on information that comes to them from the armed military robot system. Although operators are legally responsible, one could wonder to what extent they can reasonably be held morally responsible for their actions.

In the case of geolocation, for example, the drone operator's decision relating to life-and-death situations seems to depend more on artificial intelligence (triggered by an electronic signal) rather than on human intelligence. The drawback of this kind of mediation is that operators will no longer make moral choices, but will simply exhibit influenced behavior—because subconsciously they will increasingly rely, and even over-rely, on the military robot (Cummings, 2006). This could even blur the line between nonautonomous and autonomous systems, as the decision of the human operator is not the result of human deliberation, but is mainly determined or even enforced by a military robot.

7.5.4 Man-out-of-the-Loop

Although operating an autonomous car on public roads is far more complex than flying an airplane, automotive industry consultant IHS Automotive (2014) predicts that the share of autonomous cars will grow from about 1% to 9% of the cars sold worldwide in 2025 and 2035, respectively. According to these predictions, more than 50 million driverless cars will be driving around the world in 2035.

Before such a scenario can become reality, a lot of testing needs to be done, not only on closed test circuits but also on the public road. Various governments are setting up laws to enable experiments with self-driving cars that would otherwise be illegal and uninsurable. Pleas for the establishment of free experimental zones can be found with respect to various types of robots. In Japan, Tsukuba City has been designated an experiment zone for the development of mobile and service robots. This is linked to the establishment of a Robot Safety Center for the development of safety standards and a certification system for service robots (Flatley, 2010).

Unlike car drivers, people who are transported by these autonomous robot cars are released from their responsibility to avoid accidents. That responsibility has shifted completely to the developers and operators of the robotic system: the car manufacturer and road authorities. Manufacturers of autonomous cars become types of common carriers, such as railways and taxicab companies, that are held to owe passengers the highest standard of care and are liable for even the slightest negligence.

When, in the future, the army employs fully autonomous armed robots, there will still be people involved who "plan the operation, define the parameters, prescribe the rules of engagement, and deploy the system" (Quintana, 2008). If a robot does not operate within the boundaries of its specified parameters, it is the manufacturer's fault. Moreover, the ancient Doctrine of Command Responsibility implies that the commander is responsible for both the legal and the illegal use of the autonomous robot.

7.6 Robot Systems as Dehumanizing Systems

The basic intuition from which the capability approach starts, in the political arena, is that human capabilities exert a moral claim that they should be developed.

Nussbaum (1999, p. 43)

Delegating a considerable degree of control to robots also raises the fundamental normative question of what kind of decisions and actions we want or do not want to leave to a robot. Do we want activities with a strong social and moral dimension, such as killing people or looking after children or elderly people, to be decided solely by computers? The deployment of robots in social practices outside the factory raises these type of questions because, in contrast to industrial robots, service and social robots are *intimate technologies* that nestle themselves close to and between people, collect information about us, and have human traits, in the sense that they may exhibit intelligent behavior or social capabilities or may touch us with their outward appearances (Van Est, Rerimassie, van Keulen, & Dorren, 2014). As a result, such intimate robotic technologies have the power to radically change the conditions of our humanity. What it means to be human has therefore become one of the central moral and bio-political issues of this century (Broadbent et al., 2013). Whereas Section 7.2 described how robots might improve the techno-human condition, this section deals with the existential concern that robots might cause dehumanization. This can happen when robotization as rationalization overshoots its mark and leads to socio-technical systems that become antihuman or even destructive to humans.

The debate on what is human or antihuman usually circles around the concept of human dignity. Human dignity forms the basis of human rights—such as the right to autonomy and self-determination, privacy, and physical and mental integrity—as set out in the 1948 Universal Declaration of Human Rights. When we reflect on the techno-human condition, we also believe that the notion of human sustainability (Van Est, Klaassen, Schuijff, & Smits, 2008) has added value. Human sustainability concerns the preservation of what makes human beings truly human: which aspects of human beings and being human do we see as makeable and what kinds of aspects would we like to preserve? For example, the concern about whether the use of robots will lead to social de-skilling links to the notion of human sustainability. Nussbaum (1999, pp. 39–40) touches upon similar topics when she asks: "[W]hich changes or transitions are compatible with the continued existence of … [a human] being as a member of the human kind and which are not?" and "[W]hat do we believe must be there, if we are going to acknowledge that a given life is human?" Her capabilities approach (cf. Nussbaum, 1999, pp. 39–47) and the related list of central functional capabilities of humans that all citizens should have presents an interesting mix of human rights and human abilities that are required for humans to have a life in which a kind of basic human flourishing will be available. In addition to referring to basic human rights (such as bodily integrity, freedom of speech, and the right to affiliation), the list of central human capabilities refers to topics such as senses, imagination, thought (e.g., "being able to use the senses; being able to imagine, to think, and to reason—and to do these things in a 'truly human' way") and emotions (e.g., "being able to have attachments to things and persons outside ourselves; being able to love those who love and care for us") (Nussbaum, 1999, p. 41). In the following, we describe various ways in which the employment of robots might lead to dehumanization as it affects either human dignity or human sustainability.

7.6.1 Undermining Human Dignity

Large numbers of thinkers point out that the utilization of robotics in the fields of care, entertainment, and the military could undermine human dignity in various ways. Especially with regard to the utilization of robotics in care, there is a fear that this will eventually lead to an objectification of the care receivers. Pitiless circumstances in current health care fuel this discussion; think, for example, of the discussion surrounding the use of adult nappies to decrease the work it takes for caregivers to take elderly people to the bathroom. The quality of life of patients should be the leading principle of telecare. Independence and social interaction with other people are central to this.

From the viewpoint of human dignity, the utilization of tele-controlled armed drones provides a rather multifaceted picture. The story starts with the observation that a war without robots can also have a dehumanizing effect on both soldiers and citizens. Consider the fact that less than half of the soldiers that participated in the military operation Iraqi Freedom think that civilians should be treated with respect, one-third find that torture is allowed to save the life of a colleague and 10% admit to having maltreated Iraqi citizens (Office of the Surgeon Multinational Force-Iraq & Office of the Surgeon General United States Army Medical Command, 2006). Advocates of tele-controlled armed military robots state that their use will improve that situation. They believe that creating an emotional and thus moral distance between the drone operators and the ethical implications of their action will decrease the psychological suffering among military staff and lead to more rational decisions. Critics fear, however, that creating more distance between an action and its consequences involves a danger that operators will make life-or-death decisions as though they are playing a video game.

In practice, the use of drones enables a certain mix of both emphatic bridging and distancing (Coeckelbergh, 2013). Almost half of the drone pilots on active duty report high levels of stress, and about one-third report emotional exhaustion or burn out.* This might relate to the fact that drone pilots—contrary to plane pilots—regularly watch potential targets before eventually attempting to kill them and have the possibility of seeing the results of their actions in detail. On the other hand, using geolocation to target the enemy clearly has a moral distancing effect and gives rise to the danger of numbing the operator's feelings and objectifying the enemy. The use of armed drones also gives rise to anxiety and psychological trauma among civilian communities, because "[t]hose living under drones have to face the constant worry that a deadly strike may be fired at any moment" (Stanford Law School & NYU School of Law, 2012, p. vii).

7.6.2 Undermining Human Sustainability

Entering into meaningful social relations, dealing with violent conflicts, and caring for other people each requires complex skills. The use of robots to assist in such situations leads to the question of to

---

* https://forums.gunbroker.com/topic.asp?TOPIC_ID=554910.

what extent this utilization will contribute to strengthening such complex skills or actually undermining them. In the latter case, one speaks of *de-skilling*. Turkle (2011) is afraid that people will lose their social competencies—such as dealing with rejection or making up after fights—if they spend too much time with socially capable robots. Sparrow (2002) is even afraid that at some point people will no longer realize that they are dealing with robots (socio-emotional machines) rather than with people. There is a fear not only that individual skills will be affected but also that the cultural way we interact, the way we deal with each other, emotions, violence, love, and how we relate to nature may eventually be eroded. Therefore, a crucial question is: do robots have a positive or negative effect on social skills and cultures we consider to be truly human and would like to strengthen?

Section 7.2 showed some opportunities for utilizing robots for entertainment and therapeutic ends. Robots can be utilized to teach autistic people to recognize certain human emotions in small, regulated steps, in a safe environment, and sex robots could teach people the art of making love. Kloer sees the advantages of robots as ideal sex partners, but fears that this will further rationalize sex.* A sex culture dominated by robots will devalue sex so much that it will not matter anymore if the sex is human. Accordingly, Kloer puts forward an intriguing question: "Has the commercial sex industry made sex so mechanical that it will inevitably become ... mechanical?"

This remark is thought provoking because it reminds us about the rationalization processes that took place within the factory during the first half of the nineteenth century and paved the way for the mechanization of the production process. Robots then caused the next step in this ongoing attempt to rationalize production, ultimately by replacing people with machines. Kloer's observation thus forces us to reflect on how rationalization processes are currently shaping and changing various social practices outside the factory and what role robots as well as humans are playing therein. It also puts forward the issue of what stage of rationalization a certain social practice is in now and where rationalization might lead. For example, has war already been mechanized to the extent that killing people can be outsourced to robots? In order to reflect on a question like this, one needs to

---

* http://news.change.org/stories/are-the-robots-the-future-of-prostitution.

unravel the various stages of robotizing a certain social practice. For example, initially, plane pilots dropped bombs. The introduction of the drone led to the cubicle warrior. The first cubicle warriors were mostly pilots with flying and war experience. The new generation of cubicle warriors, however, are trained, from the start, as operators of tele-controlled airplanes. These observations already give a flavor of how the required skills of pilots have changed and how the culture of flying may have shifted.

Sociocultural practices both enable and constrain the development of certain individual skills. Vallor (2011), for example, refers to the benefit of the current culture of care. Through caring for others, a caregiver develops certain capacities that are indispensable in human relations, such as empathy. The utilization of care robots takes away the opportunity to develop these qualities. In the field of military and police robots, too, this matter is at stake. Hambling (2010) wonders whether, through teleoperations, the police will eventually lose the skills that are necessary for dealing with serious situations—skills that require long training and experience. Here, again, not only individual skills but also cultures of dealing with conflict and violence are at stake. The experience of Nordholt, former chief superintendent of the Amsterdam police force, was that technology had a big impact on the way police work was organized and politically debated (De Jong & Schuilenburg, 2007). He found that a focus on technology has a tendency to strengthen the supervising role (law enforcement and criminal investigation) of the police at the cost of its social role (supporting citizens).

7.6.3 Current Relevance

Robotization as rationalization thus leads to concerns about dehumanization. These concerns present the antithesis of the dream of autonomous and socially and morally capable machines. Since that engineering dream is speculative, one might argue that the concerns it provokes are also speculative and should not be taken too seriously. The influence of robotics on our individual human rights, skills, and sociocultural capital, however, is not only a matter for the long term; it deserves our attention today. For example, systems that support driving tasks are already leading to less alert and reckless driving behavior. And Turkle's (2011) fear of the antisocial influence of social robots is based on her long-term research on the influence of social media and mobile telephones on the communication between young people. Turkle states that the youngest generation is much less empathic than preceding generations, because intimacy can be avoided and relations through the Internet or devices are much less binding. The core concern about dehumanization actually forces us to continually address the question of what it means to be human in a robot society.

In debating this issue, we should consider the use of robots from the view of the long-term process of the rationalization of society. In Section 7.1, it was concluded that by enveloping our daily environment in an ICT-friendly infosphere, we are laying the basis for the application of service and social robots in various social practices. This enveloping is a socio-technical process, which is about building technical infrastructures, such as GPS, 4G mobile networks, and data management systems, and also about social shaping. For example, the effective use of a robot vacuum cleaner depends on "roombarization" (Sung et al., 2007): the rationalization of the home environment to make it robot-friendly. As said earlier, rationalization presents all kinds of opportunities. The concern about dehumanization reminds us about the risk we are running, as Floridi (2014, p. 150) explains: "[B]y enveloping the world, our technologies might shape our physical and conceptual environments and constrain us to adjust to them because that is the best, or easiest, or indeed sometimes the only, way to make things work."

We should try to avoid building robotic systems that push us into an irritating straightjacket of efficiency. This is certainly a topical subject, especially because subtle forms of systematism easily escape our attention (Van't Hof, Van Est, & Daemen, 2011, pp. 142–143). For example, facial recognition in digital passports compels every citizen not to laugh upon having their passport photograph taken. In this way, we quietly lost something beautiful: a moment of freedom of expression. The dehumanization perspective keeps us aware of the potential cumulative effect of these easily unnoticed ways in which rationalized robotic systems may constrain us. In particular, the effect of drones on our privacy and autonomy calls for a timely public discussion. Living under drones in war zones not only gives rise to anxiety and psychological trauma among civilian communities, but may also have a chilling effect on the public's behavior in peaceful times. Purposely echoing Bentham and Foucault, Canadian Information and Privacy Commissioner Ann Cavoukian (2012, p. 1) states that "the increased use of drones ... has the potential to result in widespread deployment of panoptic structures that may persist invisibly throughout society." In a society with both surveillance—Big Brother who is watching us from above—and sousveillance—Little Brothers who are watching us from below—people will anticipate the fact that they can be observed anywhere at any time, and autonomy is lost.

7.7 Governance of Robotics in Society

The discourse on how to enact democratic governance of innovation properly is currently dominated by the notion of responsible innovation. Responsible innovation aims to achieve ethically sustainable and, from a societal point of view, acceptable types of innovation (Von Schomberg, 2011). Achieving responsible innovation requires the timely involvement of the opinions and capabilities of relevant societal actors. With respect to the organization of the required interactive processes, we will restrict ourselves to two remarks. First of all, it is important that the introduction of robotics in society is brought to the attention of policy makers, politicians, and the broader public. Governments have a responsibility to ensure that the debate on the application of new robotics is started early. For example, the debate on the utilization of military robots has clearly begun way too late. Second, this section starts with a strong plea for user involvement. The remainder of this section then reviews a number of political and regulatory issues that need to be tackled in order to introduce robotics into society in a responsible manner.

7.7.1 Putting Users at the Center

With innovation endeavors, it is important to bridge the gap between the expert views of (technological) developers and the wishes of users. Therefore, the views, needs, and worries of future users must be addressed as early as possible in the developmental process. In the current R&D climate, users are often only consulted at a late stage, or not at all (Van der Plas, Smits, & Wehrman, 2010). Every subject, and thus also the subject of robot systems, has an assigned value. It is important that during the design process the various views of stakeholders are taken into consideration. This can be done by disciplines known as "constructive technology assessment" (Schot & Rip, 1997) and "value sensitive design." With this last approach, specific attention is given to the ethical issues and the broad set of values that play a role in the development and application of technology.

Users must be educated early in the process about the possibilities and impossibilities of robotics systems. The technologies change the various social practices, and this often happens stealthily. This is a concern in all of the fields of application. In health care education, serious attention must be given to telecare and the consequent changing role of caregivers. A similar problem can be seen in the shift from airplane pilot to airplane operator. In the field of mobility, it appears important that drivers become familiar with, for example, Adaptive Cruise Control. A first step could be to make driving using task-supporting systems an obligatory element of passing one's driving test.

7.7.2 Political and Regulatory Issues

Worldwide investment in research on and the utilization of military robot technology has rapidly accelerated over the last decade. This novel arms race calls for a broad international debate on the consequences of using military robotics. An important goal for such a debate is to curb the proliferation of armed military robots and to keep these machines out of the hands of fundamentalists and terrorists. A second goal is to develop common ethical and legal principles for a responsible utilization of armed military robots. An essential condition of international humanitarian law is that at all times someone can be held responsible for an undesirable consequence. It is, therefore, important that humans always stay "in the loop" and take decisions on life-and-death issues. This principle is under threat because of the trend of "man-in-the-loop" heading to "man-on-the-loop" and ultimately to "man-out-of-the-loop." Therefore, an international ban on autonomous armed military robots within the foreseeable future is desirable. The start of a worldwide discussion on that topic was witnessed in May 2014, when the UN Convention on Certain Conventional Weapons hosted an informal meeting about autonomous killer robots.

The gradual but steadily persistent robotization of traffic, too, calls for timely governmental action and the formation of views on its future. For robot cars, the comparative assessment between safety and privacy plays a central role. In several countries, self-driving cars are being promoted to increase traffic safety. Discussing safety in terms of such a long-term option sounds like a clever maneuver to evoke a political debate about the use of technical measures that already exist. A debate should already be taking place on the use and necessity of the introduction of a system for automatic speed adaptation. The expected safety gain, meaning the decrease in the number of traffic casualties, makes a serious political assessment obligatory. As indicated in Section 7.2, the core ethic at stake here is the principle of unnecessary risk (cf. Strawser, 2010). In essence, this concerns the political question of whether we, as a society, accept that traffic accidents cause casualties. If we choose to take this view, like the Swedish parliament in its *Vision Zero*, we thereby choose to make traffic systems optimally safe by means of the best available (robotics) technologies.

Cooperative driving may begin to play a role in the medium term. While it will still take years before it is safe enough, it is already high time that governments, industries, knowledge institutes, and relevant societal organizations begin talking about the technical and legal aspects, which deserve attention because of the potential effects of cooperative driving, such as system safety and standardization, and liability if there is a malfunction. Good management of these aspects requires a lot of time. It is also important to anticipate the gradual introduction of autonomous cars, and thinking about different scenarios in order to get to grips with the potential radical consequences for public transportation, car ownership, road usage and urban planning, and so on, is important. The autonomous car also forces regulators to rethink the relationship between the car and the driver. In the near future, maybe the car will be obliged to take a driving test. Finally, autonomous cars will encounter situations in which they have to make life-and-death choices. Encoding moral decisions into software is a technological and ethical challenge, but also a political and regulatory one.

In the field of home robots, societal issues that are raised by sex robots deserve attention. This concerns ethical and legal questions surrounding the possible use of child robots for sex and the question of whether sex robots could eventually become an acceptable substitute for human prostitution. In various countries, the solicitation of or possession of virtual child pornography is punishable. The legislation used for that purpose, however, is not sufficient for making sex with child robots punishable. If politicians wish to forbid this kind of behavior, lawmakers must develop a legal framework for this. At the moment, the prostitution industry knows about very distressing abuse, ranging from the trafficking of women to unpaid labor (slavery). This justifies research on the question of whether sex robots could be reasonable alternatives to human prostitutes in the future.

The subject of care, especially the expected breakthrough of telecare, or home automation, calls for a formation of views. How can we deal with privacy, patient autonomy, and informed consent? A first challenge is to search for a good balance between privacy and health and patient safety. Robotics and home automation could be utilized to increase the autonomy and independence of patients. However, it is important that such technological support is always assessed from a patient perspective. In many cases, outside care is experienced as an infringement of autonomy, but it does help to fight loneliness. Loneliness is a large social problem and we must watch out that home automation does not make it worse.

The demand among law enforcement agencies, citizens, and businesses to use drones in the city is on the rise. The deployment of public and civil drones, however, raises various safety, security, and privacy issues that are in need of regulatory guidelines. Safety is the overriding concern among European and U.S. policy makers. Fragmented rules exist to authorize the ad hoc use of drones. A more comprehensive regulatory framework is needed, however, that safely integrates a wide variety and use of drones into the aviation system. Currently, the progress in regulating the safe use of drones in both the United States and Europe is slow. Although basic privacy rules apply to all drone operators, a coherent way to address some key issues with regard to privacy does not exist. In particular, the exploding use of civilian drones is a threat to privacy. To deal with the privacy issue, it would be a good idea to set up a baseline consumer protection law that details permissible use of drones in domestic airspace by both law enforcement agencies and private parties (Schlag, 2013).

7.7.3 Balancing Precaution and Proaction

Policymaking can be guided by both precautionary and proactive attitudes toward risk governance: the regulatory focus of precautionary policy makers is on the prevention of worst outcomes, whereas proactive ones seek to promote the best available opportunities (Fuller, 2012). Regulation should aim to balance precaution and proaction. Robotics is developing within a network of existing national and international legal frameworks. On the one hand, existing regulatory structures may pose unnecessary barriers to the utilization of robots and may hamper innovation. On the other hand, regulation is needed to introduce robotics into society in a responsible manner. Where this concerns privacy, think of the European Data Protection Directive, or of military robots that must meet the requirements of international humanitarian law. However, we saw that new robotics often requires the adaptation of existing laws or even entirely new types of regulatory frameworks. For example, many new robot systems lack a detailed legal–social system for dealing properly with the issue of liability. This lack of regulation, too, can hamper innovation. In such an uncertain situation, it is important to create space for innovation, and at the same time to, step-by-step, build a legal–social system. Our plea is to have experimental zones in which experiments with robotics can take place as long as certain rules are met. The government should take responsibility for various risks. This space for experiments ought to be linked to the stimulation of the setting of standards and the development of safety standards. Think of the Japanese Robot Safety Center, which is working on a certification system for service robots.

Finally, besides challenging us to make legal frameworks future proof, robotics also challenges us to reflect on our existing norms and laws, and sometimes obliges us to reconsider what is currently seen as feasible and desirable. This in turn may stimulate debate and lead to the formation of new policy goals and the instruments to meet them. For example, the prospect of a car robot that makes autonomous ethical decisions raises complex questions about which types of ethical-decision profiles should be used, and whether the user should be free to select the car's ethics setting. Lin (2014) thinks that allowing cars to have an adjustable ethics setting is a bad idea. But that is exactly how we have organized the current situation, in which users are free to buy a heavyweight car with a strong safety bumper that especially diminishes the damage done to its passengers. Thus, in an interesting manner, the difficult moral questions that are raised by the prospect of robotic systems may actually stimulate us to reflect on the norms and values underlying our existing socio-technical systems. Making such norms and values visible and debatable and, where deemed necessary, reconsidering them and adjusting current systems accordingly, may be the best way to prepare for the responsible introduction of robotic systems in our society.

7.8 Epilogue

> When I lost my faith in people, I put my trust in things
> To avoid the disappointment trusting people brings…
> I tried to do it all myself then, surrounded by my stuff
> All I found were limitations I could not rise above
> There are gadgets and contraptions, immaculate machines
> There's a program you can download now that will even dream your dreams
> It'll even dream your dreams for a monthly fee
> Clear up your complexion, you get a hundred hours free
> Possessions cannot save you the way some body can
> When I learned to care for others then the boy became a man
>
> **John Gorka (2001)**

The introduction of robotics into society is paired with an enormous human challenge. Making use of opportunities and dealing with their social, legal, and ethical aspects calls for human wisdom. Trust in our technological capabilities is an important part of this. But let us hope that trust in technology will not get the upper hand, because that would be a perfect formula for creating a technocracy, or more aptly here, robocracy. Trust in humans and the acceptance of human abilities, but also human failures, ought to be the driving force behind our actions.

Like no other technology, new robotics is inspired by humans. This technology aims to copy and subsequently improve human characteristics and capacities. It appears to be a race between machines and us. Bill Joy (2001) is afraid of this race, because he fears that humans will ultimately be worse off. His miserable worst-case scenario describes "a future that does not need us." In such a hyper-rationalized world, the robots and the elite owning them have come to control us or even replace us. Rationalization through robot-ization, however, should never be a goal in itself. The ultimate goal of robotics should not be to create an autonomous and socially and morally capable machine. This is an engineer's dream. Such a vision should not be the leading principle. Robotics, namely, is not about building the perfect machine, but about supporting the well-being of humans.

Our exploratory study shows that there is a balance in social practices between "hard" and "soft" tasks. The police take care of law enforcement and criminal investigation and also offer support to citizens. The war must be won, but also the "hearts and minds" of people. Care is about "taking care" (washing and feeding) and "caring for" through a kind word or a good conversation. We enjoy making love, but we especially want to give and receive love. Robotics can play a role in the functional side of social practices. But we must watch out that the focus on technology does not erase the "soft" human side of the practice in question. We should, therefore, take heed of insights from people like Nordholt, former chief superintendent of the Amsterdam police force, who states: "The current thinking in terms of control is strongly motivated by technology" (De Jong & Schuilenburg, 2007). Such a trap can easily lead to appalling practices: inhumane care, a repressive police force, a hardening of our sex culture and cruel war crimes.

Robotics does not exist for itself, but for society. Robotics ought to support humankind, not overshadow it. Our objective should not be to build a high-tech, robot-friendly world, but to create a human-friendly world. This begins with the realization that robotics offers numerous opportunities for improving our lives and the insight that how we envelop the world in a robot-friendly environment will define whether we seize those chances or whether that leads to a dehumanized society, guided solely by a strong belief in efficiency. This implies that sometimes there simply is no place for robots. A robot exists that can play the trumpet very well (see Figure 7.1).

Figure 7.1   A robot that can play the trumpet, developed by Toyota. (Photo courtesy of Rinie van Est.)

And yet, it would be disgraceful if the daily performance of the "Last Post" in Ypres in Belgium (see Figure 7.2), in memory of millions of victims of the World War I, were to be outsourced to a robot. Furthermore, we must watch out that trust in technology does not lead to technological paternalism. Even if, in the very distant future, there are robots that are better at raising our children than we are, we must still do this ourselves. An important aspect of this is the notion of personal responsibility and the human right to

Figure 7.2   In Ypres, the "Last Post" is played every day in honor of the liberators of Ypres. (Photo courtesy of Joost van den Broek/Hollandse Hoogte.)

make autonomous decisions and mistakes. Even if robots can carry out some tasks better than humans can, it might still make more sense for humans to carry on doing those tasks, despite doing them less well.

References

Arkin, R. (2009). *Governing lethal behaviour in autonomous robots*. Boca Raton, FL: CRC Press.

Arkin, R., Ulam, P., & Duncan, B. (2009). *An ethical governor for constraining lethal action in an autonomous system* (Technical Report No. GIT-GVU-09-02). Atlanta, GA: Georgia Institute of Technology.

Borenstein, J., & Pearson, Y. (2010). Robot caregivers: Harbingers of expanded freedom for all? *Ethics and Information Technology, 12*(3), 277–288.

Broadbent, S., Dewandre, N., Ess, C., Floridi, L., Ganascia, J.-G., Hildebrandt, M., … Verbeek, P.-P. (2013). *The online manifesto: Being human in a hyperconnected era*. Brussels, Belgium: European Commission.

Canning, J. S. (2006). A concept of operations for armed autonomous systems. Presented at the *third annual disruptive technology conference*, September 6–7, Washington, DC. www.dtic.mil/ndia/2006disruptive_tech/2006disruptive_tech.html (accessed January 23, 2014).

Cavoukian, A. (2012). *Privacy and drones: Unmanned aerial vehicles*. Toronto, Ontario, Canada: Information and Privacy Commissioner. http://www.ipc.on.ca/images/Resources/pbd-drones.pdf (accessed November 16, 2014).

Coeckelbergh, M. (2013). Drones, information technology, and distance: Mapping the moral epistemology of remote fighting. *Ethics and Information Technology, 15*(2), 87–98.

Cummings, M. L. (2006). Automation and accountability in decision support system interface design. *Journal of Technology Studies, 32*(1), 23–31.

De Jong, A., & Schuilenburg, M. (2007). Een cultuur van controle. Interview met Eric Nordholt. *Gonzo* (circus), *79*, 12–15.

Duffy, B. R. (2006). Fundamental issues in social robotics. *International Review of Information Ethics, 6*, 31–36.

European Commission. (2010). *Towards a European road safety area: Policy orientations on road safety 2011–2020*. Brussels, Belgium: European Commission. http://ec.europa.eu/transport/road_safety/pdf/road_safety_citizen/road_safety_citizen_100924_en.pdf (accessed September 4, 2014).

Flatley, J. L. (2010). *Robot safety center opens up in Japan: Crash test dummies still an unfortunate name for a band*. http://www.engadget.com/2010/12/28/robot-safety-center-opens-up-in-japan-crash-test-dummies-still/ (accessed August 1, 2014).

Floridi, L. (2014). *The fourth revolution: How the infosphere is reshaping human reality*. Oxford, UK: Oxford University Press.

Fogg, B. J. (2003). *Persuasive technology: Using computers to change what we think and do*. San Francisco, CA: Morgan Kaufmann.

Fong, T., Nourbakhsh, I., & Dautenhahn, K. (2003). A survey of socially interactive robots. *Robotics and Autonomous Systems, 42*(3–4), 143–166.

Friedman, B., & Kahn, P. (1992). Human agency and responsible computing: Implications for computer system design. *Journal of Systems and Software, 17*(1): 7–14.

Fuller, S. (2012, May 8). Beyond left and right: The future of ideological conflict. *ABC Religion and Ethics*. http://www.abc.net.au/religion/articles/2012/05/08/3497973.htm (accessed March 17, 2014).

Goldberg, M. (2014, August 8). Swedish prostitution law is spreading worldwide—Here's how to improve it. *The Guardian*. http://www.theguardian.com/commentisfree/2014/aug/08/criminsalise-buying-not-selling-sex (accessed January 23, 2015).

Gorka, J. (2001). When I lost my faith. *CD The company you keep*. St. Paul, MN: Red House Records.

Hambling, D. (2010, February 10). Future police: Meet the UK's armed robot drones. *Wired*. http://www.wired.co.uk/news/archive/2010-02/10/future-police-meet-the-uks-armed-robot-drones (accessed September 4, 2014).

Hill, K. (2014, July 14). Are child sex-robots inevitable? *Forbes*. http://www.forbes.com/sites/kashmirhill/2014/07/14/are-child-sex-robots-inevitable/ (accessed October 26, 2014).

Hughes, T. P. (1983). *Networks of power: Electrification in Western society, 1880–1930*. Baltimore, MD: The Johns Hopkins University Press.

IHS Automotive. (2014). *Emerging technologies: Autonomous cars—Not if, but when*. http://orfe.princeton.edu/~alaink/SmartDrivingCars/PDFs/IHS%20_EmergingTechnologies_AutonomousCars.pdf (accessed October 26, 2014).

IJsselsteijn, W. A. (2013). *Psychology 2.0: Towards a new science of mind and technology*. Eindhoven, the Netherlands: Eindhoven University of Technology.

International Labour Organization. (2012). *ILO global estimate of forced labour: Results and methodology*. Geneva, Switzerland: International Labour Organization.

Joy, B. (2001). Why the future doesn't need us. *Wired, 8*(4), 238–262.

KLPD. (2008). *Schone schijn: De signalering van mensenhandel in de vergunde prostitutiesector*. Driebergen, the Netherlands: Korps Landelijke Politiediensten—Dienst Nationale Recherche.

Lau, Y. Y., Van't Hof, C., & Van Est, R. (2009). *Beyond the surface: An exploration in healthcare robotics in Japan*. The Hague, the Netherlands: Rathenau Institute.

Levy, D. (2007). *Love + sex with robots: The evolution of human–robot relationships*. New York: Harper Collins.

Lewis, J. (2005, November 13). Robots of Arabia. *Wired*. http://www.wired.com/wired/archive/13.11/camel_pr.html (accessed October 26, 2014).

Lin, P. (2014, August 18). Here's a terrible idea: Robot cars with adjustable ethics settings. *Wired*. http://www.wired.com/2014/08/heres-a-terrible-idea-robot-cars-with-adjustable-ethics-settings/ (accessed September 4, 2014).

Mori, M. (1970). Bukimi no tami [The uncanny valley]. *Energy, 7*(4), 33–35.

Nielsen, M. (2011). *Reinventing discovery: The new era of networked science.* Princeton, NJ: Princeton University Press.

Nussbaum, M. C. (1999). *Sex & social justice.* Oxford, UK: Oxford University Press.

Office of the Surgeon Multinational Force-Iraq, & Office of the Surgeon General United States Army Medical Command. (2006). *Mental Health Advisory Team (MHAT) IV. Operation Iraqi Freedom 05-07* (final report). http://www.armymedicine.army.mil/reports/mhat/mhat_iv/mhat-iv.cfm (accessed October 14, 2014).

Quintana, E. (2008). *The ethics and legal implications of military unmanned vehicles* (occasional paper). London, UK: RUSI. Retrieved from http://www.rusi.org/downloads/assets/RUSI_ethics.pdf (accessed October 26, 2014).

Raymond, J. G. (2008). Ten reasons for not legalizing prostitution and a legal response to the demand for prostitution. *Journal of Trauma Practice, 2*(3–4), 315–332.

Ritzer, G. (1983). The McDonaldization of society. *Journal of American Culture, 6*(1), 100–107.

Schlag, C. (2013). The new privacy battle: How the expanding use of drones continues to erode our concept of privacy and privacy rights. *Pittsburgh Journal of Technology Law & Policy, 13*(2), 1–23.

Schot, J., & Rip, A. (1997). The past and future of constructive technology assessment. *Technological Forecasting and Social Change, 54*(2–3), 251–268.

Schreienberg, A., Van't Hof, C., & Koffijberg, J. (2011). Street images: A glance behind the scenes of camera surveillance. In C. van't Hof, R. van Est, & F. Daemen (Eds.), *Check in/check out: The public space as an internet of things* (pp. 83–93). Rotterdam, the Netherlands: NAi.

Schuurman, G., Moelaert El-Hadidy, F., Krom, A., & Walhout, B. (2009). *Ambient intelligence: Viable future or dangerous illusion?* The Hague, the Netherlands: Rathenau Institute.

Sharkey, N. (2008). 2084: Big robot is watching you. Report on the future for policing, surveillance and security. http://staffwww.dcs.shef.ac.uk/people/N.Sharkey/ (accessed March 17, 2014).

Sparrow, R. (2002). The march of the robot dogs. *Ethics and Information Technology, 4*(4), 305–318.

Stanford Law School, & NYU School of Law. (2012). Living under drones: Death, injury, and trauma to civilians from US drone practices in Pakistan. http://livingunderdrones.org/ (accessed October 5, 2014).

Strawser, B. J. (2010). Moral predators: The duty to employ uninhabited aerial vehicles. *Journal of Military Ethics 9*(4), 342–368.

Sung, J.-Y., Guo, L., Grinter, R. E., & Christensen, H. I. (2007). "My Roomba is a Rambo": Intimate home appliances. In J. Krumm et al. (Eds.), *LNCS: Vol. 4717. UbiComp 2007* (pp. 145–162). Berlin, Germany: Springer.

Thompson, M. (2007, September 26). V-22 Osprey: A flying shame. *Time.* http://content.time.com/time/magazine/article/0,9171,1666282,00. html (accessed November 16, 2014).

Tingvall, C., & Haworth, N. (1999). Vision Zero: An ethical approach to safety and mobility. Paper presented to *the sixth international conference road safety & traffic environment: Beyond 2000,* September 6–7, Melbourne, Victoria, Australia.

TNO. (2008). *Moving forward: To safer cleaner and more efficient mobility.* The Hague, the Netherlands: TNO.

Turkle, S. (2011). *Alone together: Why we expect more from technology and less from each other.* New York: Basic Books.

Vallor, S. (2011). Carebots and caregivers: Sustaining the ethical ideal of care in the twenty-first century. *Philosophy and Technology, 24*(3), 251–268.

Van der Plas, A., Smits, M., & Wehrman, C. (2010). Beyond speculative robot ethics: A vision assessment study on the future of the robotic caretaker. *Accountability in Research Policies and Quality Assurance, 17*(6), 299–315.

Van Est, R., Klaassen, P., Schuijff, M., & Smits, M. (2008). *Future man—No future man: Connecting the technological, cultural and political dots of human enhancement.* The Hague, Netherlands: NWO, Rathenau Institute.

Van Est, R. with the assistance of Rerimassie, V., van Keulen, I., & Dorren, G. (Translation Kaldenbach, K.) (2014). *Intimate technology: The battle for our body and behaviour.* The Hague, the Netherlands: Rathenau Institute.

Van't Hof, C., Van Est, R., & Daemen, F. (Eds.). (2011). *Check in/check out: The public space as an internet of things.* Rotterdam, the Netherlands: NAi.

Van Wynsberghe, A. (2013). Designing robots for care: Care centered value-sensitive design. *Science and Engineering Ethics, 19*(2), 407–433.

Von Schomberg, R. (2011). Prospects for Technology Assessment in a framework of responsible research and innovation. In M. Dusseldorp & R. Beecroft (Eds.), *Technikfolgen abschätzen lehren: Bildungspotenziale transdisziplinärer Methoden* (pp. 39–61). Wiesbaden, Germany: Springer.

Wallach, W., & Allen, C. (2010). *Moral machines: Teaching robots right from wrong.* Oxford, UK: Oxford University Press.

Whitelegg, J. & Haq, G. (2006). *Vision Zero: Adopting a target of zero for road traffic fatalities and serious injuries.* Stockholm, Sweden: The Stockholm Environment Institute.

Index

A

ABS, *see* Anti-lock braking system (ABS)
Adaptive cruise control (ACC) system, 194
Advanced driver assistance systems (ADAS) support, 191, 233
AI, *see* Artificial intelligence (AI)
Ambient-assisted living, 95
Ambient intelligence (AmI)
  artificial social intelligence, 19–20
  care domotics, elderly, 96–97
  ICT, 303
Anti-lock braking system (ABS), 186, 191, 193, 217, 234
Armed military drones
  autonomy, 270–271
  commanding officer responsibility, 276–277
  DARPA, 254, 258
  Department of Defense, 257
  ELROB, 258
  Future Combat Systems program, 249, 254, 257

human operators responsibility
  AI, 275
  cubicle warrior, 273, 281
  in-the-loop, 276
  moral responsibility, 274
  on-the-loop, 275–276
  operator, 274–275
  potential targets, 274
  residual stress, 273–274
IED, 258
international humanitarian law (*see* International humanitarian law)
joint robotics program, 254
manufacturer responsibility, 271–272
National Research Council, 256–257
proliferation and security
  dual-use technology, 278
  international regulations, 278
  legislation, 278–279
  spoofing, 280
  state and non-state actors, 278

regulation
  broad international debate, 285
  curbing proliferation, 284–285
  international ban, autonomous
    robots, 283–284
  robotics revolution, 256–257
  self-learning systems, 252–253
  social and ethical issues
    autonomy, 282–283
    counterproductive nature,
      281–282
    humanization *vs.*
      dehumanization, 282
    proliferation and abuse, 281
  tele-guided, 252–253, 257
  Unmanned Systems Roadmap
    2009–2034, 249
  U.S. budget, 254
  uses, 249–250
  U.S. Predators and Reapers,
    255–256
Artificial intelligence (AI), 275
  brute computational power, 17–18
  ethical agents, 304–306
  influence, 299–300
  Leo, two-legged robot, 16–17
  morality, 21–23
  networked robots and human-
    based computing, 23–25
  past predictions, 15–16
  social intelligence, 303–304
    AmI, 19–20
    chatbot, 20–21
    FACS, 21
    HRI, 19
    research and development, 19–20
    robotic moment, 19
  strong and weak vision, 14–15
  successful and pragmatic weak
    approach
    automotive industry and
      military, 300–301
    domestic drones, 302

    low-technology practice, 301
    monomaniacal robots,
      300–301
    sex industry and care,
      300–301
    socio-technical environment, 302
Artificial moral agents, 19, 21–23
Autonomous car, 238–239
  automatic highways, 207
  cars crash, 239
  car's system, 214
  connected autonomous car,
    186–187
  cyber security, 238
  DARPA, 207–208
  driverless taxi mobility, 214–215
  equipment/luggage, 215
  Google (*see* Autonomous
    Google car)
  IHS Automotive, 238
  job-eliminating technology, 239
  Made in Germany, 213–214
  Motion of Mercedes-Benz,
    209–210
  pedestrians, 240–243
  private car owner, 215
  public roads, 237–238
  reducing ownership and parking
    costs, 216
  smooth driving, 214
  societal and economic benefits,
    208–209
  software reliability, 238
  U.S. savings, 208–209
  V2V communication, 236–237
Autonomous Google car, 210–211
  Google engineers, 211
  Google Street View, 212
  legal fines, 211
  self-steering car, 212–213
Autonomous robots, 252–253
  civilians attack, 306
  DARPA, 258

firing loop, 283
international ban
  enforcement, 288–289
  Europe, 287–288
  ICRAC, 285–287
  regulation, 283–284
  remote-controlled armed
    drones, 286
  terrorist arsenal, 288
  United States, 287–288
robotic vacuum cleaner, 302
semiautonomous robots, 54

B

B-52 bomber, 260

C

Care domotics, elderly
  ambient-assisted living, 95
  care staff, 91–92
  care tasks, 93
  definition, 92–93
  ethical issues, 94
    average toddler, 109
    caregivers and care recipients,
      99–101
    companion technology (*see*
      Companion robots, ethical
      issues)
    design, 107–108
    human contact and quality of
      care, 99
    physical appearance, 108
    privacy, 97–99
    safety, 105–107
  infrastructures, 94–95
  KOALA, 95–96
  long-term care, 93
    companionship, 119
    dehumanization, 120
    ethical framework, 121

innovation, 119
    misleading relationships,
      119–120
    risk of paternalism, 119–120
    staff shortages, 118–119
    tele-technologies, 120–121
  paradigmatic shift
    AmI, 96–97
    CompanionAble project,
      101, 104
    emotions, 103
    HOBBIT, 101, 104–105
    KSERA, 101, 104
    Mobiserv, 101, 104
    RIBA II, 101–103
    user-centered design
      approach, 101
  people's independence, housing
    policy, 94
  remote care, 95
  Roessingh R&D division
    betrayal, 124–125
    coaching role, 122
    rehabilitation center, 121–122
    subjective perception, 123
    wheelchair, preferring, 123–124
  senior citizens, 93–94
  technological issues, 94
  telecare, 95
Caregiver
  care domotics, 93
  competences, 99–101
  empathy, 326
  ethical issues
    dehumanization, 115–116
    human contact, 99, 118
    quality of care, 99, 116–117
  home automation, 312
  in-the-loop, 101
  monitoring and privacy, 311
  out-of-the-loop, 101
  RIBA II, 102
  telecare, 96

Car robotization, 227–228
  autonomous car
    automatic highways, 207
    cars crash, 239
    car's system, 214
    connected autonomous car,
      186–187
    cyber security, 238
    DARPA, 207–208
    driverless taxi mobility,
      214–215
    equipment/luggage, 215
    Google (see Autonomous
      Google car)
    IHS Automotive, 238
    job-eliminating
      technology, 239
    Made in Germany, 213–214
    Motion of Mercedes-Benz,
      209–210
    pedestrians, 240–243
    private car owner, 215
    public roads, 237–238
    reducing ownership and
      parking costs, 216
    smooth driving, 214
    social impact, 238–239
    societal and economic benefits,
      208–209
    software reliability, 238
    U.S. savings, 208–209
    V2V communication, 236–237
  driver assistance systems, 233
    ABS, 191, 193
    ACC, 194
    application, 191
    blind spot information system,
      191, 193
    cyclist airbags, 195
    drive alert, 191–192
    eSafety Forum, 233
    ESC, 191, 193–194

    European Commission,
      234–235
    HSA, 191, 193
    intelligent speed adaption,
      191–192
    lane departure warning, 191–192
    lane keeping aid, 191–192
    parking space, 191, 193
    PCS, 196
    pedestrian airbags, 195
    policy makers, 235
    stop-and-go system, 194–195
    traffic sign recognition, 191–192
    types, 191
  GPS, 187
  limited self-driving automation,
    196–197
    cooperative driving (see
      Cooperative driving)
    cooperative systems (see
      Cooperative systems)
    traffic management, 197–199
  National Highway Traffic Safety
    Administration, 186
  road traffic, negative effects
    costs, 188
    pollution, 190
    traffic congestion, 189–190
    traffic victims, 188–189
  social and ethical issues
    acceptance, 216–218
    conscious/unconscious human
      error, 223–224
    cyber security, 223
    decision-making system,
      225–226
    drivers liability, 229–231
    liability of manufacturers,
      228–229
    life-and-death choices, 227
    limited legislation and full
      self-driving, 231–232

negative behavioral adaptation, 222–223
personalized ethics setting, 226–227
Poisson distribution, 224–225
privacy, 218–221
reliability, 221–222
road authorities, 229
speed limits, 224
SUV, 225
TNO, 187–188
CCW, *see* Convention on Certain Conventional Weapons (CCW)
Certificate of Authorization (COA), 138–139
Closed-circuit television (CCTV), 311–312
Cognitive assistance
autonomy, medication, 111, 113
cognitive prosthesis, 111
Kompai, 111, 114–115
seal robot Paro, 111–113
Collaborative robots, 5
CompanionAble project, 101, 104
Companion robots
AIBO, 60–61
android robots (*see* Sex robot)
anthropomorphism, 56, 81
authentic and inauthentic interaction, 84–85
childish/abstract life forms, 56
ecstatic happiness/deep sadness, 82–83
ethical issues
cognitive assistant, care recipient (*see* Cognitive assistance)
deception, 110–111
dehumanization, 115–116
human contact, 118

physical assistance, 109–110
quality of care, 116–117
types, 109
Furby, 57–59
hearing-impaired children, sign language, 83–84
HRI
face-to-face conversation, 65
human emotions, 66–67, 83
natural language, 65–66
perfect friendship, 67–68, 83
social de-skilling, 68–69
JIBO, 62–63
NAO, 61–62
nonverbal communication, 56, 59
Paro, 59
Repliee Q2 robot, 63–64
social competence, 55–56, 82
societal expectations, 80
speech recognition, 82
uncanny valley, 63–64
verbal communication, 56
Convention on Certain Conventional Weapons (CCW), 284, 287
Convergence, 24–25
Cooperative driving, 319, 330
advantages, 202
road haulage, 205–206
road train system, 202–203
SATRE project, 202, 204–205
self-propelling, 206
Cooperative systems
eCall system, 200–201
platoon, 235–236
social, ethical, and regulatory issues, 236
V2V and V2I communication, 199–200
Co-robots, 5, 300
Cyber security, 25, 223, 238, 313–314

D

Defense Advanced Research
  Projects Agency (DARPA),
  207–208, 254, 258
Degree of autonomy
  armed military drones (*see* Armed
    military drones)
  man-in-the-loop
    driver assistance systems, 320
    ISO standards, 318
    NHTSA, 319–320
    road authority, 319
    V2V and V2I
      communications, 319
  man-on-the-loop, 320–321
  man-out-of-the-loop, 321
  responsibility and liability,
    317–318
  unarmed military robots, 250–251
Dehumanization
  care domotics, elderly,
    115–116, 120
  human dignity, 323–324
  human sustainability
    cubicle warriors, 326
    military and police robots, 326
    preservation, 322–323
    rationalization process, 325
    sex culture, 325
    socio-emotional machines,
      324–325
  intimate technologies, 322
  relevance, 326–328
  sex robot, 75–76
Digitizing, 24–25
Driver assistance systems, 233
  ABS, 191, 193
  ACC, 194
  application, 191
  blind spot information system,
    191, 193
  cyclist airbags, 195

drive alert, 191–192
eSafety Forum, 233
ESC, 191, 193–194
European Commission, 234–235
HSA, 191, 193
intelligent speed adaption, 191–192
lane departure warning, 191–192
lane keeping aid, 191–192
parking space, 191, 193
PCS, 196
pedestrian airbags, 195
policy makers, 235
stop-and-go system, 194–195
traffic sign recognition, 191–192
types, 191
Drones, 14
  Amazon Prime Air, 132
    ambulance drone, 134
    commercial applications, 133
    delivery service, 131
    emergency services, 134
    GPS navigation and electric
      motors, 131
    India, 132–133
    parcelcopters, 134
    public and civil drones, 135
    R&D testing, 131–132
    TacoCopter, 133
    thermal sensor, 134
  armed military (*see* Armed
    military drones)
  autonomous/remote-controlled
    vehicles, 180
  civilian market, 176–177
  drone journalism, 138–140
  false positives, 179
  flying cameras, 180
  infrared surveillance, 175
  law enforcement
    legal and ethical issues, 152–154
    police drones (*see* Police drone
      technology)
    Robocop, 144–146

Little Brother, 176
load-carrying robots, 179
long-term vision, 181
panopticon, 175–176
precision farming
    commercial drone
      applications, 141
    crop survey, 141
    definition, 140
    InventWorks, Inc. and
      Boulder Labs, Inc., 141
    multispectral images, airborne
      cameras, 141
    remote sensors, 140
    reveal patterns, crop, 140
    RMAX, Japan, 142–143
privacy
    Big Brother, 164–165
    chilling effect, 165
    civilians problem, 161
    domestic drones, 162
    fixed surveillance cameras, 161
    reasonable expectation, 162
    Senate Commerce
      Committee, 160
    sousveillance, 160
    visual surveillance, 160
    voyeurism, 163
recreational use
    Amazon's Drone Store
      section, 136
    Citizens of Vancouver, 138
    DIYDrones, 135
    DJI Phantom, 137
    Hubsan X4, 137
    model aircraft use, 137
    Parrot AR.Drone 2.0,
      136–137
    TU Delft, 137
regulations
    European Commission, 170–175
    local and state governments,
      169–170

    proliferation, 173–174
    U.S. Federal Aviation
      Administration, 166–169
robot helicopters, 178
safety
    aerial, 156
    drone hunting, 158–159
    drones hacking, 157–158
    FAA, 156
    improper operations, 156–157
    Washington Post, 156
surveillance activities, 177, 180
thermal-imaging security
    camera, 175

E

EATR, *see* Energetically
    Autonomous Robot
    (EATR)
eCall system, 200–201, 218–219,
    228, 236, 312
Electrocardiograms (ECGs), 95
Electronic stability control (ESC),
    189, 191, 193–194, 217
ELROB, *see* European Land
    Robotic Trial (ELROB)
EMI, *see* Experiments in Musical
    Intelligence (EMI)
Energetically Autonomous Robot
    (EATR), 13
Epilogue
    exploratory study, 334
    human characteristics and
      capacities, 333–334
    human wisdom, 333
    "Last Post" in Ypres, 335
    robot-friendly environment,
      334–335
ESC, *see* Electronic stability control
    (ESC)
European Land Robotic Trial
    (ELROB), 258

European Robotics Technology
    Platform (EUROP), 3–6
Expected social gains
    benefits, 306–307
    camel races, Qatar, 307–308
    human suffering, 308
    investing public money, 307
    ISA system, 309
    personal responsibility, 310,
        313–314
    principle of unnecessary risk,
        307–309
    Swedish sex workers, 308
    Vision Zero, 309
Experiments in Musical Intelligence
    (EMI), 18

F

FAA, *see* Federal Aviation
    Administration (FAA)
FAA Modernization and Reform
    Act (FMRA), 167
Facial Action Coding System
    (FACS), 21
Federal Aviation Administration
    (FAA)
    drone journalism, 138–139
    European Commission, 171, 177
    law enforcement agencies, 153
    privacy, 161
    Qube's flight, 148
    RMAXs, 143
    safety, 156
    UAS, 133
    U.S. Federal Aviation
        Administration, 166–169
FMRA, *see* FAA Modernization
    and Reform Act (FMRA)
Future Combat Systems program,
    249, 254, 257

G

Global positioning system (GPS),
    23, 131, 134, 187, 199, 201,
    212, 260, 302
Governance
    political and regulatory issues
        armed military robots, 329
        autonomous cars and thinking,
            330
        child robots, 330–331
        cooperative driving, 330
        privacy and health and patient
            safety, 331
        public and civil drones, 331
        safety gain, 330
        worldwide investment, 329
    precautionary and proactive
        attitudes, 332–333
    responsibility, 328
    user involvement, 328–329
GPS, *see* Global positioning system
    (GPS)

H

High tech automotive systems
    (HTAS), 202
Hill start assist (HSA), 191, 193
Household robots, *see* Vacuum
    cleaner
Human-based computing, 23–25
Humanoids, *see* Companion robots
Human–robot interaction (HRI)
    care domotics, elderly, 106–107
    face-to-face conversation, 65
    friendship, 67–68, 83
    human emotions, 66–67, 83
    natural language, 65–66
    social de-skilling, 68–69
    social intelligence, 18–19

I

IFR, *see* International Federation of Robotics (IFR)
Improvised explosive device (IED), 258
Information and communication technology (ICT)
  AmI, 303
  blood glucose and blood pressure, 96
  long-term goal, 80
  physical and mental living space, 32
  precision farming, 140
  R&D efforts, 303
  telecare, 96
  TNO, 187–188
Information technology (IT), 25–26
  monitoring and privacy
    CCTV, 311–312
    citizens, 311
    eCall system, 312
    home automation, 312
    utilization, 310–311
  safety, cyber security, and misuse, 313–314
Intelligent agent, 17, 20
Intelligent Speed Assistant (ISA), 309
International Federation of Robotics (IFR), 4, 9
International humanitarian law
  aspects, 259
  autonomous drones
    adequate artificial ethical conscience, 268
    combatant, 265–266
    demilitarized zone, 266–267
    ethical decisions, 269
    ICRAC, 269
    kill boxes, 266
    non-reducibility, 268

robotic sensors, 267
    technological imperative, 268
  jus ad bellum, 259
  tele-led drones
    discrimination principle, 262–263
    proportionality principle, 260–262
    targeted killing, 263–265
International Organization for Standardization (ISO), 9, 106, 272, 318–319
ISA, *see* Intelligent Speed Assistant (ISA)
IT, *see* Information technology (IT)

J

Jus in bello, *see* International humanitarian law

K

Knowledgeable Service Robots for Aging (KSERA), 101, 104

L

Lifelike appearance
  childlike appearance, 314–315
  claimed effect, 316–317
  KASPAR, 316
  nonverbal communication, 314
  persuasive robot technology, 317
  sex robot, 315–316
  social interaction robots, 316

M

Mechanoid robots, *see* Vacuum cleaner
Ministry of Economy, Trade and Industry (METI), 6

Mobile robot, 12–13, 101,
    103–104, 179
Montgomery County Sheriff's
    Office (MCSO), 150
Moore's Law, 18

N

Nano Hummingbird, 150–151
National Aeronautics and Space
    Administration (NASA),
    13, 37, 156
National Aerospace Laboratory
    (NLR), 147
National Highway Traffic
    Safety Administration
    (NHTSA), 319
National Robotics Initiative (NRI), 5
NATO, *see* North Atlantic Treaty
    Organization (NATO)
Neo mechatronic society, 6
Networked robots, 10, 23–25
New Energy and Industrial
    Technology Development
    Organization (NEDO), 206
NHTSA, *see* National Highway
    Traffic Safety
    Administration (NHTSA)
NLR, *see* National Aerospace
    Laboratory (NLR)
North Atlantic Treaty Organization
    (NATO), 260

P

Panopticon, 175
Parcelcopters, 134
PCS, *see* Pre-crash system (PCS)
Persuasive robot technology, 317
Police drone technology
    flight times and wind force, 151
    intelligent camera systems, 151–152
    Nano Hummingbird, 150–151

Qube drone, 148
    RMAX, Japan, 142–143
    ShadowHawk, 150
Pre-crash system (PCS), 196, 234

Q

Qube's flight, 148

R

Radio police automaton, 144–145
Remote care, 95–97, 100
Robotic trading, 22
Robots, 10
    Amazon, 32–33
    artificial life
        artificial senses, 13–14
        from automata to robots, 6–8
        physical activity, 12–13
        physical appearance, 10–12
        robot brain (*see* Artificial
            intelligence (AI))
        robot-friendly environments,
            8–10, 33
    autonomy, 26–27
    drones, 14
    hugging robots, 1
    humanoid robots, 1–2
    IT, 25–26
    laws of Asimov, 1–2
    life-like appearance, 26, 31
    long and short term, 32
    mental presence, 33–34
    nuclear energy program, 36–37
    rationalization
        bureaucracy, 27–28
        environments, 28–29
        iron cage of rationality, 28
        irrationality, 30–31
        The McDonaldization of Society,
            29–30
        production process, 28

RUR, 6
sex robot, 35
social and ethical issues, 32
social media, 34–35
technical promises and societal
    expectations, 32
vision, 2
    EUROP, 5–6
    human skills, 4–5
    IFR, 4
    From Internet to Robotics, 5
    lights-out factory, 4
    METI, 6
    NRI, 5
    penetration, 4
    service robots/serve-us
        robots, 5
    Unimate, 3–5
Wired for War, 1
working envelopes, 32
Roomba
    commercial market, 46–47
    costs, 47
    fixed base, 47
    roombarization, 48–50
    sales ratio, 46–47
Rossum's Universal Robot (RUR),
    6, 45

S

Safe Road Trains for the
    Environment (SATRE)
    project, 202, 204–206, 237
Sex robot
    benefits
        Amsterdam, 73
        ethical justification, 73–74
        prostitution and sex
            trafficking, 72–73
        unlimited private lessons, 72
    child robots, 76–78
    cultural acceptance, 74–75

ethical issues, 81
human performance, 69–70
Internet, 69
Rocky, 71
Roxxxy, 70–71
sex slavery, 81
trafficking of women, 81
Westworld thriller, 69
ShadowHawk, 150
Social intelligence, 303–304
    AmI, 19–20
    chatbot, 20–21
    FACS, 21
    HRI, 18–19
    ISO 13482 standard, 106–107
    research and development, 19–20
    robotic moment, 19
Softbot, 20–21
Sousveillance, 160, 328
Special Weapons and Tactics
    (SWAT), 150
Spoofing, 157, 280
Sport utility vehicle (SUV), 225

T

TacoCopter, 133
Telecare, 95–96, 301, 312, 323, 329
The Netherlands Organisation for
    Applied Scientific Research
    (TNO), 187–188
TomTom traffic, 198–199
Turing test, 15–16, 20

U

Unarmed military robots, 250–251
Uncanny valley theory, 11–12,
    63–64, 71, 314
Unmanned aerial vehicles (UAVs),
    250–251
Unmanned aircraft systems
    (UASs), 133

Unmanned combat aerial vehicles
    (UCAVs), 251–252
Unmanned military aircraft, 14

V

Vacuum cleaner, 44–45, 79
    household tasks
        cooking complexity, 51–52
        degree of difficulty, 51
        frame problem, 50–51
        household tasks, 45
        Lego box, 52–53
        liability, 53–55
        non-monomaniacal weekly
            cleaner, 53
    lawnmower robot, 46
    low power capacity, 47
    monomaniacal household robots,
        45–46
    Roomba
        commercial market, 46–47
        costs, 47
        fixed base, 47

roombarization, 48–50
    sales ratio, 46–47
    RUR, 45
    virtual walls, 47
V-A1 robots, 146
Vehicle-to-infrastructure (V2I)
        communication
    automatic highways, 207
    electronic safety system, 200–201
    European Commission, 235–236
    producer liability complex,
        319–320
    traffic management, 196–197
    transmission and control
        apparatus, 199
Vehicle-to-vehicle (V2V)
        communication, 196–197,
        199–200, 204, 235–237, 319
Voyeurnalism, 139–140

Y

Yamaha Attitude Control System
    (YACS), 142